Canines, page 212

Hoofed mammals, page 274

Bears, page 228

Birds, page 300

Raccoon and ringtail, page 240

Reptiles, page 334

Weasels and their kin (Mustelidae), page 244

Skunks, page 268

Amphibians, page 334

Insects and other invertebrates, page 341

WILDLIFE *of the* PACIFIC NORTHWEST

Tracking and Identifying Mammals, Birds, Reptiles, Amphibians, and Invertebrates

DAVID MOSKOWITZ

TIMBER PRESS FIELD GUIDE

Photographs and other illustrations are by the author unless otherwise indicated.

Inside front cover illustrations by Jenn Wolfe. Page 1: A Rocky Mountain red fox pauses on a rainy day. Canadian Rockies, Montana. Frontispiece: Tracks of a galloping black-tailed deer cross a coastal sand dune. Northwest Coast, Oregon. Facing page: An American marten peers out from the safety of its perch in a spruce tree. Middle Rockies, Idaho. Page 6 (opposite contents): A great egret hunts in the still water of a coastal lagoon. North Coast, California.

Published in 2010 by Timber Press, Inc.

The Haseltine Building
133 S.W. Second Avenue, Suite 450
Portland, Oregon 97204-3527
timberpress.com

Printed in China
Designed by Susan Applegate
Fourth printing 2015

Library of Congress
Cataloging-in-Publication Data

Moskowitz, David, 1976–
Wildlife of the Pacific Northwest: tracking and identifying mammals, birds, reptiles, amphibians, and invertebrates/David Moskowitz.
p. cm.—(Timber Press field guide)
Includes bibliographical references and index.
ISBN 978-0-88192-949-2
1. Animal tracks—Northwest, Pacific—Identification. 2. Tracking and trailing—Northwest, Pacific. 3. Animals—Northwest, Pacific—Identification. I. Title.
QL768.M675 2010
591.9795—dc22 2009039066

For Keith O'Rourke and Benayshe Titus,
who sparked my interest in the natural world,
and Mount Diablo,
my first classroom as a naturalist.

CONTENTS

Female bighorn sheep browse after a spring snow squall. East Cascades, Washington.

PREFACE

It's a sunny winter day in the Cascades. Since close to first light I have been following the fresh trail of a bobcat over a perfect blanket of gleaming white snow. As the day has begun to warm, the bobcat's trail winds its way up through open timber toward a small cliff band and angles toward a narrow slot through the middle of the rock. I approach slowly to see if I, too, will be able to pass through the opening in the cliff where the trail is clearly leading. As I get closer, I realize that what at first appeared to be passageway is actually an overhanging recess, and I understand where these tracks have taken me. Cautiously, I take a few steps toward the base of the rock and peer into the fissure's shadowed depths. I see nothing in the dark but two green eyes staring back at me.

Again and again while writing this book, I have been overwhelmed at the magnificent diversity of wildlife in the Pacific Northwest and the captivating signs animals leave across the landscape. For me, wild animal tracks and signs have a compelling, magical quality. This book is my attempt to inspire others to discover that magic and to deepen and enrich their personal relationship with the natural world.

Wild animals are a universally fascinating feature of the natural world, yet they are often difficult to observe directly. Many species of wildlife are shy, nocturnal, or occur in such low densities that they are rarely seen. An ability to interpret the tracks and signs animals leave behind is a vital part of studying wildlife, especially mammals, in the field. Even when we do have the opportunity to observe wildlife directly, we need a foundational understanding of wildlife behavior and ecology to grasp the depth of what we are seeing and to understand why we saw that particular animal, in that particular place, do that particular thing.

This guide is designed to do more than help you attach names to things. My hope is that it will help you discover and understand the fascinating stories of wildlife of the Pacific Northwest. Based on my years of field work and teaching, I have tailored the content and layout of this guide for ease of use in the field and to increase the depth and richness of your understanding of wildlife in our region. Photographs, descriptions, and measurements of tracks and signs are representative of what you will typically encounter when you venture out into the forests, mountains, shores, and deserts of the Pacific Northwest.

Introductory material provides context for species accounts that follow, illuminating how topography, climate, and other ecological factors shape the distribution of wildlife in our region and introducing basic wildlife biology concepts. "Wildlife Tracking Basics" and "Finding Tracks, Signs, and Wildlife" cover terminology used to describe tracks, track patterns, and field techniques and tracking concepts that will help you develop your observational skills related to wildlife field studies.

Visual keys provide a starting point for identifying wildlife, tracks, and signs. Tracks and signs of mammals can at times be confused with those left by birds, reptiles, amphibians, and invertebrates. Short sections on each of these other groups of wildlife help you distinguish other animal signs from those of mammals and provide a window into their fascinating world.

We humans are not just passive observers of the world around us; we are active participants. Through our choices, we affect not only our own lives and those of other people, but also the existence of flora and fauna in wildlands that provide the services and natural resources on which we depend for our survival and quality of life. The forests on which endangered woodland caribou depend for shelter and food also protect watersheds and provide carbon sequestration that humans require for drinking water, electricity, and a stable climate.

Our culture's interest in, and affinity with, wild animals is not new or unique (though our lack of awareness of our direct impacts may be). In any hunter-gatherer culture, a detailed understanding, respect, and appreciation for wild creatures is integral to the material survival and safety of its people. Though we may think that we in this modern world have moved far from our roots in subsistence cultures, in actuality our survival and safety still depend on understanding, respect, and appreciation for our natural environment. It is my sincere hope that reading and using this book furthers your understanding, deepens your respect, and awakens your appreciation of animals, their signs, and the natural world upon which we all depend.

ACKNOWLEDGMENTS

Without the unending love, support, encouragement, and critical eye for details of my partner, Darcy Ottey, I would never have been able to translate my passions into this book. I am tremendously grateful for the support and inspiration of my family. The value of curiosity, critical thinking, and deductive reasoning was instilled in me from a young age by my parents, Johanna Goldfarb and Ralph Moskowitz, and my extended family. This paved the way for my career as a naturalist. I am also grateful for years of mentorship, and feedback I received during the writing process, from Charles Worsham. His ideas, more than anyone else's, have shaped the way in which I teach wildlife tracking and how these concepts have been presented here.

A number of people took a keen interest in this project, and I am grateful for their support. Jeff Rosier helped me locate rare species in the field and helped me formulate how to present several tricky topics in the book, and he spearheaded the production of the track measurement graphs in the back of the book. Jenn Wolfe and I shared many adventures in the field, and she collaborated with me on several illustrations in the opening chapters. Dr. James Halfpenny (of A Naturalist's World) provided access to unpublished track measurement data, guidance in creating reliable measurement parameters and track descriptions for several species in the book, and instruction on how to handle small mammals for photography. The Nature Conservancy, through the diligent work of Erica Simek, created and donated the ecoregions map. Cristina Eisenberg gave me amazing opportunities to collect material for the book in the field and reviewed my entire manuscript. Brian McConnell spent many hours in the field with me searching for tracks and wildlife, and his keen eyes spotted several subjects of my photos. Mark Elbroch, through his books and time in the field, provided a great deal of insight, particularly in the realm of small mammal and bird tracking. Sue Morse (of Keeping Track, Inc.) reviewed several sections of the book and contributed to my understanding of the fields of wildlife biology and tracking in general. Joe Kiegel, Casey McFarland, Mallory Clarke, Aaron Goldfarb, and Mark Darrach all contributed to the production of this book in unique and substantial ways. I am grateful for the support and flexibility of my employer, Wilderness Awareness School, and my colleagues and students there.

I would also like to thank the following mentors, colleagues, friends, and institutions for assisting in the production of this book: Jeff Bradley; the Burke Museum of Natural History and Culture; Northwest Trek; Roger Bean; Hart Mountain National Antelope Refuge; James Begley (Western Transportation Institute); Steve Smith; Fiona Clark; Jonathan Goff; Roger Christophersen (North Cascades National Park); Kurt Jenkins (U.S. Geological Survey); Patti Happe (Olympic National Park); Jeff Lewis (Washington Department of Fish and Wildlife); Karston Heuer; Alan St. John; Brandon Sheely; John Rohr (Okanogan National

Forest); Mark Hebblewhite; Glenn Lorton (Bureau of Land Management; Lakeview field office); Jeff Copeland; Paul Rezendes; Jason Knight (Alderleaf Wilderness College); Paul Houghtaling; Shannon Kachel; Paul Bannick; Conservation Northwest; Kevin Mack; the Progressive Animal Welfare Society; Kainoa, Leif, and Amara Oden; Al Thieme; Mark, Heather and Bea Timken; Julie Perry; Dan Corcoran; Emily Gibson; Dan Gardoqui and Dan Hanshe (White Pines Programs); Emily Pease; Linda Bittle; Matt Monjello; Shannon Kachel; Patricia A. Garvey-Darda and John Agar (Wenatchee National Forest); Susan Applegate; and Lisa Theobald.

The tracking funnel diagram is an elaboration of a teaching method developed by Jon Young. Eve Goodman and others at Timber Press helped refine my original idea into this unique field guide. Many individuals already mentioned reviewed sections of the book related to their areas of expertise. Numerous other people, too many to name, supported me in writing this book and I am grateful to them all.

In writing this field guide, I drew from a wide variety of published sources and personal communications, along with my own field work. I am grateful to all the naturalists, biologists, wildlife trackers, and writers upon whose shoulders I have perched to write this book. However, any mistakes in the pages that follow are mine alone.

HOW TO USE THIS BOOK

This book is divided into three parts. Part One, "Wildlife Tracking Basics," covers the fundamentals of wildlife habitat, tracks, track patterns, signs, foot morphology, locomotion, and track and sign measurements. Part Two, "Mammals," focuses specifically on habitat, tracks, track patterns, and signs of the order Mammalia. Part Three, "Other Wildlife," offers habitat, track, and sign information on birds, reptiles and amphibians, and insects and other invertebrates.

Use the visual keys and brief descriptions to find specific group or individual species accounts. Keys are not comprehensive, but they are good starting points in your quest to identify typical presentations of a variety of tracks and signs.

The key to mammal tracks provides life-sized or exactly scaled drawings of the front and hind feet of individual species. Flip through the drawings until you find the closest match in terms of size and shape to tracks you have found, and then go to the page noted next to the drawings for detailed species accounts. You'll also find a key to bird tracks in the relevant section.

Species or families of wildlife leave distinct signs other than footprints, including scats, feeding signs, marks on trees, and predation signs on other animals. Use the photographs and descriptions in the respective keys to narrow down the possible maker of those signs and follow page references to find more information.

Use the silhouettes (inside the front cover) of mammals and other groups of wildlife to help identify the type of animal you have discovered.

Families or Groups

Species accounts are divided into groups according to a combination of taxonomy, appearance, and similarities in tracks and signs. Group accounts provide details on life history and track and sign patterns that are shared among all group members. Details that distinguish individual orders, families, genera, or species within the group are also noted. The pages for each group are color coded.

Species Accounts

Each species or group is presented in a specific format to help you use the text most efficiently. Some closely related species, such as shrews (*Sorex* species), are difficult

Before You Head into the Field

Reviewing accounts of groups, families, or species before you go into the field can help you develop search images of what to look for and where to look. Learning about the habitat and types of signs left by a species will help you determine where to look for wildlife and recognize their signs when you find them. Use information regarding tracking techniques to define a plan for discovering and studying an interesting species in the field. After time afield, you will find the material in this guide helpful in interpreting your findings.

to distinguish in the field, even when you have an animal in hand, and you may find it impossible to differentiate them through their tracks and signs. Such groups are described together as a genus, with notes on distinguishing species where relevant. Contextual guidelines such as habitat preferences are included to help you narrow down what you may have found in the field.

DESCRIPTION FORMAT

COMMON NAME or GROUP NAME

Species

Important life-history, behavioral information, and tracking tips are provided for each species or group.

SIZE Body length (including tail) and weight. Size differences between male and females are noted when applicable.* **FIELD MARKS** Distinctive appearance attributes and markings. **REPRODUCTION** Reproductive behavior notes designed to help you interpret field observations. **DIET** Seasonal shifts in diet with a focus on details that aid with interpretation of feeding signs in the field. **HABITAT & DISTRIBUTION** Helps in your search for a particular species, with a focus on distinguishing similar species. Primary sources for range maps are Csuti et al. (2001) and Johnson and Cassidy (1997), with other sources consulted for areas outside Washington and Oregon and for specific species. **SIMILAR SPECIES** Other species, or groups of species, that might be confused with the species at hand either through tracks and signs or actual visual appearance. **CONSERVATION ISSUES** When relevant, issues associated with the conservation of a species are mentioned. Impacts on native wildlife and ecosystems are included for introduced species.

TRACKS

Detailed description of tracks that can help you identify the species at hand. Measurement ranges are specified for front and hind feet. **TRACK PATTERNS** Common and/or distinctive track patterns and measurement parameters left by a species and notable behavioral interpretations of track patterns.

SIGNS

Relevant signs such as scats, feeding signs, or scent-marking are described to help you identify and interpret what you find.

* The source for the majority of these parameters is Kays and Wilson (2002), supplemented with other sources for some species.

A beaver has felled a large black cottonwood tree (*Populus balsamifera*) to access the inner bark of the trunk and branches, a primary food source for beavers. East Cascades, Washington.

Upper Napeequa River valley. North Cascades, Washington.

THE PACIFIC NORTHWEST AND ITS WILDLIFE

I follow the trail of a wolverine that has lithely crossed a vast and snow-covered alpine basin, skirted the edge of the tree line, and scampered up a series of ledges. I can see that it slid down steep snow slopes like an otter and returned several times to a colony of marmots along the edge of a large cliff band just below the ridgeline. Around the basin, steep, glacier-carved mountain faces rise up and active glaciers spill down from the peaks. Below, where the basin drops into a deep U-shaped valley, miles of dense conifer forest broaden as the valley winds its way through the heart of the mountains.

One of the rarest species of mammals in the Pacific Northwest, wolverines are wide-ranging animals, and their large, heavily furred feet are well adapted to travel in deep snow. It is a lucky backcountry skier or mountaineer who crosses the trail of a wolverine in the high mountains of the Northwest. I discovered these tracks while on a weeklong trip to a remote part of the Glacier Peak Wilderness in the North Cascades. I was crossing a trailless mountain pass at the highest elevation and most remote point in my trip.

Other reports of wolverine tracks come from similarly remote and stunning locales. Unfortunately, some who have encountered wolverine trails haven't recognized them or understood how valuable such a sighting could be to our understanding of this rare mammal.

Wolverines have been considered habitat generalists because of their broad and ecologically diverse geographic range. Their distribution in the high, remote mountains of western North America is often attributed to their need for large tracts of wildlands and perhaps their aversion to humans. Studies of natal den site selection suggest, however, that wolverines may require specific features in their rugged habitats.

All documented natal dens have been excavated into deep snowpack (Magoun and Copeland 1998). Wondering whether the need for adequate snow cover during the denning season might be a defining habitat requirement for wolverines, researchers (Aubry et al. 2007) overlaid satellite data for snow cover in the month of May with known current and historic wolverine locations globally. The overlap of the two data sets was clearly significant, far more so than any other habitat parameter the researchers tested. Based on this work, it appears that wolverines may be an excellent example of a species whose distribution is intimately connected to the climate of its environment.

Abiotic Influences on Wildlife

A complex mosaic of weather, climate, topography, and latitude defines Pacific Northwest ecology. These abiotic factors play a large role in regional wildlife distribution. On the range maps provided for individual species, you will notice that some species are restricted either to the western or eastern portions of our region, while others inhabit only mountainous sections or lower elevations. The wildlife communities of our region change drastically from the coast, to the mountains, to the arid interior. Predictable weather shifts and large elevation ranges in our temperate climate zone influence seasonal weather and wildlife seasonal patterns of use.

The East-West divide. The Pacific Ocean and the Cascade Mountains are the most important geological features for defining our weather and climate. The Cascades split the region into two distinct climates: the maritime-influenced western section and the arid interior to the east. Temperatures are more moderate west of the Cascades, affected by the proximity of the Pacific Ocean. East of the Cascades, the landscape experiences greater fluctuations in temperature—hotter in the summer and colder in the winter.

Most of our storm systems come from the Pacific Ocean. The Cascades cause storms to back up against the Coast Range and west slope of the Cascades, dumping large amounts of precipitation. As clouds are pushed up and over the mountains, precipitation turns to heavy snow in winter. By the time storm systems have crossed the Cascades, they have lost most of their moisture, creating the rain shadow that is responsible for the arid lands and true desert regions east of the Cascades.

Flora and fauna that thrive only on the west side are associated with moist forest environments and are largely intolerant of the prolonged absence of water. East of the Cascades, various species have adapted to the arid environment, and many demonstrate fascinating adaptations for conserving water.

Latitude. In summer, the weather-influencing jet stream usually shifts well to the north, carrying the Pacific's storms with it and leaving the region dry for months. When the jet stream shifts south again in the fall and winter, storms bring deep snows to the high mountains and overcast skies and rainy conditions in the lowlands. East of the Cascades, arctic air masses move south, bringing consistently subfreezing temperatures.

Located in a temperate zone (the 45th parallel, the midpoint between the equator and the North Pole, falls in central Oregon), our region experiences four distinct seasons, particularly east of the Cascades. The diverse suite of regional plant communities in the Pacific Northwest are assembled in part to reflect this range of seasonal variation. In addition, you can find microhabitats similar to areas hundreds of miles to the north or south, such as the bogs and cold-air drainage valleys of northern Washington more typical of boreal settings well into Canada, and the arid playa lake salt flats of southeastern Oregon more typical of desert settings farther south.

Some species reach the northern end of their range along the southern boundary of our region, and, conversely, the southern limit for several species lies in the northern section of our region.

Elevation and topography. The Cascades and other Northwest mountain ranges are essentially southern peninsular or island-like extensions of the boreal life zone. As a result, some species that occupy expansive areas much farther north can eke out an existence at the higher elevations of our mountains. In this environment, deep winter snow and short summers mimic boreal conditions. Conversely, lower elevations east of the Cascades are far more arid and prone to extreme seasonal temperature fluctuations. These areas are home to desert-adapted species such as the Great Basin pocket mouse and other members of the mammal family Heteromyidae.

ECOREGIONS

The abiotic influences on a landscape combine to create conditions that support unique biological communities. Ecoregions, zones of broad ecological similarity based on their climate and topography, present unique assemblages of associated plant and wildlife communities and are an excellent framework for understanding regional wildlife distribution. Ecoregion descriptions and associated maps are based on the work of Robert Bailey (1995), with the U.S. Geological Survey, and The Nature Conservancy. Location references included in photograph captions throughout the book refer to these ecoregions.

Northwest Coast. The wettest ecoregion in the Pacific Northwest is defined by temperate rain forests and a coastal mountain range that reaches its apex in the Olympic Mountains of Washington and along the spine of Vancouver Island in British Columbia. The mouths of large rivers have created extensive estuary systems, and intermittent sand dunes that extend from northern California through much of Oregon create another distinct habitat for wildlife. Because of their isolation from other mountain ranges and landforms, the Olympic Mountains and Vancouver Island have several endemic species of mammals.

Willamette Valley, Puget Trough, and Georgia Basin. This ecoregion sits in a low-elevation trough between the Coast Range and Cascades and includes the cities of Portland, Seattle, and Vancouver, British Columbia. With a moderate marine-influenced climate, this is the most human-populated and developed part of the Pacific Northwest. It is characterized by tracts of agricultural land and urban sprawl interspersed with ecologically fragmented and modified naturally occurring habitats. Native ecological communities such as oak woodlands and prairies are threatened and reduced in extent relative to presettlement conditions. Much of the ecoregion's fauna is dominated or at least influenced by introduced and invasive species of mammals, amphibians, and invertebrates. The construction of extensive road systems and developed areas have severed large portions of the Coast Range from the Cascades, affecting dispersal of some wildlife species between ecoregions.

North Cascades. Including both the North Cascades and British Columbia's Coast Range north of the Fraser River, this ecoregion reflects a history of recent extensive glaciation. These are the most rugged mountains in our region and are the most intensely glaciated mountains in the contiguous United States. Ice age glacial scouring created rocky peaks and alpine tundra at the highest elevations, steep-sided mountain canyons, and heavily forested valley bottoms. Many species found here have adapted to the rigors of deep snowfall and steep terrain, including several whose primary distributions are in the boreal forests farther north on the continent. The rugged country of the North Cascades includes

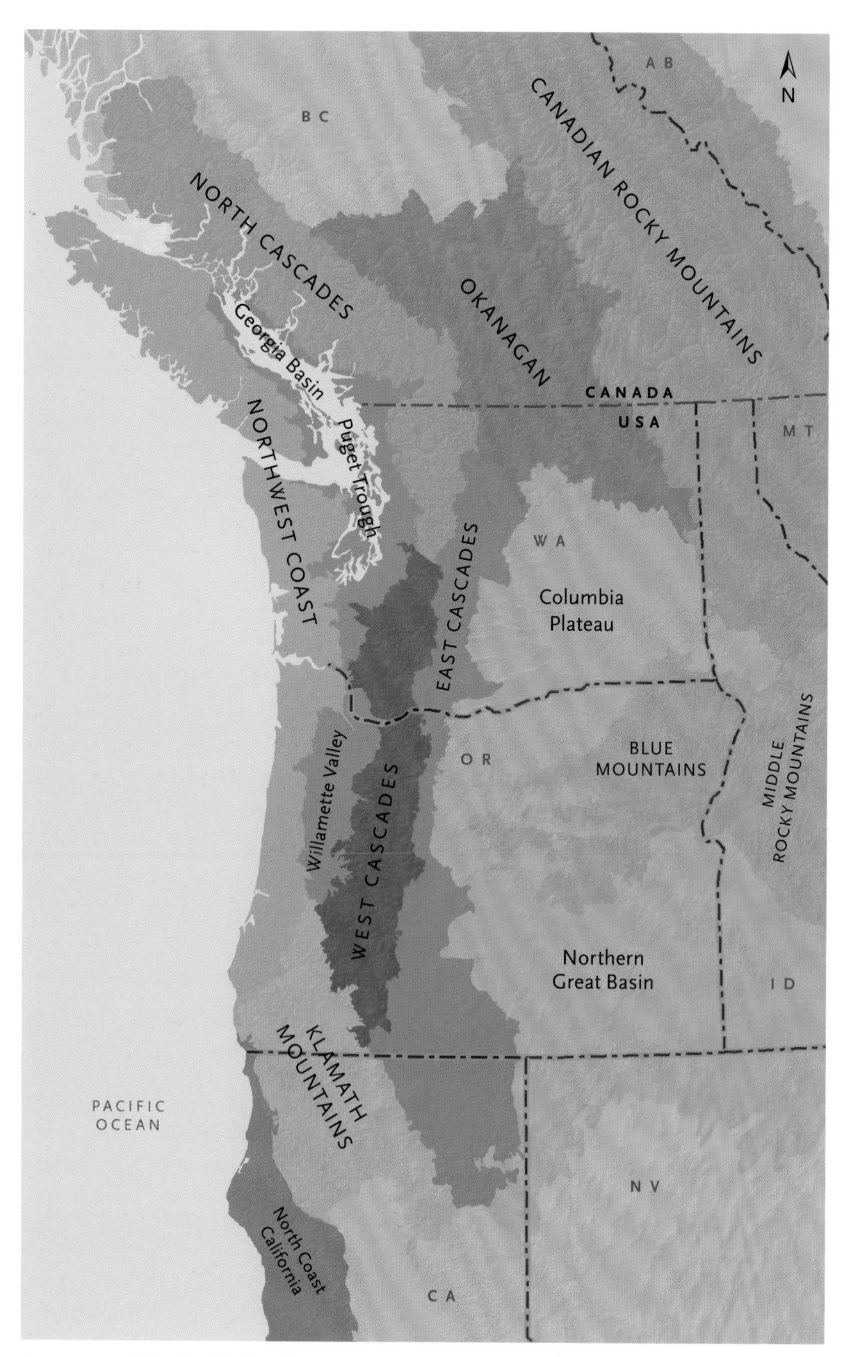

Ecoregions of the Pacific Northwest. Illustration by Erica L. Simek, The Nature Conservancy.

extensive tracts of remote wilderness. Accelerating climate change is impacting this ecoregion, with glacial recession occurring rapidly. The portion of this ecoregion that falls within the United States is a federally designated recovery area for endangered grizzly bears.

West Cascades. South of Snoqualmie Pass, starting in the middle of the state of Washington, the Cascades change in character. Slightly lower elevations and more gentle topography punctuated by towering stratovolcanoes (such as Mount Rainer and Mount Hood) define this part of the Cascades. West of the Cascade Crest, the ecoregion is heavily forested and is similar to the coastal mountains to the west, in terms of flora and fauna.

East Cascades. Colder in winter and drier than the West Cascades, this ecoregion blends some elements of Rocky Mountain flora with westside species. Fire strongly drives the ecoregion's ecology, with hot summer temperatures, regular seasonal droughts, and occasional "dry lightning" thunderstorm activity. At the southern terminus of the Cascades in northern California, both sides of the mountain range share the arid climate and plant communities of this ecoregion.

Okanagan. This broad mountainous region connects the Cascades and Rocky Mountains, with wildlands of less dramatic topography than the North Cascades. Large tracts of dry conifer forests are dominated by ponderosa and lodgepole pine (*Pinus ponderosa* and *P. contorta*) and are subject to regular summer wildfires. Because of its latitude and location east of the higher North Cascades, the Okanagan has a blend of northern boreal and arid-adapted species more common farther to the south.

Canadian Rocky Mountains. This vast and wild ecoregion is part of an inland, continent-spanning mountain range. The climate is colder and drier than the Cascades. It includes the Selkirk and Columbia mountain ranges to the west of the Rocky Mountains proper. The collection of fauna is defined by boreal species, including woodland caribou.

Blue Mountains and Middle Rocky Mountains. This ecoregion includes the Blue Mountains of Oregon and Washington, the Salmon River Mountains of central Idaho, and ranges to the east. The Blue Mountains are considered the Rocky Mountains' westernmost extension, and the flora and fauna found here reflect this. These mountains are drier than the Cascades, with bitterly cold winters. Forests are dominated by fire-adapted conifers. Lower elevations are considerably drier, and species common in the Columbia Plateau and the Great Basin are found here. The mountains of central Idaho contain the largest contiguous wilderness area in the lower 48 states.

Columbia Plateau and Northern Great Basin. The Cascades' rain shadow is the most important defining ecological feature of this ecoregion. It includes the northern terminus of the vast Great Basin desert and is dominated by desert-adapted shrubs and fauna. The ecoregion's northern portion falls within the Columbia and Snake River watersheds. The southern portion is colder in winter and drier, with characteristic basin-and-range topography typified by dramatic north-south–oriented mountain ranges whose watersheds terminate in the broad valleys (basins) that separate the ranges. Numerous desert-specialist wildlife species occur only in this ecoregion, including several whose range barely extends into the southeastern corner of Oregon.

Klamath Mountains. This complex jumble of mountains in southwest Oregon and northwestern California is dominated by mixed hardwood and conifer forests. Though these mountains are not particularly tall, wild rivers have cut steep canyons through them, and the area's diverse geologic and climatic settings make for a broad range of habitats that support perhaps the highest plant diversity north of Mexico. Several species common farther south find the northern end of their range here. Extensive stands of oak trees and scrub (*Quercus* spp.) support a significant number of wildlife species that use acorns as an important food source.

North Coast California. This ecoregion's coastline includes sand dunes, rocky headlands, river estuaries, salt marshes, and several north–south mountain ranges. Some coastal locations receive up to 120 in. (3 m) of precipitation per year, almost all during the winter and some as snow at higher elevations. Interior mountains and valleys are considerably more arid with greater temperature fluctuations. Coastal summers are characterized by cool ocean winds and fog.

This ecoregion is known for its towering redwood trees (*Sequoia sempervirens*), one of the largest and longest living tree species in the world. Redwoods occupy areas within reach of summer fog that drifts in from the ocean at low elevations. Higher, drier, and interior regions' forests are dominated by oak trees and Douglas-fir. Most rivers and large streams have been historically salmon-bearing, but anthropogenic (human-caused) influences have diminished salmon runs in this area. Invasive plant species have also impacted coastal and grassland plant communities.

Biological Influences on Wildlife Distribution

In general, all animals share the same basic needs: food and water, shelter from adverse environmental conditions, refuge from predation, and access to other members of their species for reproduction. The tactics and physical adaptations employed by animals for meeting these needs are amazingly diverse and are often evident in an animal's body shape, behaviors, and tracks and signs. As you read about a particular species, ask yourself this: How does this animal's unique morphology and behavior enable it to meet its basic survival needs?

The wide range of adaptations—both physical and behavioral—of different animals allows a diversity of food sources and habitats to be exploited, thus reducing competition between similar species. The unique way an organism meets its survival needs is called its ecological niche. Specific biological influences on distribution are covered in the introductory section to mammals and in individual animal accounts.

Human Influences on Wildlife

Regionally, human influences on wildlife distribution, population size, and behavior are similar in scale to the effects of geography and general biotic factors. We need to understand the impacts and influences of anthropogenic changes to the region's natural systems and wildlife before we can gain a complete picture of wildlife behavior and distribution.

Wolverines, for example, were hunted, trapped, and poisoned for decades, leading to their extirpation from large sections of their historic range in the western United States, including parts of the Pacific Northwest. During the second half of the 20th century, as human sources of mortality abated, wolverines began reestablishing in areas where they had previously existed. This wide-ranging species exists in low densities and reproduces slowly, and it faces a new threat to its survival in our region, even with complete protection from hunting and trapping. The research and models that link wolverine distribution with abundant spring snowpack suggest that the current changes in our region's (and the planet's) climate may prove challenging for wolverines. Earlier melt-off of mountain snowpack is already apparent, and these trends are predicted to continue with global climate changes. How these changes will affect wolverine populations in the Northwest remains to be seen, but it is a cause for concern relative to their persistence in our region.

Native habitat destruction, and fragmentation. Since Europeans began settling in the Pacific Northwest, native landscapes have changed dramatically after being converted to uses such as agriculture, urban and suburban development, and industrial wood production. Even many so-called "wildlands" bear little resemblance to the habitats that existed prior to settlement.

Low-elevation forests west of the Cascades have been particularly impacted. In the Puget Trough ecoregion, prairies and oak woodlands have almost completely disappeared, leaving some animal and plant species on the brink of extirpation from this part of their traditional range. In the northern portion of the Columbia Plateau, large-scale conversion of native arid bunchgrass prairies and shrub-steppe habitat to agricultural uses has left disjointed fragments of this originally vast habitat. Most locations not converted to cropland are infested with invasive plant species as a direct result of long-term livestock overgrazing. However, habitat destruction has not been evenly distributed across all parts of our region, and although some wildlife has suffered immensely from these changes, others, such as coyotes and crows, have adapted well to the new habitats.

Where high-quality native habitats still exist, habitat fragmentation, such as a mature forest segmented by clearcuts, can decrease its value for wildlife. Animals such as the American marten, which tend to avoid large clearings, are not able to use small chunks of mature forest surrounded by clearcuts. Increased forest edge habitats prevalent in fragmented settings create ready inroads for invasive and generalist species that would not otherwise be present.

Invasive species. Humans have also introduced many species of mammals, birds, insects, amphibians, and plants that have had significant effects on native wildlife through competition for resources, disease introduction, direct predation, and habitat degradation. Invasive species covered in this guide include the opossum, nutria, and Norway rat.

Invasive plant species also impact wildlife, such as aggressive alien plants that out-compete native species and eradicate plant communities on which wildlife species rely for food or shelter. Cheat grass (*Bromus tectorum*) is a typical example of such a species in arid, shrub-steppe landscapes of the Columbia Plateau and is responsible for widespread ecological degradation. Himalayan blackberry (*Rubus discolor*), while providing an abundant and nutritious food source, also creates impenetrable thickets that can stunt the succession of native forests and ultimately reduce the diversity of habitats and food sources available to wildlife.

Reduced permeability of wildlife corridors. In our topographically diverse landscape, wildlife with specific needs often have narrow corridors of preferred habitat in which to live and travel to access other members of their species for reproduction. This issue is of particular concern for wide-ranging, slowly reproducing species such as some large carnivores. It also poses a problem for smaller animals that do not disperse far and therefore can be isolated from other portions of their population. What was already a challenging landscape has become infinitely more difficult because of

Interstate 90 passes over Snoqualmie Pass at 3019 ft. (920 m) in the Cascades. Eight lanes of high-speed traffic, lift-access ski resorts, a large reservoir, an extensive forest road system with industrial logging operations, and housing subdivisions on private lands have created extensive habitat fragmentation and major obstacles to north–south passage of wildlife along the Cascade Crest.

human-made features such as highways, large blocks of human development in biologically critical locations, and agricultural fields that were once key wildlands. Species whose travel has been documented to be deterred by roads range from moles to grizzly bears.

Hunting, trapping, and poisoning. During the 19th and first half of the 20th centuries, systematic carnivore eradication programs included trapping and poisoning, with hunters paid bounties by the government. Commercial trapping of fur-bearing animals such as beavers and fishers was widespread. These activities had a devastating impact on several species in our region, resulting in the population crash or complete disappearance of many carnivores.

Climate change. We are just beginning to understand the long-term impacts of global climate change on wildlife populations. As with the wolverine, the southern limit of boreal species will likely shift north as temperatures increase, but many effects may not be so linear. Researchers have linked recent massive infestations of

The Tieton River Canyon, in the East Cascades ecoregion in Washington, is home to a diverse array of flora and wildlife and is important winter range for elk, deer, and bighorn sheep. Conservation of the canyon's biodiversity and restoration of healthy forest and river habitats has been the focus of the Nature Conservancy and a number of other organizations.

bark beetles in the arid conifer forests of the eastern parts of our region with increased stress from higher summer temperatures, drier conditions, and warmer winters (Williams and Liebhold 2002). Exactly how this landscape-level impact will affect wildlife remains to be seen, but we can assume that the climate change factor will place numerous wildlife populations under significant, if not catastrophic, stress.

Conservation and restoration. Humans play a critical role in threats to and survival of our region's wildlife biodiversity. However, we can also effect positive change to support the survival and persistence of wildlife. Conservation groups advocate for protection of wildlife habitat, and informed citizens play an important part in conservation. Citizen scientists and amateur naturalists have improved our understanding of wildlife biology on many fronts, with reports of many sightings of rare animals. Helping young people develop a personal connection to the natural world through direct experiences in nature allows them to appreciate and understand the wild heritage of our region and ensures that future generations continue to protect these resources.

PART ONE
Wildlife Tracking Basics

Mountain goats feed on an alpine mountainside. West Cascades, Washington.

"One of the most remarkable attributes of humans is our ability to recognize relationships between disparate phenomena or events—to discover order, pattern, symmetry, predictability, and beauty in an apparently disordered world."

T. A. Vaughan et al., *Mammology*

On a snowy day in the Washington Cascades, during a class on wildlife tracking, my students and I came across the trail of a small group of hoofed mammals. I assumed at a quick glance that these were tracks of black-tailed deer. One of my students followed the trail of one animal off the old railroad grade where we were walking. The tracks led across a rocky ledge, which the animal crossed. We thought this path was a bit odd for a deer. I closely inspected the tracks and noticed that the cleaves' tips were more blunt than those of a deer. The cleaves were spread consistently and the bottom of the track was flat or slightly rounded. These were the tracks of a mountain goat. Following the trail, we found where the animal leaped from one rock ledge to another before scampering over rock slabs lightly covered with snow, a place where no deer would choose to travel.

We knew that the tracks were fresh, as snow had started falling only a few hours earlier. From the various track sizes and the number of trails, we estimated that several adult and two young goats had traveled here. I had seen a large cliff along the ridge above us from a bridge over a ravine we had just crossed. Knowing that mountain goats prefer cliffs and steep terrain, we used binoculars to scan the cliffs and quickly located the small band of goats traveling along the ridge.

Wildlife tracking involves finding, identifying, interpreting, and following the tracks and signs left behind by wildlife. Tracking is a valuable tool that can help you understand the often mysterious lives of the animals with whom we share this corner of the world. The ability to track requires physical observation skills, keen analytical skills, and an openness to intuitive insights. Studying tracks and signs can help you identify the maker, tell you about its behavior, and lead you to the creature itself in the wild.

Solving the Mystery

To identify and interpret an animal's story from the tracks and signs it leaves behind, you must use both deductive reasoning and creative, imaginative thought. Often many details of a track have disappeared over time, or conditions have allowed only part of a track to register. As a tracker, you must be able to make reasonable deductions about what you observe and create a hypothesis about how a clue came to be there.

After you create a hypothesis, natural corollaries follow, which you can test for validity. The route chosen by the animals we were tracking led me to hypothesize that these were indeed mountain goats rather than deer. I reasoned that further study of their trail would show more evidence of route selection indicative of mountain goat rather than deer, and that the tracks themselves would show details that also resemble those of mountain goats rather than deer. Both of these turned out to be true, so I hypothesized that because mountain goats tend to linger in exposed and cliffy areas, and because their tracks seemed to be heading in the direction of just such an area, we might actually be able to see the animals there. This turned out to be true as well. Getting final confirmation from an actual sighting of an animal is rare, however. You must be open to conflicting information and not become overly attached to an opinion that is unsupported by further study. Ask yourself questions about what you have found to spur deeper thinking, deductions, and hypotheses.

As a tracker, you must be able to create detailed images in your mind—to recall, for example, a detailed image of a known footprint to compare with what you see on the ground. For interpreting track patterns or understanding an unusual sign or disturbance created by an animal, you can imagine the animal moving to create the pattern or sign at hand. Watching animals move—your pet, a squirrel in your yard, wildlife on television, or actual wild animals—helps build your bank of stored images from which you can draw when re-creating events in your imagination as you track.

Context: The Tracking Funnel

To understand and interpret wildlife tracks and signs accurately, it helps to be aware of their ecological context. The tracking funnel is a systematic approach to finding and interpreting tracks and signs based on an ecological perspective of the landscape. For example, you are not likely to find the tracks of a pika just anywhere in the Pacific Northwest. Pikas occupy a particular habitat, and within that habitat they have particular preferences that accommodate their basic needs. Knowing this can help you determine where to look for pikas and their signs and can help you identify pika signs in the field. Analyzing tracks and signs at all levels will help you create an ecological context that tells the animal's complete story. Start at the top and work your way down to gain a richer image of what you have found.

Bioregion and ecoregion. At the top of the funnel is the broadest view of our region and the basic ecological factors that influence wildlife here. Various topographic and climatic influences create smaller ecoregions. Study the range maps and distribution

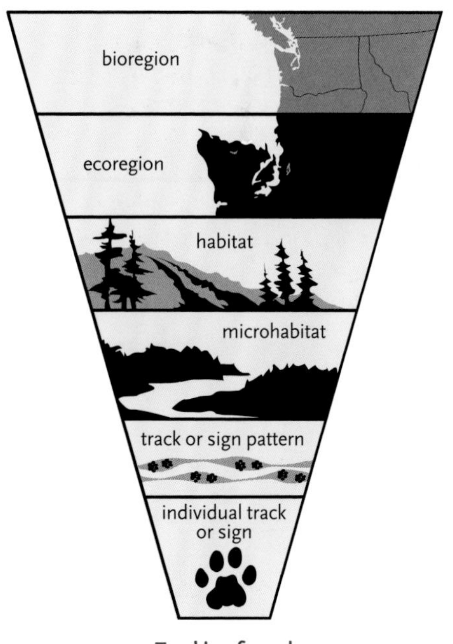

Tracking funnel
Illustration by Jenn Wolfe and David Moskowitz.

criteria in each species account, and you can create a list of what to expect in a certain area or narrow down the possibilities of what you have found.

Habitat. Within each ecoregion is a variety of habitats. For example, in the high elevations of the North Cascades, the subalpine zone is a particular habitat. Within the subalpine zone are meadows, forest edges, cliffs, and talus fields. The habitat descriptions in each species account outline the animal's basic requirements.

Microhabitat. Pay attention to the specific parts of a habitat in which the tracks or signs are located, as this can reveal important information about how an animal is using the landscape. Are you in the middle of a vast field of jumbled boulders or on the edge, where the talus gives way to a flat, grassy valley bottom?

Track and sign patterns. Consider all the tracks or signs as a whole. Do you see a pattern? Perhaps you notice that the grasses along the entire edge of the talus field are shorter than the grasses farther out in the valley.

Individual tracks and signs. After sifting through each level, return to the initial tracks or signs that drew your attention. Explore them in detail using the criteria laid out for individual track analysis and sign types.

INDIVIDUAL TRACK IDENTIFICATION

You can use clear tracks to identify the maker's family or species. Sometimes, when tracks are used in conjunction with your basic knowledge about the biology of a family or species, you can identify the age class, sex, and occasionally an individual animal through the tracks it leaves. Learning to identify the parts of animals' feet and tracks will help you make more detailed observations and develop detailed search images.

Mammal Foot Morphology

Mammals have evolved to place their feet on the ground in three basic ways, which is reflected in the structure of their feet and tracks. Plantigrade animals such as humans and bears usually place their entire foot on the ground. Digitigrade animals such as canines usually place only their toes on the ground, and unguligrade animals (hoofed animals) place only the tips of their toes and nails on the ground. Sometimes, however, a digitigrade animal places its entire foot on the ground, usually when landing with a large impact or in deep substrates. Plantigrade animals sometimes do not place their entire foot on the ground, such as a sprinting human who does not touch down with the heel. Digitigrade and unguligrade foot structures are believed to be adaptations that aid in faster movement, making these animals adept runners.

When analyzing a clear track, you should systematically look for and describe each part of the track, starting from the largest overall shape and ending with the smallest details you can find.

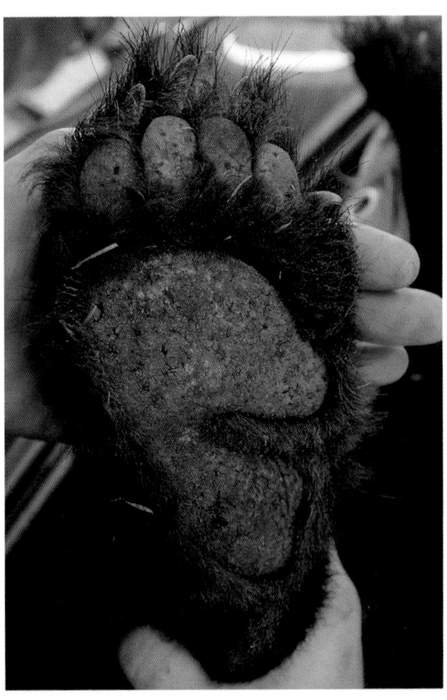

Plantigrade: Right hind foot of a female black bear. The palm and heel form one fused pad.

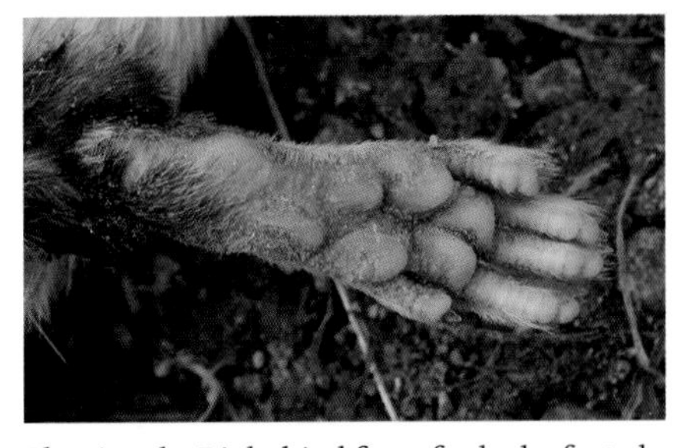

Plantigrade: Right hind foot of a dusky-footed woodrat. The palm comprises four distinct metacarpal pads, which may register as a single entity. The heel is mainly unfurred, and two distinct proximal pads appear below the metacarpal pads.

Digitigrade: Front feet of a juvenile coyote. The palm is one fused pad, but careful inspection reveals three distinct subpads within it. The proximal pad, high on the foot, rarely registers.

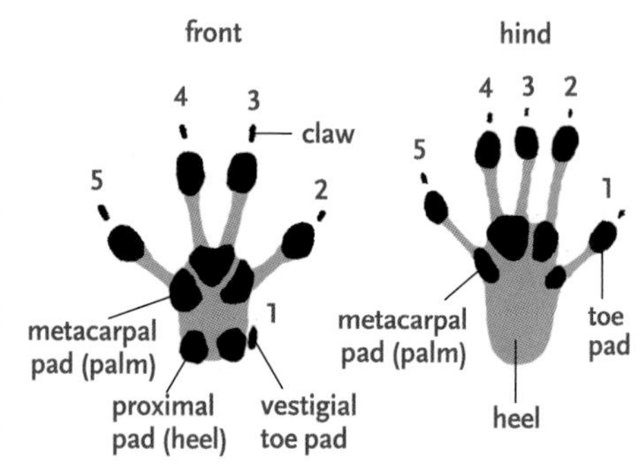

Plantigrade rodent tracks
(left tracks with digital reference numbers)

Unguligrade: Left front foot of a mature male Rocky Mountain elk.

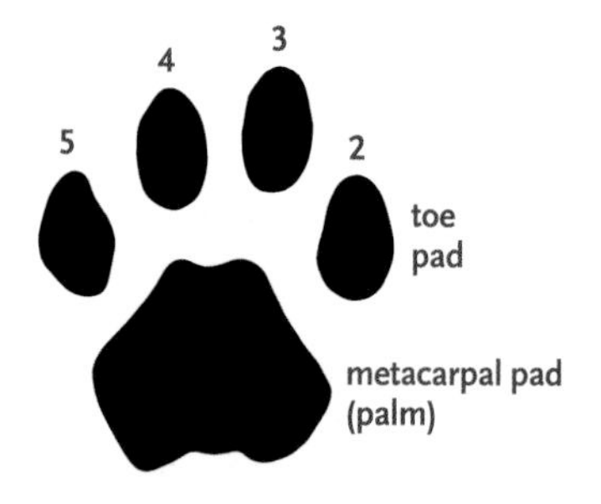

Digitigrade feline track
(left front with digital reference numbers; toe 1 rarely registers)

PLANTIGRADE AND DIGITIGRADE TRACKS

Overall shape. If you draw a line around the track's outside edge, what sort of shape does it create?

Toes. Notice the number, shape, and orientation of each toe. Digits are numbered 1 to 5 on each foot, with toe 1 on the inside of the foot and toe 5 on the outside. For most mammals, toe 1 is the smallest and registers farthest back in tracks. Animals that have only four toes on a foot have lost toe 1 over evolutionary time. For these species, the first toe on the inside is labeled as toe 2.

Palm or metacarpal pads. Notice the shape and orientation of the pad or pads between the toes. Does the palm comprise one smooth pad or several smaller, partially fused or entirely separate pads? How big is the palm relative to the rest of the track?

Heel or proximal pads. In addition to the palm pads, on some mammal feet the heel also has unfurred pads, referred to here as proximal pads. Other species' heels are completely furred. For plantigrade animals, heels often register and associated pads may be fused with the palm pad. For digitigrade animals, heels rarely register. How many proximal pads does the animal have? What are their shape and relative size? Do they create a specific shape in the heel?

Negative space. Negative space refers to the space around and between the toes and other pads. Sometimes it creates distinct and recognizable shapes, such as the distinctive star shape created by the fur between the toes and palm pad in the coyote foot. Pay attention to negative space to notice the relationship between toes and the palm.

Claws. Most mammals have a claw on each digit. Claws vary in shape and function. If claws show in the track, note their quality. Are they large and blunt or short and sharp? Do they appear deeper than the rest of the track or as small pricks in the soil? How do the claws on the front foot compare with those on the hind foot?

Webbing. Webbing appears between some animals' toes. Webbing that connects the entire length of the toes is called distal webbing. If the webbing connects the toes about midway along their length, it is referred to as mesial, and if only a small amount of webbing is present at the toes' base, it is called proximal. Webbing often does not appear consistently in tracks, which can be misleading if you expect it to be an important identifying feature.

UNGULIGRADE TRACKS

Cleaves. Digits on an unguligrade mammal are called cleaves. For all of our native hoofed mammals, toes 3 and 4 register consistently in tracks. Cleaves spread more in deeper substrates, at faster speeds, or for heavier animals. The cleaves' overall shape, size, and spread are key features in identifying the track maker.

Dewclaws. Toes 2 and 5, the dewclaws, are higher on the foot and register only in deeper substrates or for running animals. You can usually see differences between the dewclaws on front and hind feet (page 274). The pronghorn does not have dewclaws. Caribou dewclaws on the front feet usually register.

Hoof wall. The cleave's outer rim is made of a hard material called the hoof wall or unguis (the same material as a human fingernail). The hoof wall registers distinctly in clear tracks.

Subunguis. The material inside the hoof wall is softer and similar to the material under the human fingernail. The amount of subunguis varies among species and between front and hind feet, and registers as a raised area in the track.

Pad. Synonymous with a human fingertip, the pad is a convex structure that starts at the back of track and takes up more or less space within the track for different species.

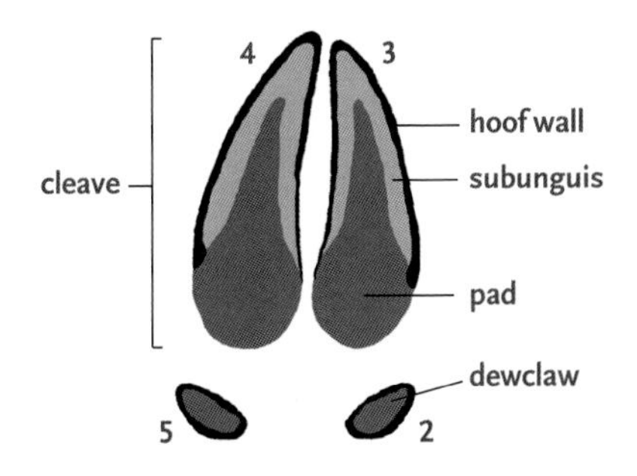

Unguligrade deer track
(left front with digital reference numbers)

PATTERNS IN TRACK SHAPES

An animal's body shape and lifestyle are often clearly reflected in the shape and features of its feet. You can often deduce a great deal about a track's maker without knowing which animal made it by analyzing the track structure. What purposes are served by the shape and structure of this track? You can use several general guidelines for correlating tracks with their maker:

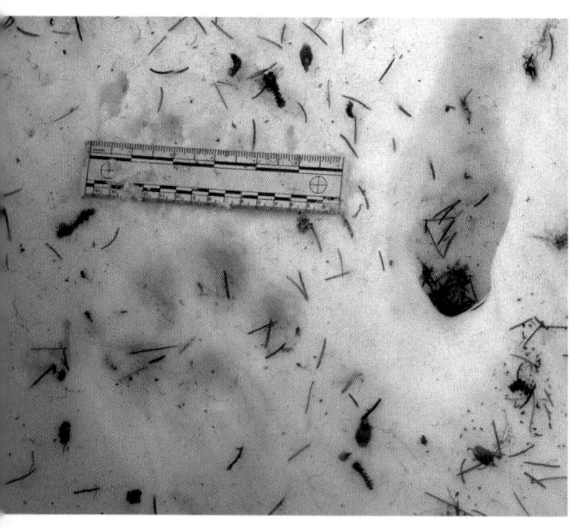

A Canada lynx (left) walked easily on top of the snow where a coyote (right) sunk in deeply. Okanagan, Washington.

Body shape. The relative size of an animal's front and hind feet often reflects how weight is distributed in its body. A canine carries most of its weight in its head and chest and has larger front feet. Many mustelids, such as weasels and fishers, carry an even distribution of weight throughout their body and have similarly sized front and hind feet.

Digging animals. Digging specialists such as badgers and pocket gophers often have long, relatively flat claws on their front feet, with smaller claws on their hind feet. Many travel with a pigeon-toed gait.

Semiaquatic animals. Animals such as otters and beavers have larger hind feet that are either webbed or have other adaptations to aid with propulsion in water.

Adaptations to snow. The feet of animals well adapted to snowy environments, such as snowshoe hares, Canada lynx, and American marten, are often exceptionally large for their body size, and fur may cover the entire bottom of the foot in winter.

ANIMAL LOCOMOTION AND TRACK PATTERNS

Animal locomotion involves several basic patterns of movement, called gaits. Each gait produces distinctive track patterns. Very different gaits can, however, produce track patterns that may appear similar at first glance. Understanding animal locomotion and the resulting track patterns may seem difficult but is ultimately a rewarding endeavor, as it reveals secrets about the track maker and its behavior.

The format I use to classify gaits and track patterns mainly follows the work of Bang and Dalstrom (2001), Halfpenny (1986), and Elbroch (2003), who base their systems on the work of Howell (1944), Hildebrand (1989), and Muybridge (1957). I have modified their system to make it more accessible for a wider audience.

Quadruped Locomotion

As wildlife observers and trackers, we can classify the motion of mammals into several gaits: walks, trots, lopes and gallops, and bounds. Gaits are defined by the order and rhythm of foot movement, not the animal's speed. Each gait can be employed at various speeds. All mammals, from a tiny shrew to a bull elk, use some or all of these basic gaits or variations of them. Some

Rocky Mountain bull elk walking

mammals use one gait almost exclusively, while others use various gaits, depending on their behavior.

Walk. Walking is an energy-efficient, though relatively slow, method of locomotion and is the most typical gait for many animals, including hoofed mammals and felines. In a walk, each foot moves independently of the others. If each footfall made a beat on a drum, the cycle would have four distinct beats. At least one foot, and often three feet, are on the ground at all times. Animals walk in a variety of ways depending on their body shape and speed. A coyote, for instance, often brings its hind foot well beyond where its same-side front foot landed, before setting it down, while a deer usually places its hind foot exactly where the front had been.

Trot. In a trot, feet on diagonally opposite sides of the body move in tandem; for instance, the left front and the right hind feet followed by the right front and left hind feet. The rhythm of this gait has two beats, as diagonal feet lift off and land at precisely the same moment. In a trot, which is usually faster than a walk, an airborne phase often occurs between each lift of one pair of feet and the landing of the next. Animals can use trot variations, such as the side trot used by all our wild canines, which accommodate different body structures.

A pace is similar to a trot but is not commonly used by any of our region's wildlife. A pace is generally faster than a walk; the front and hind feet on the same side of the body move in tandem (as opposed to diagonal feet in a trot). As with a trot, a brief airborne phase often occurs between the feet pushing off on one side of the body and landing on the other.

Gallop and lope. Gallops and lopes (a subset of gallops) involve a type of body motion that differs entirely from that of walking and trotting. These gaits require more energy than walks. A gallop is usually the fastest form

Bighorn ewes loping

of movement for a mammal, but some animals, such as several members of the Mustelidae family, use a loping gait as a primary method of locomotion.

In gallops and lopes, the animal moves each leg independently but rapidly. At some point in the gait, a "gathered suspension" occurs when all four feet are off the ground after the front feet push off and before the hind feet land. If this is the only airborne phase, the gait is a lope. For a gallop, which is faster, a second airborne phase occurs between the push off of the last hind foot and the landing of the first front foot, and this is called "extended suspension," as the animal's front feet are extended out in front of it and its hind feet are extended behind it. Occasionally only an extended suspension occurs.

Bound. Bounds are more closely related to gallops and lopes than to walks or trots. The animal pushes off strongly with both hind feet in unison, or nearly so. An extended suspension is common before the front feet land, and for faster gaits a gathered suspension may also occur.

Several variations of bounds are used; for example, the hind feet can land behind the front feet, exactly where the front feet were located, to the side of the front feet, or well beyond where the front feet landed. The latter variation is the most common form of movement for many rodents and all of our lagomorphs and squirrels. Animals such as canines and felines may use bounds when chasing prey, traveling through deep snow, or leaping over obstacles.

A pronk is a variation of the bound used extensively by mule deer and black-tailed deer. The animal pushes off with all four feet at once, entering one long airborne phase, and then lands on all four feet at once. This gait allows for rapid changes of direction and quick acceleration but requires a tremendous amount of energy. Pronking is used by an animal fleeing from danger.

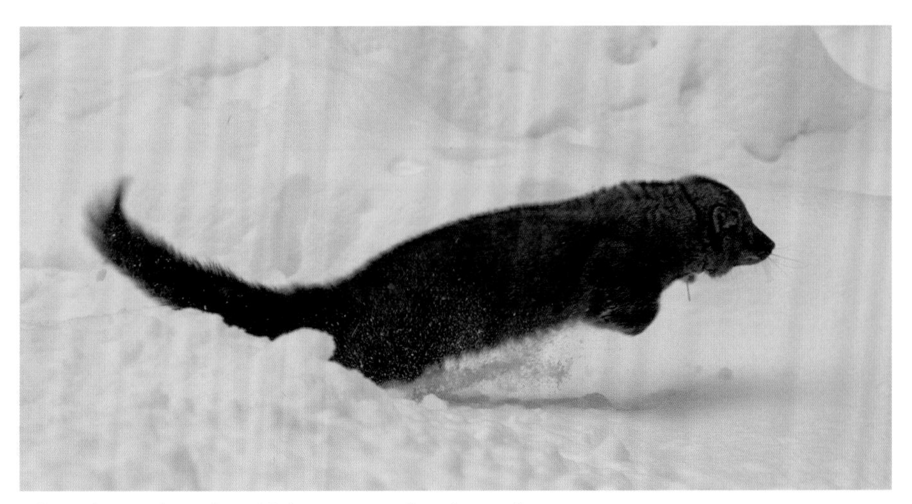

A bounding, radio-collared fisher is reintroduced into Olympic National Park, pictured here moving into extended suspension.

Track Patterns

When an animal travels using a particular gait, it leaves a distinctive track pattern. The pattern left by any galloping animal, large or small, shares characteristics with the pattern left by any other galloping animal. Once you learn these patterns, you can recognize the animal's gait from the tracks it leaves behind.

Each species also uses preferred methods of locomotion for particular activities. For instance, a coyote comfortably traveling along a trail trots, while a deer walks and an otter lopes. Some animals, such as rabbits, use one gait (bound) almost exclusively. Others, such as coyotes, use various gaits on a regular basis.

Once you recognize basic track patterns and learn to identify the maker's gait, you can often deduce the animal's behavior at the time the track was made. A startled deer will change direction and gallop away from danger. A coyote will slow from a trot to a walk to inspect an interesting scent. Sometimes the same gait is employed with more energy, such as a red squirrel traveling across the forest floor in a leisurely bound and becoming startled, increasing the stride of each bound to reach the safety of a tree.

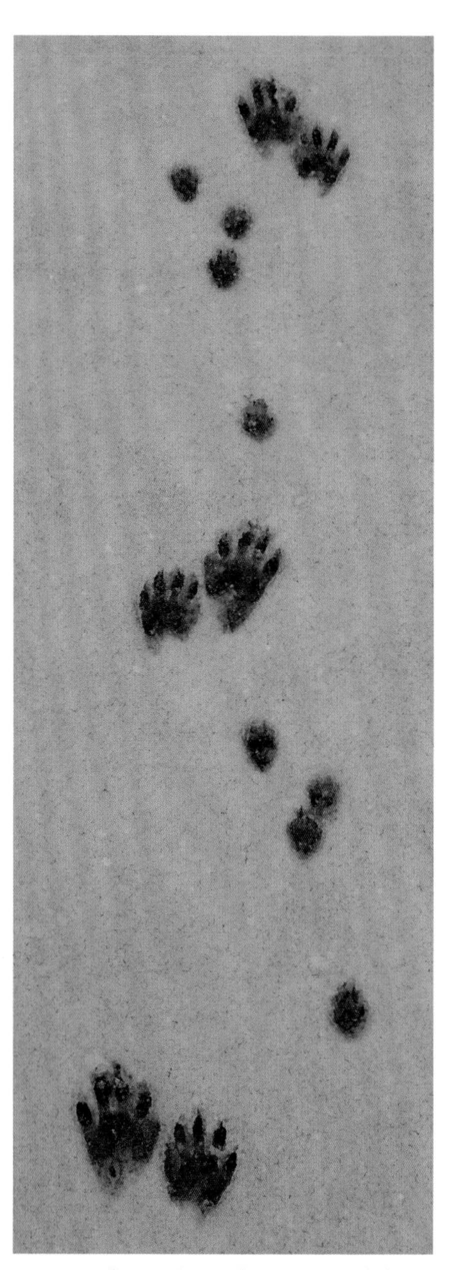

Larger tracks in groups of two were made by a walking raccoon. Smaller tracks in groups of four were made by a galloping spotted skunk. Northwest Coast, Washington.

SPECIES IDENTIFICATION

Species accounts are accompanied by illustrations of common track patterns used by each animal. Compare the pattern on the ground with the illustrated patterns. Find all the tracks on the ground, as not all feet may register equally clearly. Using track patterns (rather than, or in conjunction with, individual tracks) to identify a species is useful when individual tracks show little or no detail. Such conditions occur in dry sand, loose snow, leaf litter, or on hard surfaces where little more than scuff marks register.

GAIT IDENTIFICATION

In species accounts, each track pattern is labeled with the associated gait. Use the guidelines presented here to identify what type of movement created the track pattern you have found. The simplest way to start classifying track patterns is by distinguishing visual patterns of two tracks (created by walks and trots) and patterns of four tracks (created by lopes, gallops, and bounds).

Walks and trots: patterns of two. If the trail's primary pattern is defined by sets of two tracks, one front and one hind foot, usually from the same side of the body, it is a walk or a trot. Several variations exist.

Understep. The hind foot registers before the front foot on the same side of the body.
Overstep. The hind foot passes the front on the same side.
Direct register. The hind track is directly on top of the front track.
Indirect register. The hind track partly covers the front track.

Many animals direct register when walking or trotting, which appears as a single

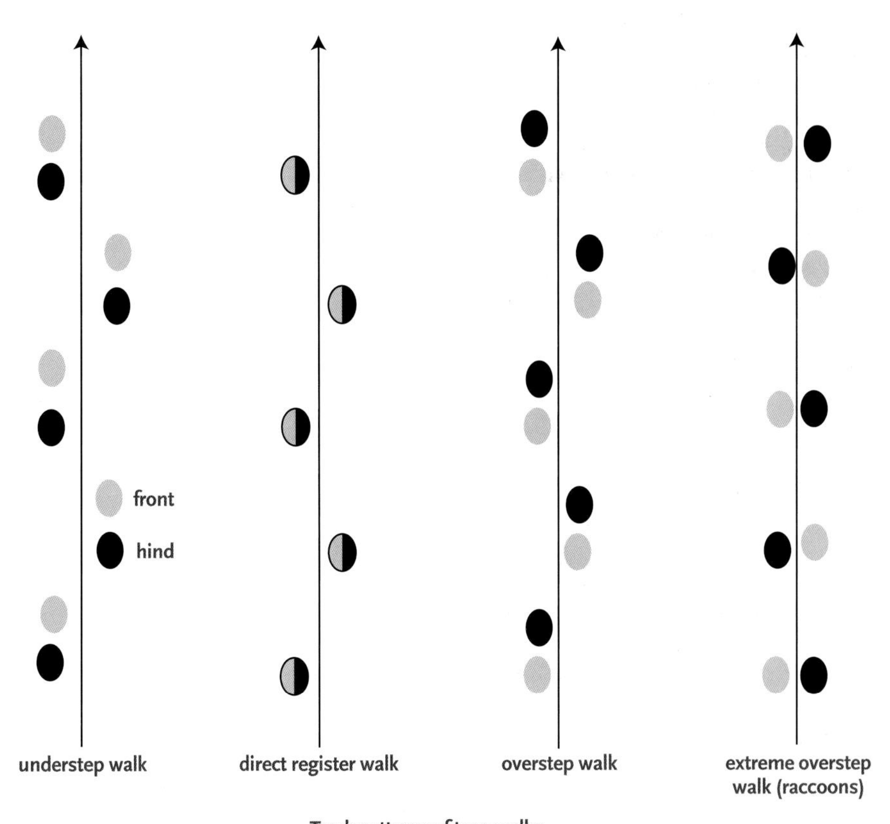

Track patterns of two: walks

track, and this can be confusing. The overall pattern appears to be an alternating pattern of one track, although it represents the footfalls of two feet (the front and hind foot from one side of the body). Look closely and you may find that some of the front track is still visible (many animals' front feet are larger than their hind feet).

After you have identified a pattern of two tracks, you can differentiate between walks and trots. Walks generally have a shorter stride and a wider trail width than trots. In a walk, the animal is never airborne, so its stride cannot be much longer than the length of its body. A trot often has an airborne phase, and the stride can be twice the length of the animal's body. The trail width is wider in walks because as an animal moves slower, it compensates by widening its stance to improve its balance. With a walk, you can usually look down a trail and distinguish the left and right tracks from each side of the body. As an animal increases its speed to a trot, its feet often fall more in line with one another, and tracks may appear almost in a straight line as you

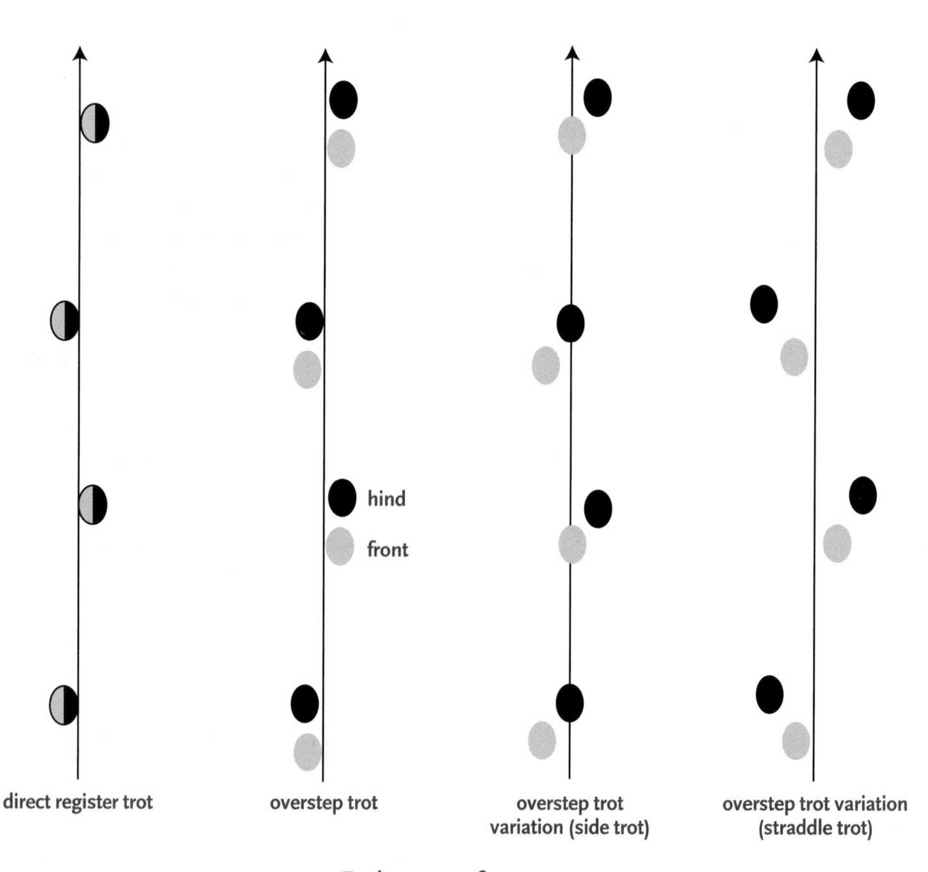

Track patterns of two: trots

look down the trail. Side trots, among others, are an exception, however, as the trail width may increase slightly as the animal switches from a walk to a trot.

In some cases, the hind foot oversteps so far ahead that it lands close to the front foot on the opposite side of the body. If the hind footprint from one side is next to the front on the other side (an extreme overstep), the overall pattern will still be that of two tracks. If the hind track is evenly spaced between the two fronts, you'll probably see no distinguishable pattern of either two or four tracks at first. To distinguish these track patterns from a pattern left by a lope, measure the distance between each front foot. If the measurement is consistent, it is likely a walk or trot. If the measurement is inconsistent (short, long, short, long), it is probably a lope. (In this case, the longer measurement is caused by the airborne phase in a lope.)

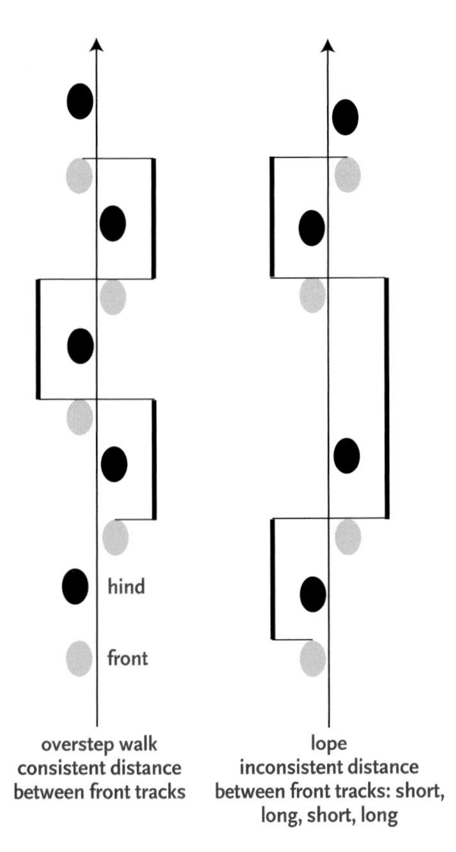

Distinguishing between patterns of two and four

Lopes, gallops, and bounds: Patterns of four. If the trail's primary pattern is made of four tracks (each of the animal's four feet), the animal was loping, galloping, or bounding. This pattern can often be clear and easily identifiable. For confusing trails with unclear patterns, use the technique already described to distinguish lopes from walks or trots.

In lopes and gallops, each foot lands and pushes off independently; whereas in bounds the hind feet push off and land in unison, or nearly so. A track pattern in which the two hind feet are placed next to each other (perpendicular to the direction of travel) is a bound. If the hind feet are offset, it is a lope or a gallop. In a bound, the front feet may be angled or paired.

Sometimes it is impossible to distinguish lopes from full gallops in a track pattern. If both hind feet are well beyond the front feet, the animal was likely galloping. If one hind foot registers before one of the fronts, it was probably loping.

As with walks and trots, an animal can sometimes place its hind foot directly over the track of its front foot, thereby leaving what appears to be a pattern of three or even two (if it places both hinds into the tracks of both fronts). Look closely to see if you can detect the front foot track beneath the hind foot. For bounds in which both hinds land where the front feet had been, look down the trail. A walk or trot usually leaves a pattern with a zigzag appearance (with the alternating two left feet followed by two right feet), while a bound will not appear so (as each set represents both lefts and rights). Look also at the stride: a bounding gait often has an airborne phase, and the stride can be considerably longer than the length of the animal's body.

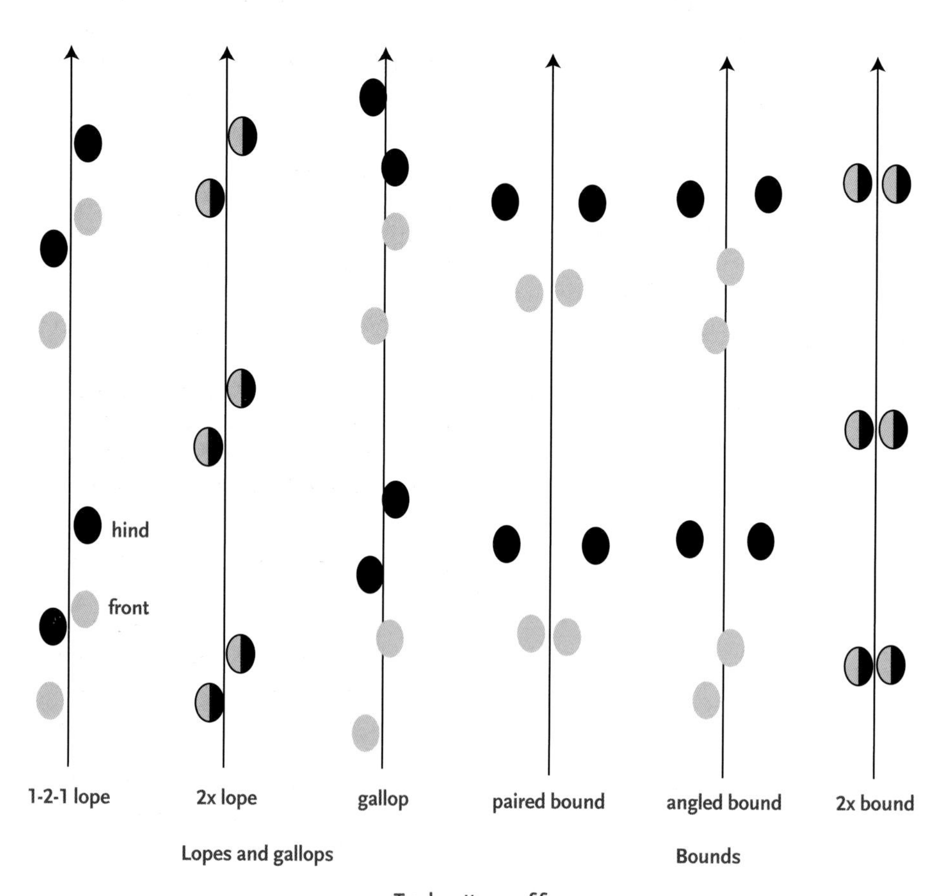

Track patterns of four

SIGN TRACKING

The lack of consistent quality substrate for holding tracks means that in many parts of the Pacific Northwest, signs rather than tracks are the most abundant evidence of wildlife. Signs can tell you more about an animal's behavior, diet, and ecology than tracks alone. Animals need food and water, shelter, refuge from predators, and intraspecies communication and access for reproduction. They leave signs associated with each of these requirements, which provide a good classification scheme for defining the types of signs to look for on the landscape. The specific ways that these signs manifest are discussed in each species account.

Feeding Signs

If you are interested in a particular animal, read about its diet and feeding habits and then look for these food sources, for feeding signs, and for the animal itself.

> *After finding a bear scat filled with salmonberries, I went searching for the largest patch of fruiting salmonberries in the vicinity, which I found along a river floodplain. There I discovered rough trails of broken branches and the stems of berries on which the bear had fed. I came back the next morning around dawn and watched the bear feeding there.*

Herbivory. Herbivores and omnivores consume plants and fungi in various ways. The sharp incisors of rodents and lagomorphs leave feeding signs characterized by clean cuts and obvious incisor grooves. None of our native hoofed mammals have upper incisors, so vegetation on which they feed has a ragged appearance from being ripped. Pay attention to the way in which seeds and cones have been eaten—were they torn or meticulously gnawed open?

Predation. Carnivores leave distinct signatures in prey remains. The prey species, the carcass location, how the carcass has been eaten, and accessory clues such as scats can provide worthwhile clues as to who fed there. For bird remains, feathers that have been sheared off are a sure sign of a mammalian predator. Both mammals and birds can pluck out their prey's feathers. Sometimes the imprint of a predatory bird's beak can be found as a paired crimp mark on a quill.

Scats. Along with examining direct signs, such as animal remains or clipped vegetation, you can inspect scats to determine what an animal has been eating. Be aware, however, that scats of some wild animals, especially raccoons and some other omnivores and carnivores, can contain parasites that can infect people. Use basic common sense to observe, measure, and analyze scat in a safe manner. In the field, you can use sticks to dissect scat to determine its contents, or for more careful analysis, use rubber gloves. When working with dry scat, avoid inhaling any dust associated with the specimen.

Shelter and Refuge

Animals need safe places to rest and sleep. Protection from the elements and safety from predators are key features of a rest site. Common shelters include underground burrows, lays in brush or at the bases of trees with low-hanging branches, and natural cavities in trees or cliffs, rocky ledges, and overhangs.

Some animals create elaborate shelters, such as the intricate burrows of kangaroo rats, which are designed to protect it from predation, retain moisture, and help it stay cool in hot, arid environments. Other animals put little or no work into their beds, such as deer that select and curl up in any spot where they can detect approaching predators. In winter, they may choose a spot with a southern exposure to maximize external warming; in summer, a north-facing slope in shade may be an ideal place to stay cool.

When you find a shelter site, ask yourself how the location and structure work to provide safety and protection from the elements.

Trails and Travel Routes

To access the resources it needs, such as bedding and feeding locations, an animal must travel. An animal may wear a trail by repetitive travel over the same terrain to reach these resources. Some trails are well worn with distinct access points on both ends, while others fade in and out of the landscape. Animals that travel widely over large home ranges may not form trails, but you can still predict travel patterns by understanding their behaviors and habitat requirements.

Trails and travel routes can be used by multiple species. When accessory clues such as tracks or scats are unavailable, identifying the trail maker can be difficult.

Trails radiate from an Ord's kangaroo rat burrow at the base of a fallen juniper log. Columbia Plateau, Oregon.

Radiating trails. These have a nexus, or starting point—often a burrow entrance but sometimes a preferred feeding area or water hole—with a network of trails fanning out from it.

Landscape-oriented trails. These are formed as animals make use of easy or safe features or adapt their movement as they encounter geographic obstacles. For example, an easy passageway across a ravine acts as a natural funnel. As the trail approaches the ravine bottom, it becomes better defined because more animals have walked in that specific spot.

Foraging trails. These occur in areas that are heavily foraged by feeding animals. Some car-

In the Selkirk Mountains, woodland caribou trails traverse forested mountainsides linking wet subalpine meadows, important feeding areas in summer. Canadian Rockies, British Columbia.

nivores hunt opportunistically, traveling long distances in hopes of flushing or catching the scent of a prey species. Coyotes often use dirt roads and human trails for this purpose. Trails often form along the edges of a wetland, a productive place to forage and hunt.

Overland trails. These can be created by semiaquatic species, such as beavers and muskrats, that move from one body of water to another. Look for these trails at the closest point between two bodies of water.

A beaver crosses over its dam to move from one body of water to the next. Puget Trough, Washington.

Intraspecies Communication: Scent and Visual Marking

These signs give us a glimpse into the mysterious, complex, and often poorly understood world of social interactions within a species. Many animals communicate with one another through scent and/or visual posts—scratches, bite marks, and rubbings on trees; scents rubbed onto rocks; dirt mounds in prominent locations; wallowing spots in dust or mud; or deliberately placed

Scarred lodgepole pine trees have been scratched, bitten, and rubbed against by numerous grizzly bears. Their prominent location at the center of a clearing where several game trails converge makes these trees an ideal place for various bears to announce their presence and check that of others in the area. Canadian Rockies, Montana.

scats or urine. These signs are used to communicate messages, including territorial boundaries, warnings to other members of the species to stay away, breeding conditions, and the presence or absence of food. Many animals seek out particular landscape features on which to create their sign posts, and once you develop a search image, you can often locate these signs in the same type of habitat in other places.

A coyote's trail veers where the animal deposited urine on a branch before continuing on its way.

MEASURING TRACKS, TRACK PATTERNS, AND SIGNS

Each species account offers measurement ranges for individual tracks, common track patterns, and common signs, such as scat or burrow diameters. A guide inside the back cover shows how to take these measurements.

Ranges are more helpful than simple averages because species include a range of different-sized individuals, and track pattern measurements can vary with the track maker's size as well as with its speed and the substrate. Most measurement ranges are based on a minimum of ten discrete track sets (five for birds). The track measurement graphs at the back of the book give sample sizes for various species for comparison.

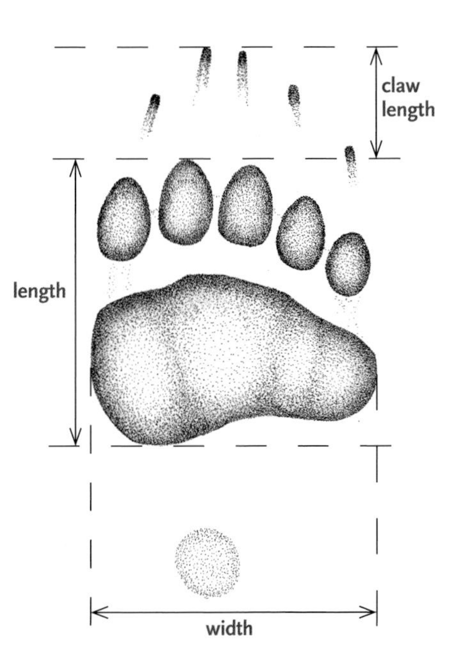

Individual Track Measurements

To maintain consistency in track measurements, measure only the weight-bearing surface of the foot, or the minimum outline. Length is measured from the front to the back of the track and includes the heel if it commonly registers for a species (as with the front feet of squirrels). For species such as bears that may or may not register a heel print, ranges include measurements with and without the heel. (The proximal pad on the front feet of members of the order Carnivora are not included in length measurements.) Width is measured perpendicular to the length, across the track's widest portion. Length and width measurements of mammals do not include claws (unless specifically noted), because many species' claws do not reliably register. Length measurements for birds include claw length because distinguishing the end of the toe from the beginning of the claw is often impossible. A claw length measurement is provided for some mammal species if this measurement is distinctive, and it is measured from the tip of the toe to the end of the claw.

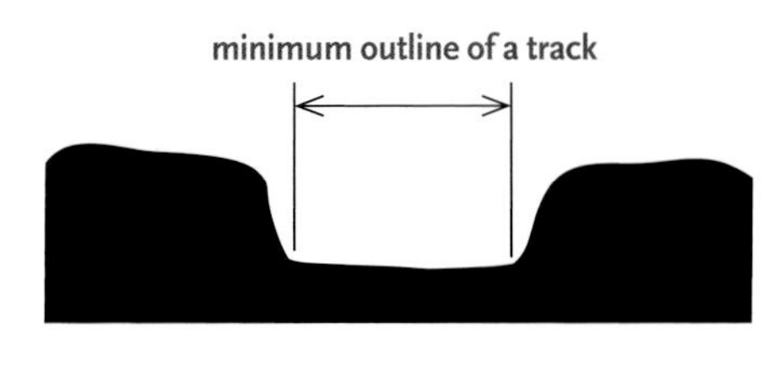

Track Pattern Measurements

Stride is measured from a place on one foot (such as the center of the left front) to the same place on the same foot the next time it appears in the trail. Trail width is the distance from the outside of the leftmost track to the outside of the rightmost track (using each track's minimum outline). These two measurements are taken for both track patterns of two and four. Group length is measured from the heel farthest back in a set of four tracks to the front of the track farthest forward. Group length is measured only for track patterns of four.

Sign Measurements

Scat diameter is measured across the widest portion. Scat length is the total length of a tubular scat. For pellets, the length of a single pellet is measured. Burrow diameter is measured at its narrowest point.

trail width

stride

Measurements for patterns of two tracks (walks and trots)

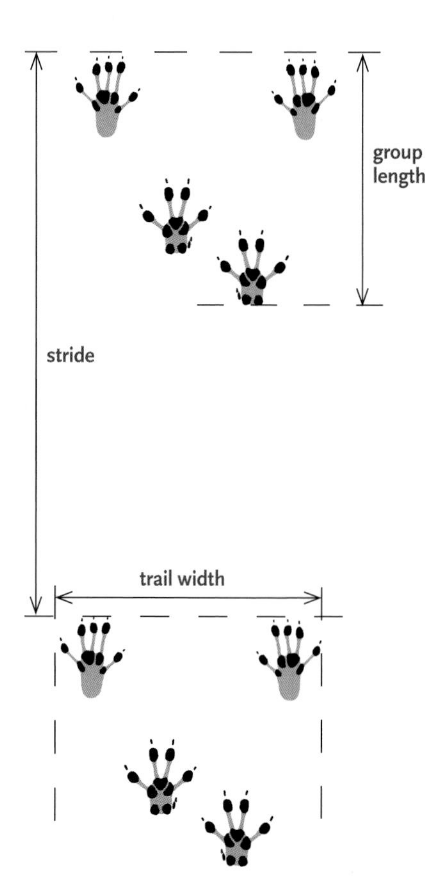

Measurements for patterns of four tracks (lopes, gallops, and bounds)

FINDING TRACKS, SIGNS, AND WILDLIFE

The wildlife tracker's first task is to hone the observation skills necessary to find clues that make identification and other tracking skills possible. This is especially important for modern people whose senses are accustomed to an urban environment that doesn't require many of the perceptual skills used in nature. For example, traffic signals and billboards are designed to get our attention. Conversely, in the natural world, many wild animals survive by their ability to blend into their surroundings and their tracks and signs are often subtle, requiring us to be more perceptive if we hope to find them. You can increase your power of observation when looking for wildlife or tracks by following a few guidelines.

Notice patterns and disturbances. Look for subtle variations from the norm. Tracks and signs can at first appear as unusual patterns on the landscape, such as repetitive lines of holes in a tree trunk made by a sapsucker or a line of vegetation bent over by a passing animal. They can also appear as variations from patterns, such as a spot on the ground that reflects light differently from its surroundings because it has been flattened by an animal's foot.

Vary your vision. Practice looking under logs or up in trees. When you are walking down the trail, notice where your eyes naturally focus and force yourself to look in new places. Our eyes are drawn to movement, edges, and contrasting or striking colors. We tend to miss details that are not distinctive in these ways.

Pay attention to details. Take a closer look. Follow a shrub's branches to their tips to look for feeding signs. Look for scratches on tree bark that might have been made by the claws of a climbing animal.

Pay attention to light. As light changes, landscape features can seem to appear or disappear. The low light of morning and evening can enhance the contrast between a track and the surrounding substrate. These times are also good for observing wildlife.

Ask yourself guiding questions. Ask questions about what you are seeing to help draw out subtleties. When examining signs in the field, I often ask myself, How did this happen and why does this look the way it does?

Develop search images. Humans have evolved to notice and recognize familiar things more quickly than unfamiliar things. Study the images and descriptions of tracks and signs in this guide to develop search images to seek out in the field.

Let go of search images. The same search images that help you recognize and identify tracks and signs can also limit your discoveries. When we construct search images, our minds want to fit what we see into categories, whether or not they belong there. Many tracks and signs are incomplete and/or subtle and don't fit a standard search image, however. Study your surroundings as an artist might—as a blend of shapes, colors, textures, shadows, and light. Looking without naming allows subtle clues to bubble up into your consciousness without the burden of familiarity.

Tracker Brian McConnell points to a false hellebore (*Veratrum viride*) that was bent over by a passing black bear. After following a string of such disturbances, we found the bear about a quarter mile beyond. East Cascades, Washington.

Be open to your instincts. People have been tracking and studying wildlife for millennia. I believe that much of our ability to do so is hard-wired, but we need to wake it up with experience. If you suspect that one spot or another might yield interesting results, check it out. In my experience, the more you do this, the more accurate your instincts become.

Take your time. Slow down and "look twice" to allow the details of what you are actually seeing to filter through any preconceived notions about what you expect.

Where to Look for Tracks and Signs

Diverse topography, climate, plant communities, and seasonal variations in parts of our region pose challenges and opportunities for finding wildlife tracks and signs. To find animal footprints in the abundant forests of the Pacific Northwest, pay attention to signs other than tracks. In arid interior regions, open substrates are ideal for finding tracks in larger areas.

Natural track traps. Track traps hold clear tracks and are likely to be used by traveling wildlife. Because of their good substrate and biological diversity, the edges of lakes, ponds, marshes, and streams are excellent spots to find tracks and signs. Other common track traps include the throw mounds of gophers, dusty game trails, and sand and mud bars along streams and rivers.

Human trails and roads. Humans have made some of the best track traps in the region, such as hiking trails and dirt roads. Timing is key on popular routes; an early start on the trail might yield the story of a host of nocturnal travelers that will be covered by daytime hikers. Old logging roads offer good opportunities for sign cutting, or picking up tracks that you can follow off the road to explore an animal's activities. Check under bridges and in dry culverts for tracks.

Snow. Trackers in the mountains and east of the Cascades benefit from snow cover that lasts for months at higher elevations. The best time to track is 24 to 72 hours after a snow that is deep enough to cover old tracks and drip marks from snow falling and melting off tree branches. In powdery snow conditions, individual track details may not register; pay attention to the track shape and pattern to identify the maker. Tracks in melting or loose snow can be distorted or enlarged.

Coastal environments. Sandy beaches, tidal mudflats, and coastal dunes offer excellent substrates for holding tracks. Coastal locations can be outstanding spots for studying the tracks and signs of shorebirds, but the tracks of many mammals and marine invertebrates can also be discovered here.

Forest environments. Although clear footprints might be scarce in the forest, other wildlife signs are plentiful. You can find signs on the trunks of trees, including claw marks from climbing, scent-marks, and incisor marks from animals feeding on the tree's cambium. Scats and feeding signs can be found under trees, in hollows, or in tree crotches. Edge habitats, ridgelines, and stream corridors are also productive places to look for signs.

Fallen logs on the forest floor are used extensively by small mammals. Rodents travel along their bases, hidden from aerial predators but attracting the attention of small predators such as weasels. Look for worn runs, scats, and feeding signs such as clipped conifer cones and gnawed seeds around logs.

Witches' broom, dense sprays of branches in conifer trees caused by mistletoe infections, create important habitat for many species, including rest spots for the American marten and nest platforms for northern flying squirrels. Check the ground around the base of a tree with a witches' broom for interesting signs.

Alpine environments. Snow in the highest elevations can provide excellent tracking conditions well into June. After the snow melts, you can find signs of animals that live a subnivean existence throughout the win-

Witches' broom are an important habitat feature for forest wildlife. This tree housed the nest of a northern flying squirrel, and the surrounding forest showed numerous small digs in duff, evidence of its foraging for truffles. East Cascades, Washington.

Trail of an American marten traveling across a snow-covered subalpine meadow. West Cascades, Oregon.

Sandbars along creeks and rivers are often excellent track traps. Here the overstep walking pattern of a grizzly bear traverses the edge of a river. Canadian Rockies, Montana.

ter. Shallow tarns (glacially carved ponds) often begin to dry up in the late summer, leaving mud that holds wildlife tracks. The fine silts produced by glaciers are an excellent substrate for holding clear tracks.

Arid environments. The deserts of eastern Oregon and along the Columbia River in Washington provide the best conditions in our area for finding tracks on a landscape level. Limited resources such as water and trees draw wildlife from long distances, and use is concentrated, making the areas excellent spots to look for tracks and signs. Desert creatures leave fascinating signs of behaviors they have developed to survive in this environment.

This juniper tree (*juniperus occidentalis*) is the only one for a couple of miles in either direction along the crest of a ridge in Oregon's Hart Mountain National Antelope Refuge. Ridgelines are common travel routes for mountain lions, which often deposit their scats at the base of prominent trees. I found six discrete mountain lion scats under this tree.

Finding Wild Animals Using Tracking

It rained hard for much of the night, but now, at first light, the sky is clear as I head down an old dirt road. I notice the tracks of a doe and her fawn. I continue down the road slowly and quietly, knowing the tracks are fresh, as they show no sign of being rained in. From reading the tracks, I can tell that the deer walked comfortably up the road for several hundred yards and then something changed. I also see coyote tracks on the road, the pattern indicating it was galloping. In the same area, the deer trails become jumbled. The road is littered with the tracks of deer and coyote, making it difficult for me to interpret what has happened. I walk around the edge of the fray until I find the tracks of the doe. Her gait now a pronk, the trail disappears into the brush on the hillside above the road. I also discover the tracks of the coyote leaving the fray, but not the tracks of the fawn.

While formulating a hypothesis of what has happened, I slowly move up the edge of the road, splitting my attention between watching the trail of the coyote, which continues up the road, and looking and listening up ahead in the direction I expect to find the coyote with its meal. Indeed, I hear scuffling in the

The edges of shallow tarns, such as this one in the North Cascades, often hold many tracks. This tarn's location at the junction of several commonly used wildlife travel routes contributes to its productivity.

brush along a small stream that crosses the road ahead of me. I crawl slowly to a vantage and peer into the direction of the sound. I see the coyote with the fawn, now dead, in a small, streamside clearing under a tree. The coyote is beginning to feed.

I watch intently for a couple of minutes until suddenly the coyote spins around and stares directly at me. After a few moments of indecision, it slinks away deeper into the brush. I move quickly to look closely at the dead fawn and then leave the area. Several hours later, I return to find the fawn gone, and drag marks on the other side of the stream indicate that the coyote had retrieved its prize and taken it to a less public location.

Early humans probably developed the ability to track animals primarily to feed themselves as hunters. Like them, you can use tracks and signs to locate wildlife in several ways.

Trailing. The most obvious, though challenging, tracking method involves following a string of tracks or signs until you catch up with the animal. Proficient trailers can infer where the animal is heading and bypass areas where tracks disappear or become time-consuming to find. You can follow a fresh trail to increase your odds of finding the actual animal. Fresh tracks on recent snow offer the best conditions for trailing.

If you suspect you are closing in on the animal, try to estimate its location and move where you can spot it without being detected—a task that requires knowledge of the animal's behavior and how it detects danger. Remember that your actions could cause the animal to flee, thus endangering it—especially in winter, when energy conservation is vital. If the trail indicates that the animal has begun to flee, consider discontinuing your trailing attempts.

Trailing can also help you learn more about wildlife biology. Try following the trail backward (backtracking). Look for changes in the track pattern that may indicate a behavior shift, and look for feeding or scent-marking signs. Backtracking can help you build search images for unfamiliar signs and does not endanger the animal.

Predicting future movements. To predict when and where an animal is likely to appear, analyze all the available signs and terrain features and review your knowledge of the animal's habits. After you have chosen the best location, return at an appropriate time, conceal yourself where you can observe the animal without being detected, get comfortable, and wait. If you're lucky, the animal will appear shortly after you do, or you may wait for hours for nothing. Either way, what you learn from the experience will hone future predictions. Remember that some animals are extremely sensitive to smells. If the wind carries your scent in the animal's direction, it may not continue toward you.

This method also works well for viewing small animals with limited home ranges, such as ground squirrels or voles. Remember that some animals experience periods of inactivity, and attempting to wait out a ground squirrel that has begun its winter dormancy would be a very long wait indeed!

Still-hunting. This method involves moving slowly, pausing, looking, and listening in areas where you expect the animal to be, especially if you come across fresh tracks and think the animal may be nearby.

KEY TO MAMMAL TRACKS

This key contains life-sized or exactly scaled drawings of the left front (left images) and left hind (right images) feet of several species. Common and scientific names for each pair and a page reference for relevant species accounts are included.

Life Size

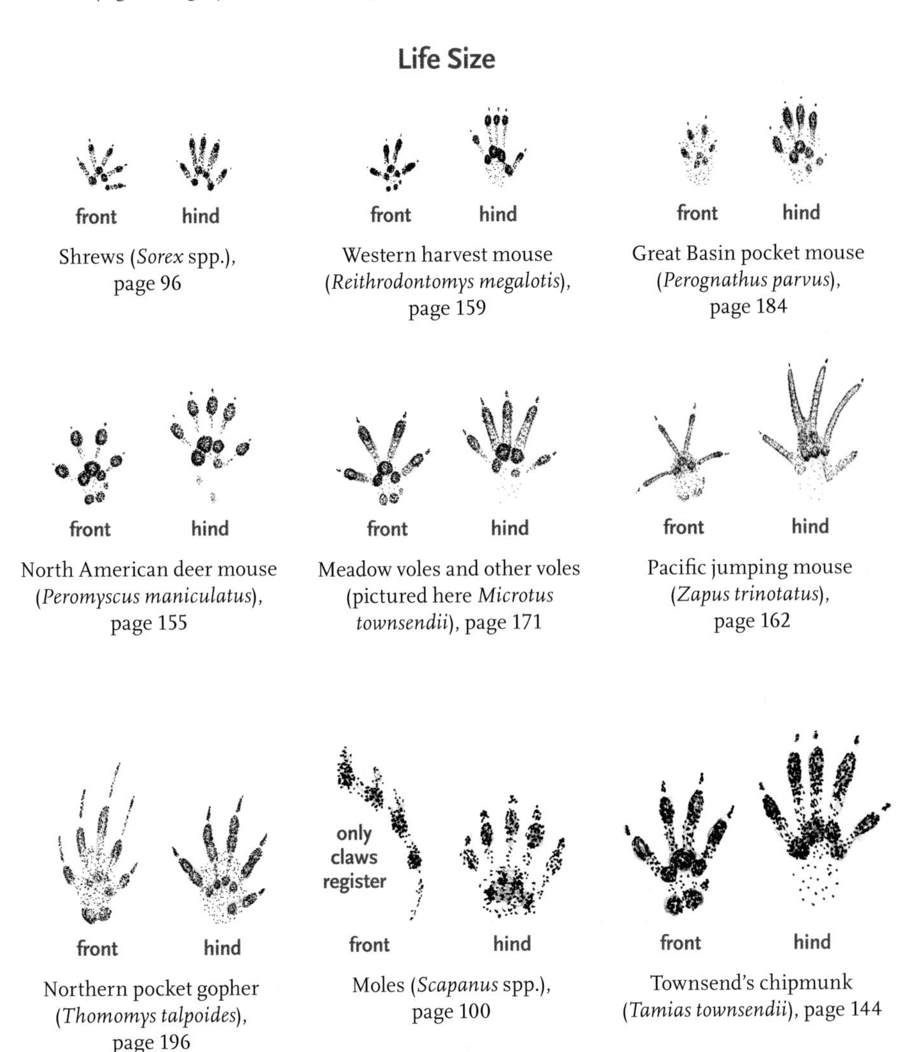

Shrews (*Sorex* spp.), page 96

Western harvest mouse (*Reithrodontomys megalotis*), page 159

Great Basin pocket mouse (*Perognathus parvus*), page 184

North American deer mouse (*Peromyscus maniculatus*), page 155

Meadow voles and other voles (pictured here *Microtus townsendii*), page 171

Pacific jumping mouse (*Zapus trinotatus*), page 162

Northern pocket gopher (*Thomomys talpoides*), page 196

Moles (*Scapanus* spp.), page 100

Townsend's chipmunk (*Tamias townsendii*), page 144

Life Size

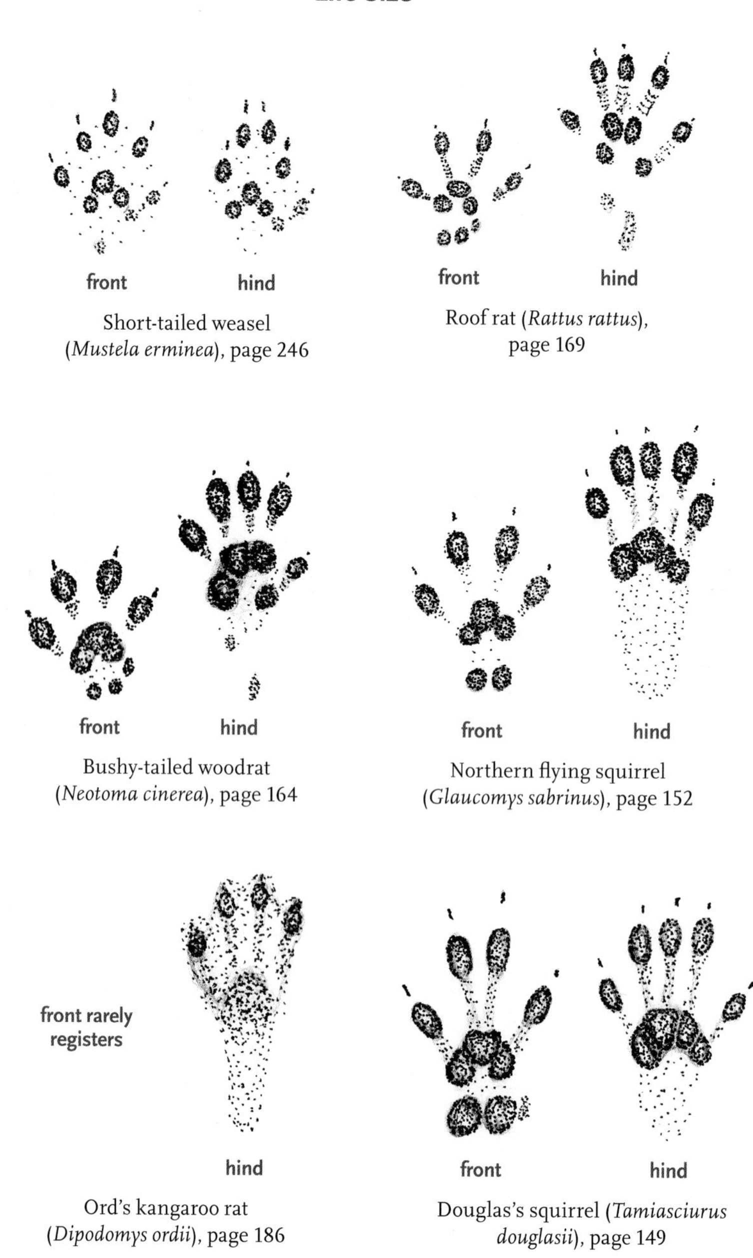

Short-tailed weasel (*Mustela erminea*), page 246

Roof rat (*Rattus rattus*), page 169

Bushy-tailed woodrat (*Neotoma cinerea*), page 164

Northern flying squirrel (*Glaucomys sabrinus*), page 152

Ord's kangaroo rat (*Dipodomys ordii*), page 186

Douglas's squirrel (*Tamiasciurus douglasii*), page 149

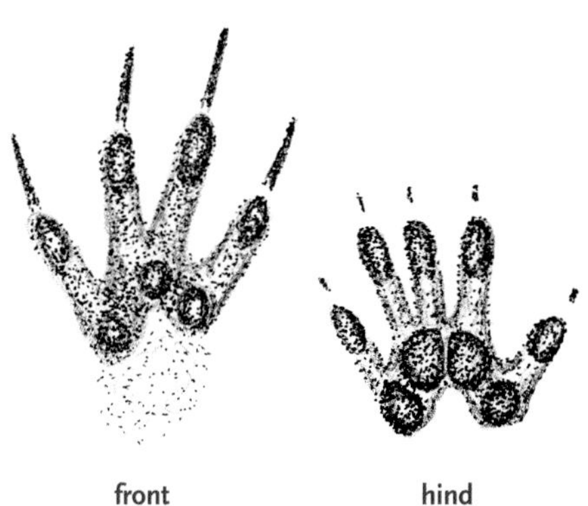

Mountain beaver (*Aplodontia rufa*), page 193

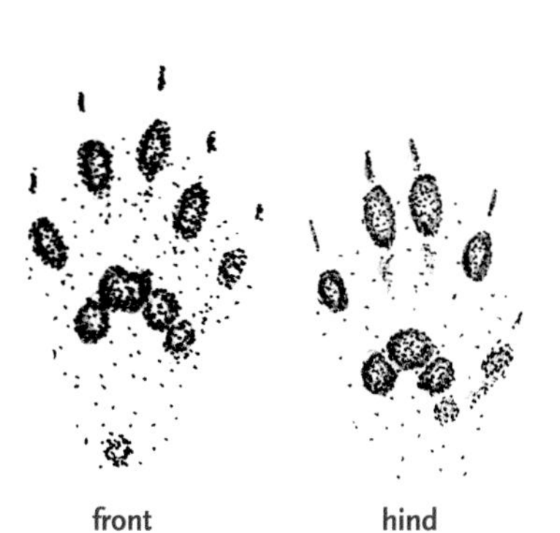

Long-tailed weasel (*Mustela frenata*), page 246

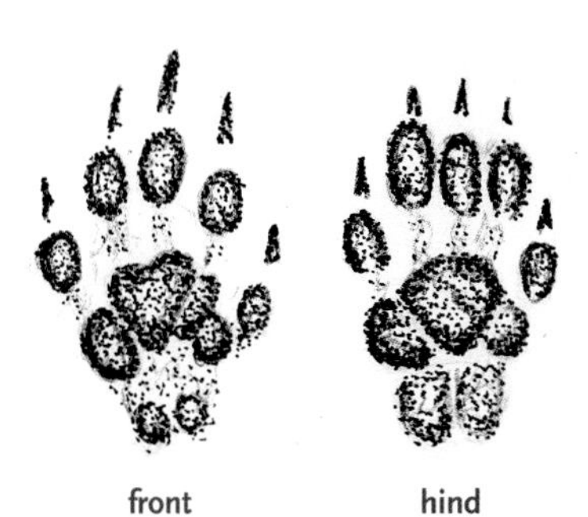

Western spotted skunk (*Spilogale gracilis*), page 271

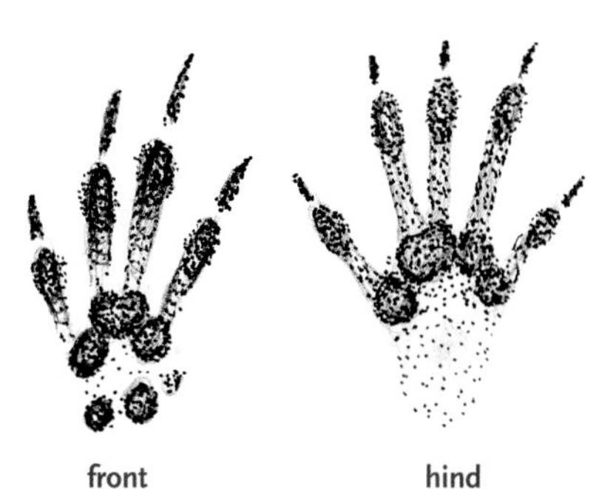

Columbian ground squirrel (*Spermophilus columbianus*), page 137

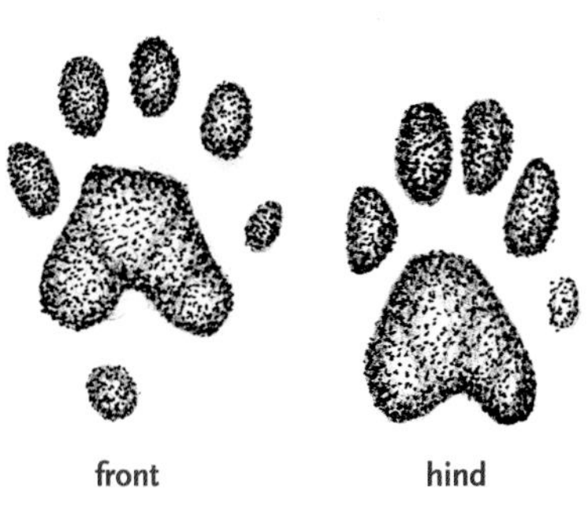

Ringtail (*Bassariscus astutus*), page 242

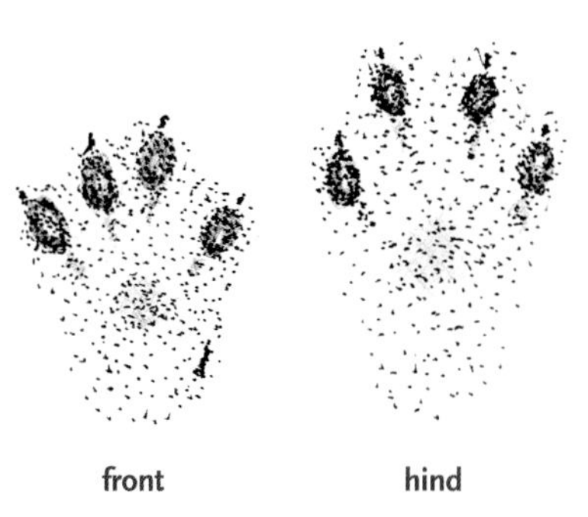

American pika (*Ochotona princeps*), page 107

Life Size

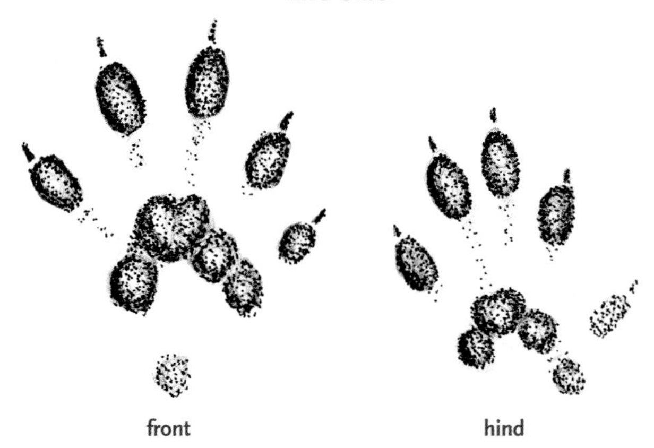

Mink (*Neovison vison*), page 250

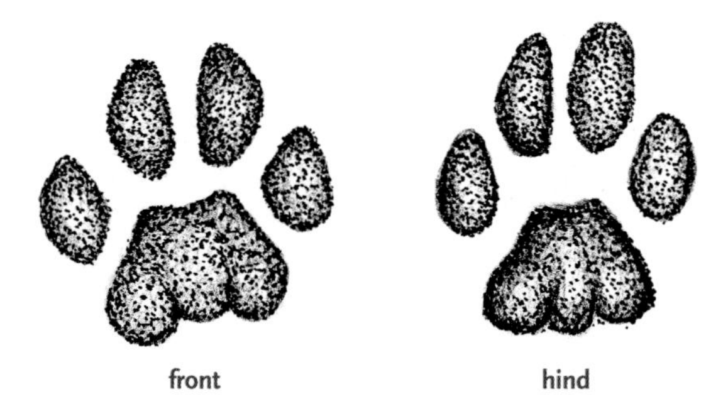

House cat (*Felis catus*), page 211

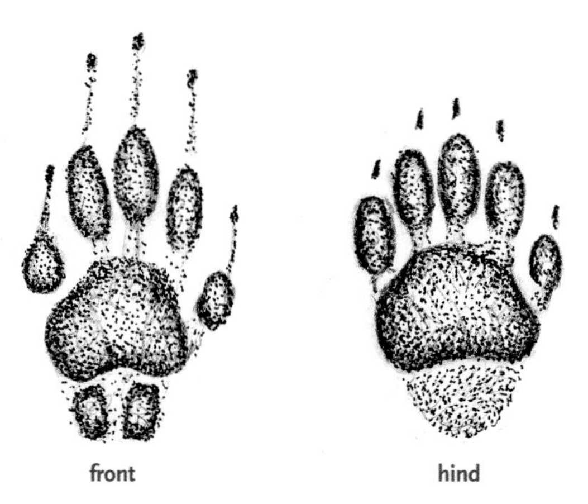

Striped skunk (*Mephitis mephitis*), page 269

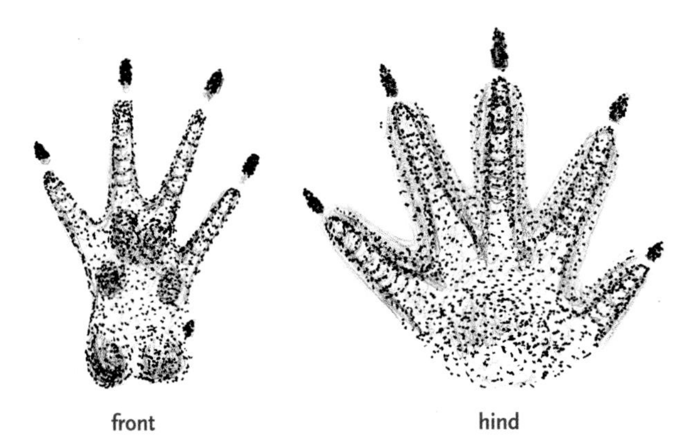

Muskrat (*Ondatra zibethicus*), page 122

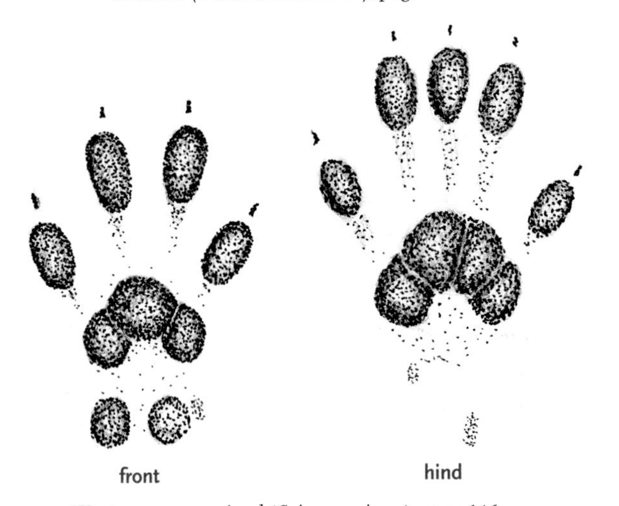

Western gray squirrel (*Sciurus griseus*), page 146

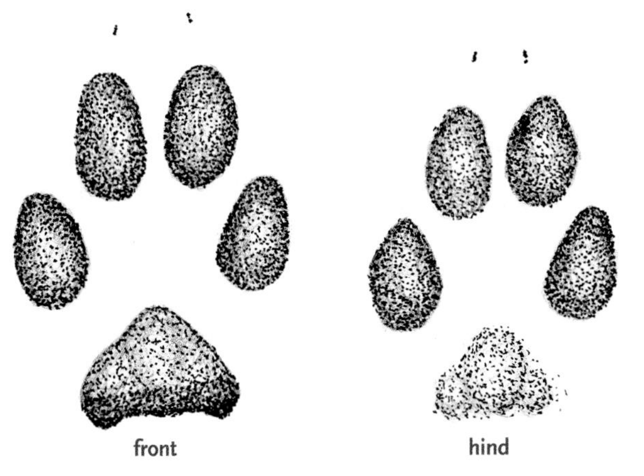

Gray fox (*Urocyon cinereoargenteus*), page 224

Two-Thirds Life Size

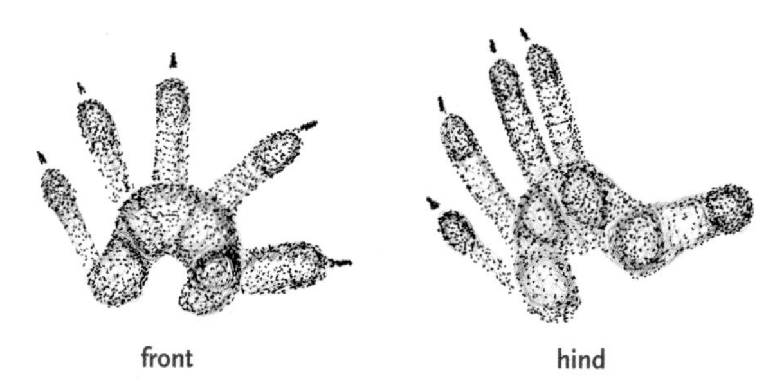

Virginia opossum (*Didelphis virginiana*), page 94

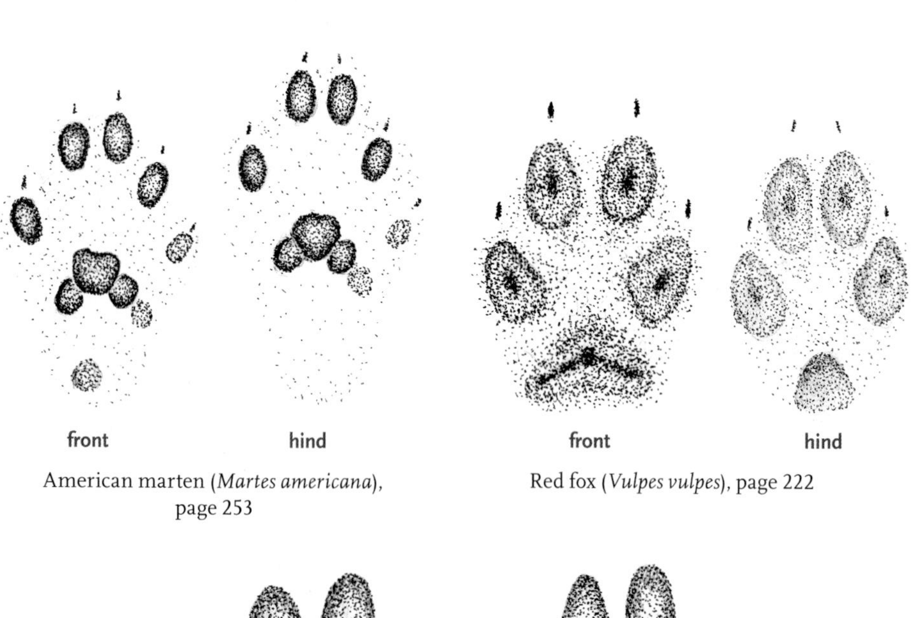

American marten (*Martes americana*), page 253

Red fox (*Vulpes vulpes*), page 222

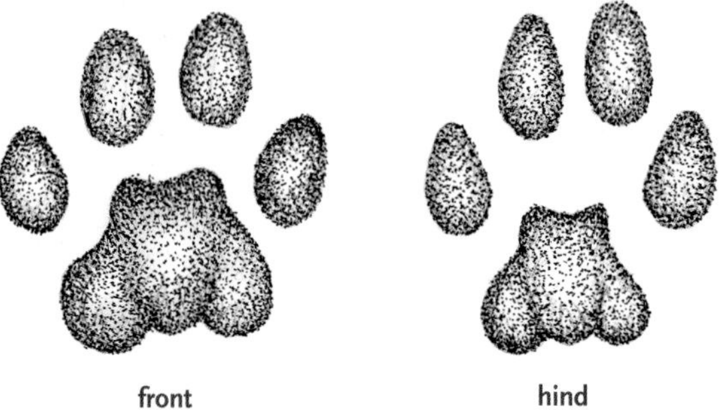

Bobcat (*Lynx rufus*), page 202

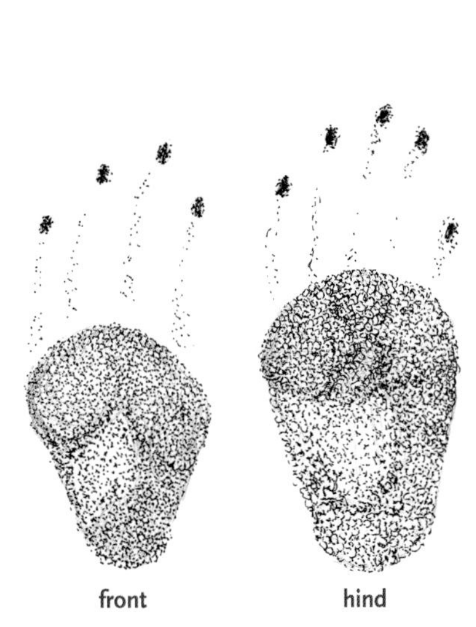

Porcupine (*Erethizon dorsatum*), page 190

American badger (*Taxidea taxus*), page 266

Nuttal's cottontail (*Sylvilagus nuttallii*), page 110

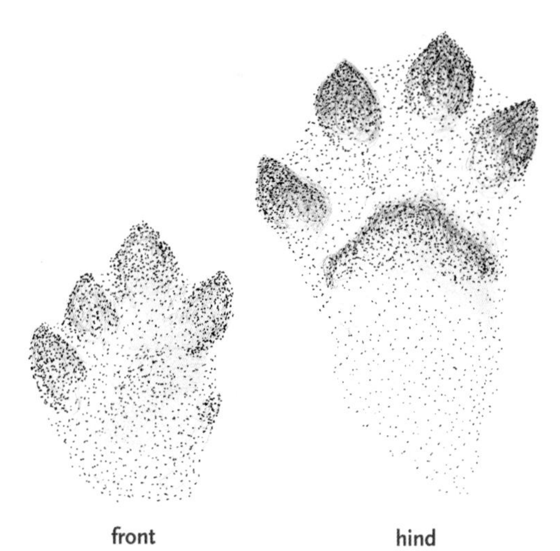

Snowshoe hare (*Lepus americanus*), page 116

Two-Thirds Life Size

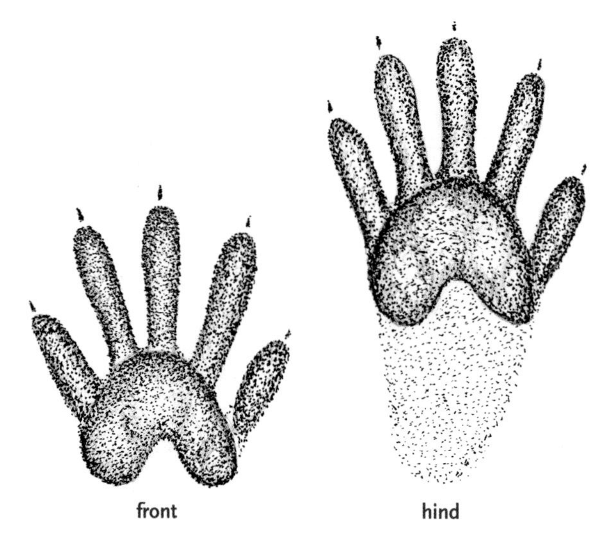

Raccoon (*Procyon lotor*), page 240

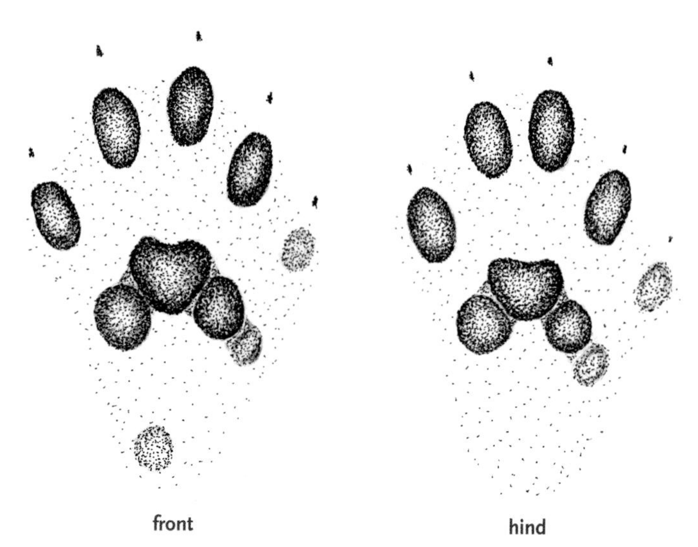

Fisher (*Martes pennanti*), page 256

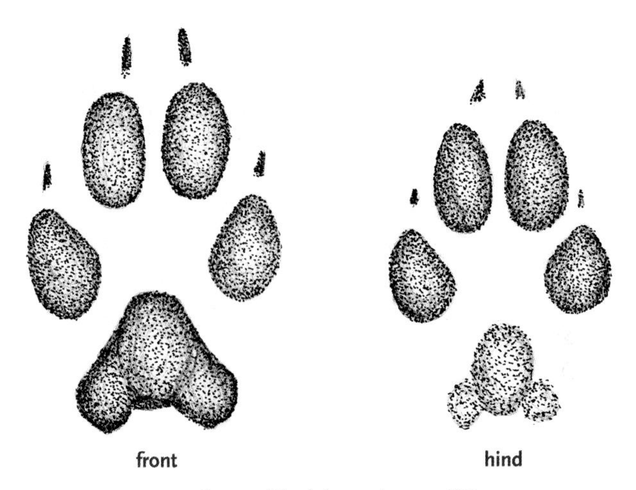

Coyote (*Canis latrans*), page 217

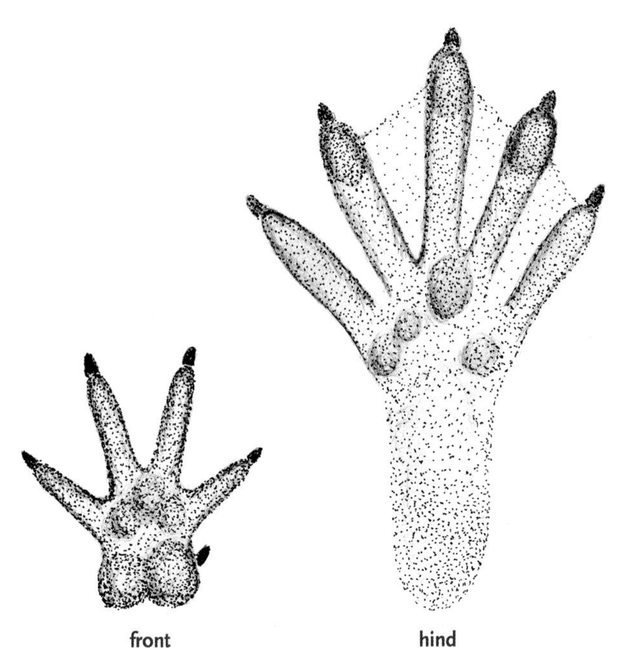

Nutria (*Myocastor coypus*), page 125

Two-Thirds Life Size

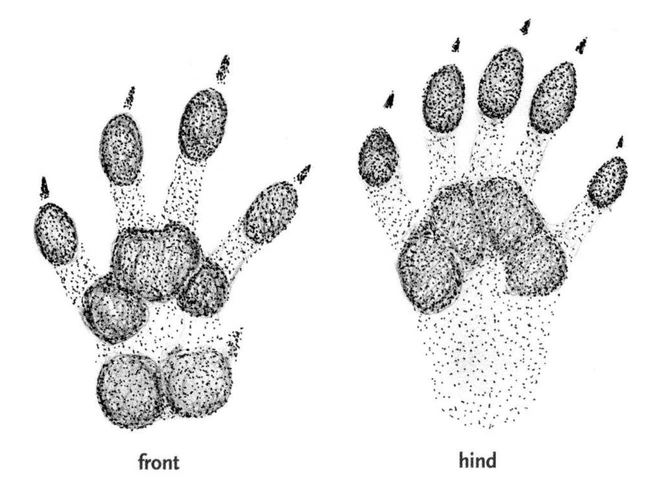

Hoary marmot (*Marmota caligata*), page 132

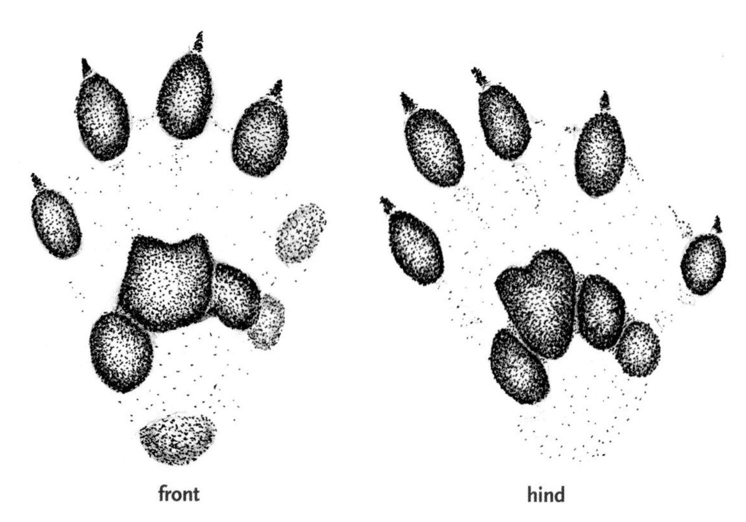

River otter (*Lontra canadensis*), page 259

Half Life Size

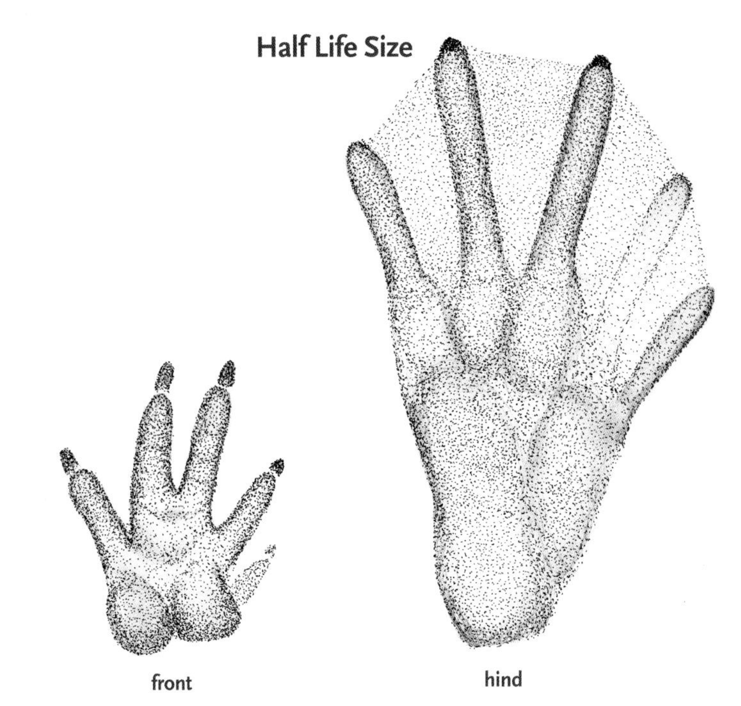

Beaver (*Castor canadensis*), page 127

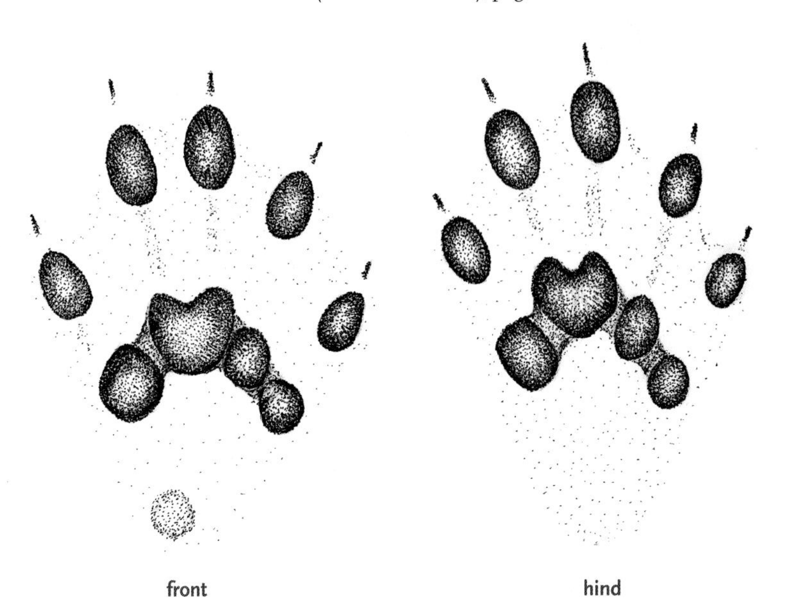

Wolverine (*Gulo gulo*), page 262

Half Life Size

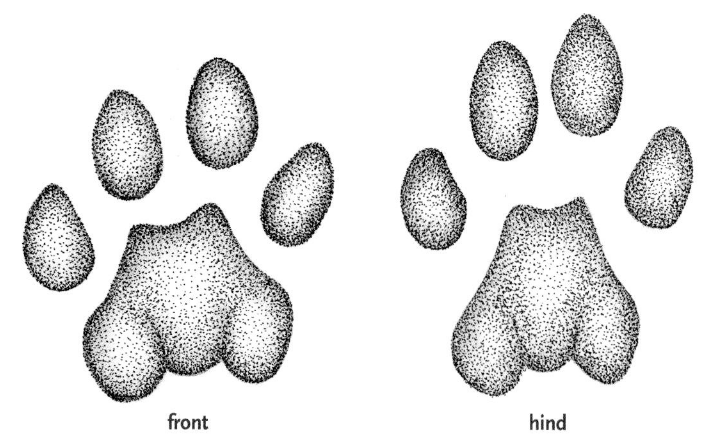

Mountain lion (*Puma concolor*), page 207

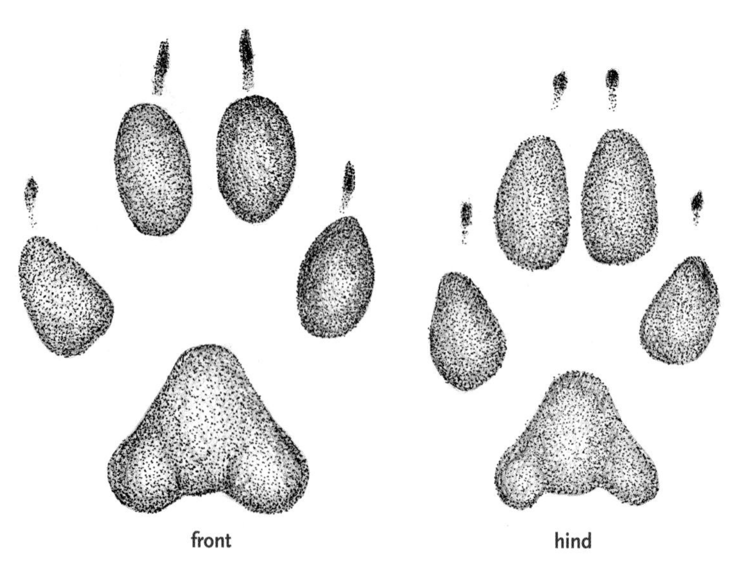

Gray wolf (*Canis lupus*), page 212

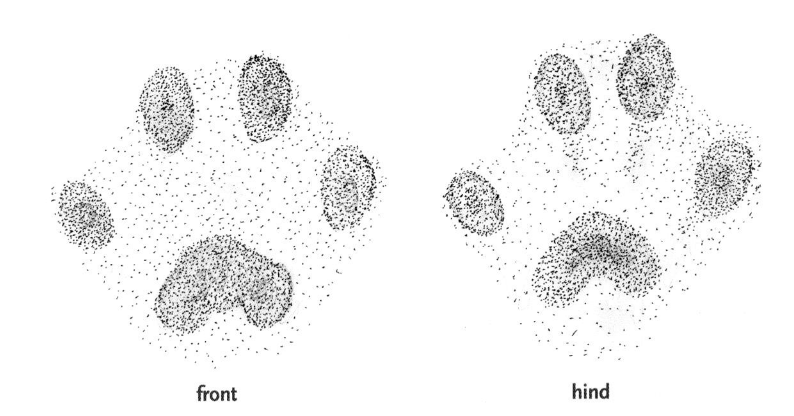

Canada lynx (*Lynx canadensis*), page 204

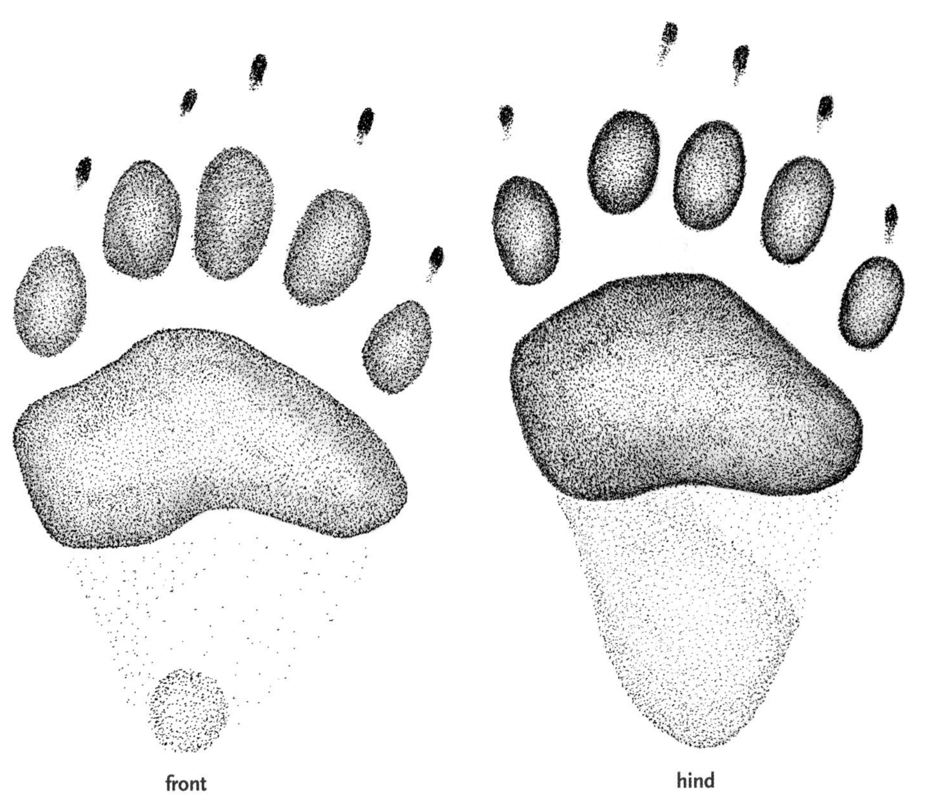

Black bear (*Ursus americanus*), page 230

Half Life Size

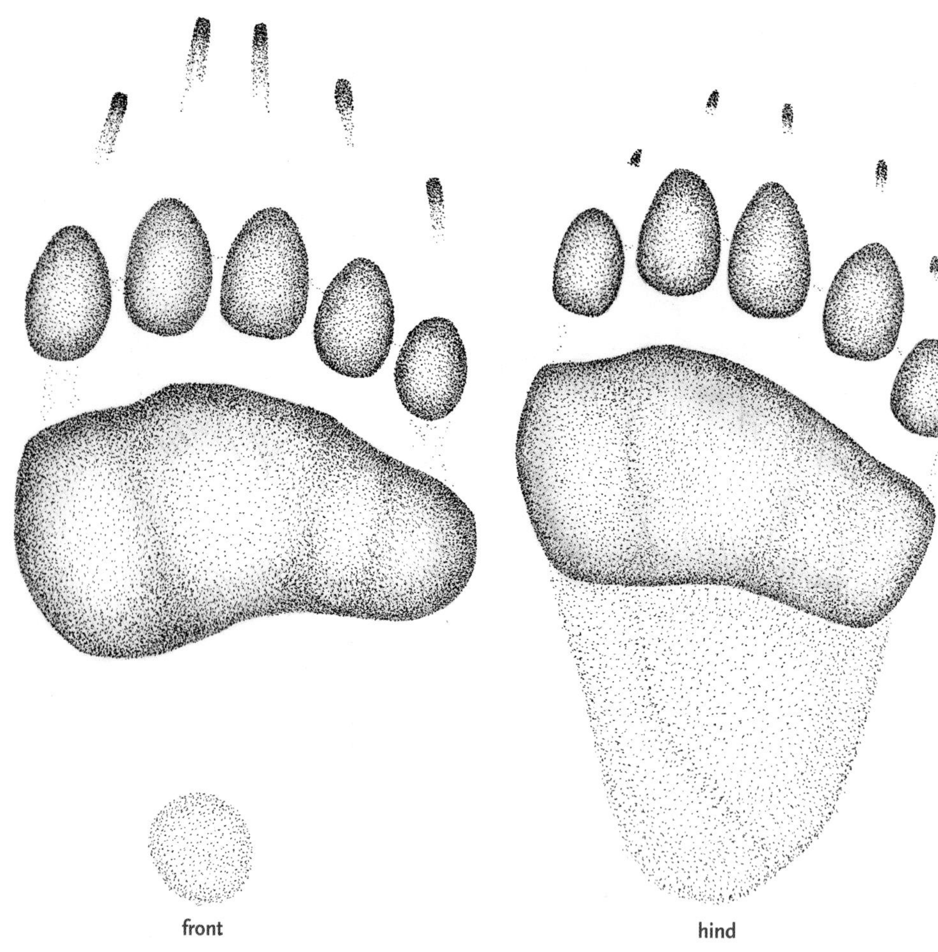

Grizzly bear (*Ursus arctos*), page 236

Life Size

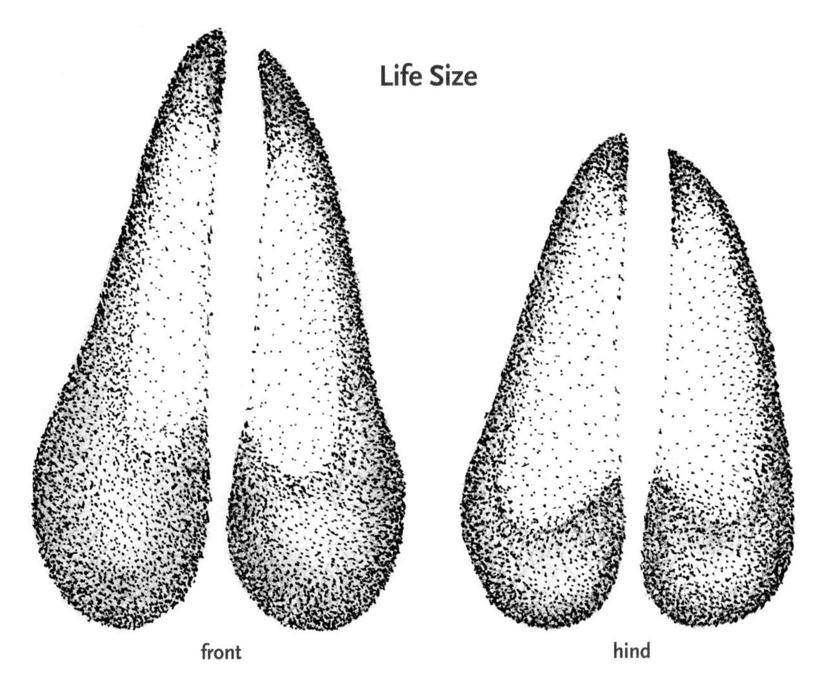

Pronghorn (*Antilocapra americana*), page 296

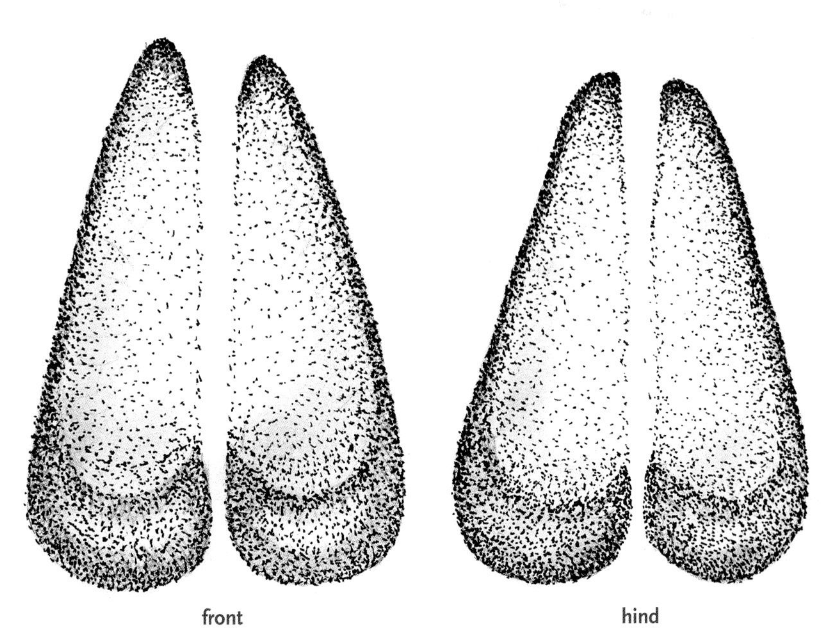

Bighorn sheep (*Ovis canadensis*), page 293

Life Size

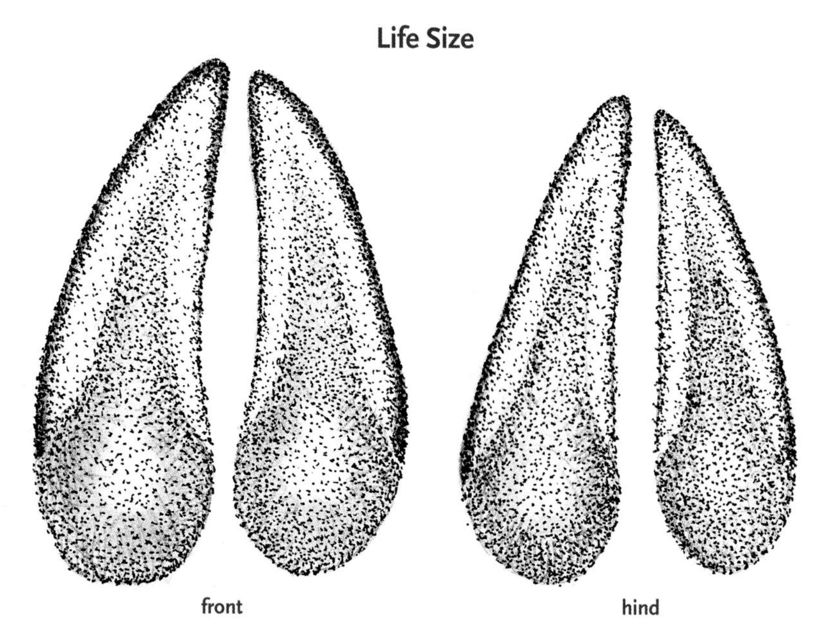

Mule deer (*Odocoileus hemionus*), page 276

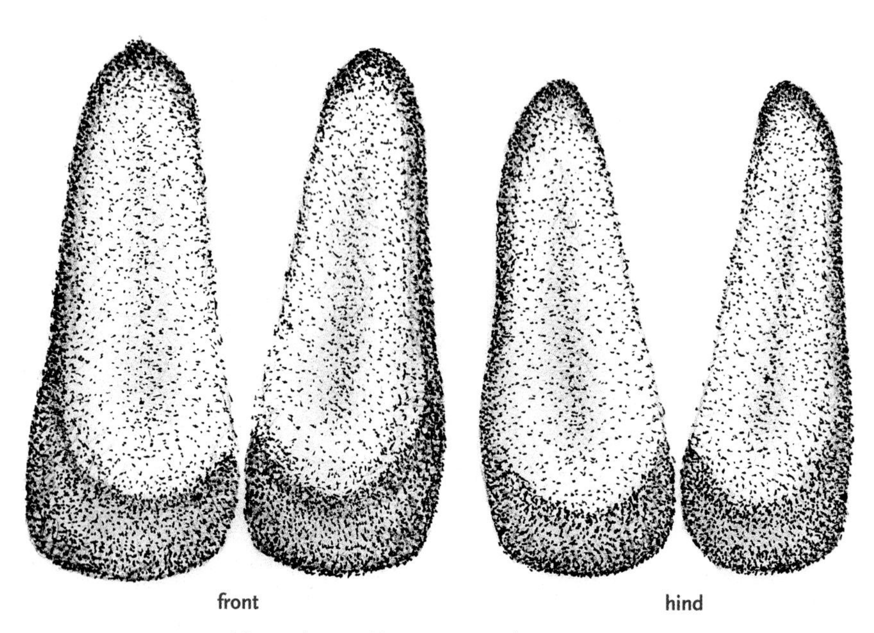

Mountain goat (*Oreamnos americanus*), page 291

Two-Thirds Life Size

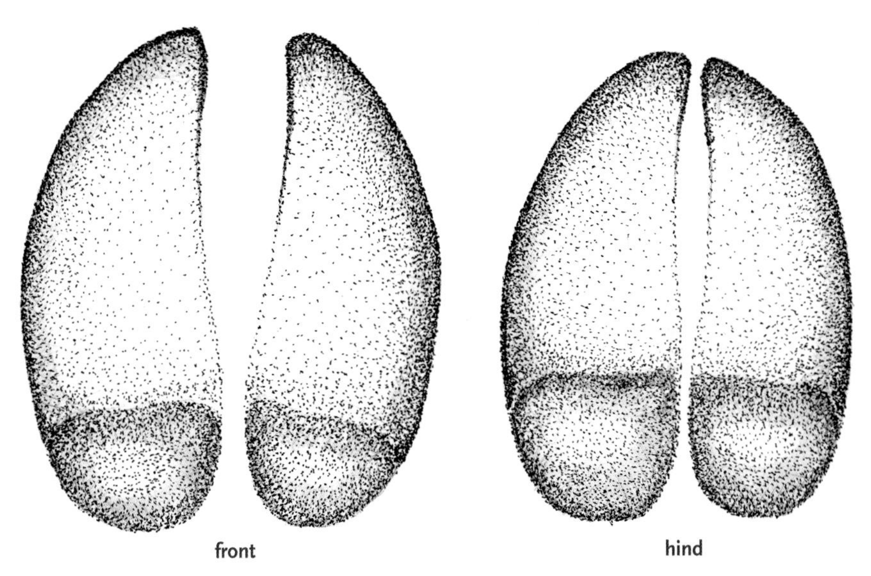

Elk (*Cervus elaphus*), page 281

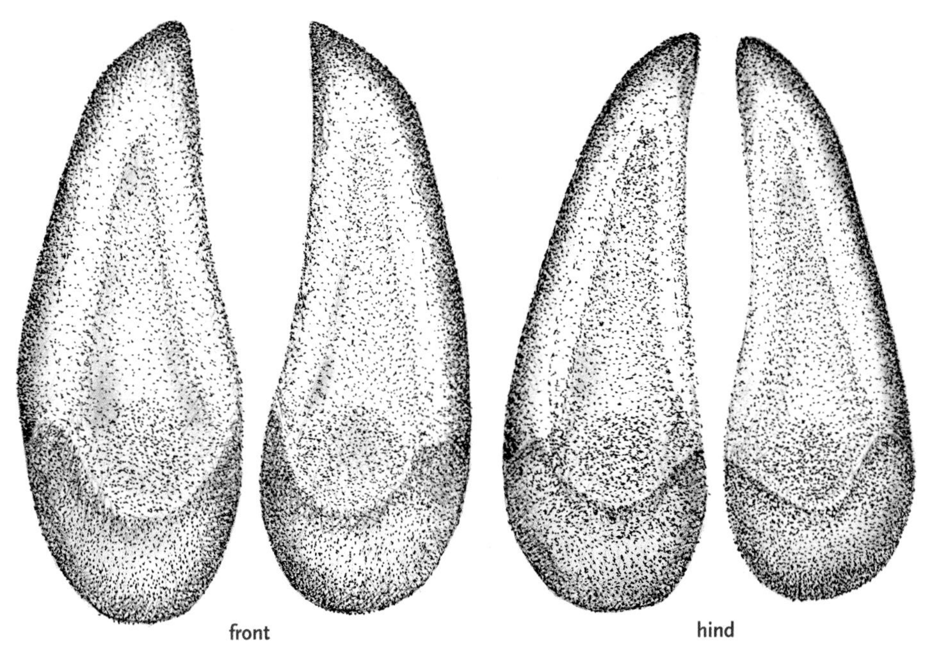

Moose (*Alces alces*), page 286

Half Life Size

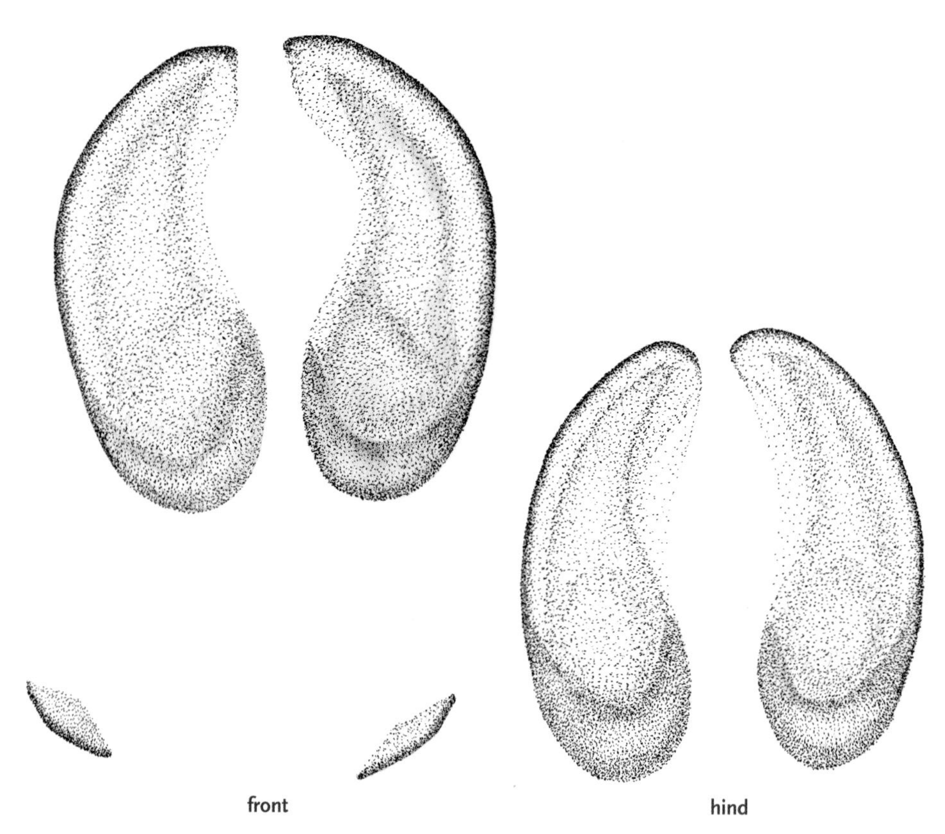

Woodland caribou (*Rangifer tarandus*), page 288

KEY TO SCATS

Lagomorph scats, page 107. Flattened spheres, from left: black-tailed jackrabbit, Nuttal's cottontail, pygmy rabbit.

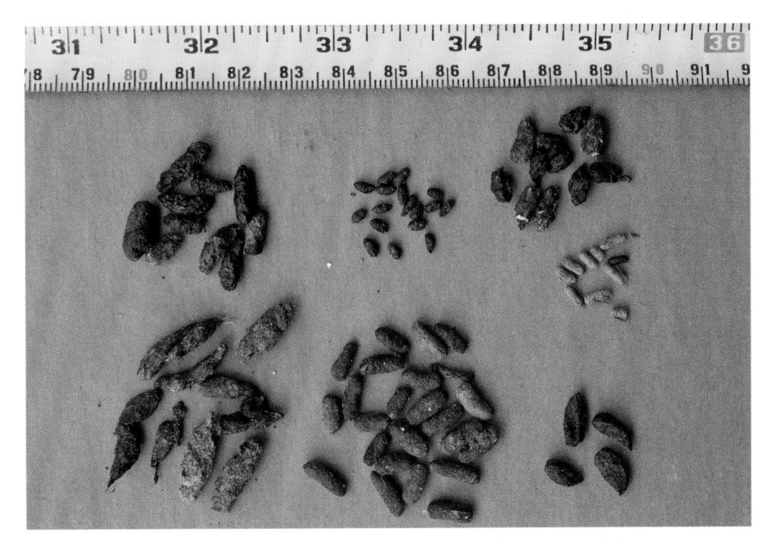

Rodent scats, page 121. Elongated pellets, clockwise from upper-left: roof rat, deer mouse, eastern gray squirrel, western heather vole, red squirrel, dusky-footed woodrat, Belding's ground squirrel.

Ungulate winter scats, page 274. Regular or irregular pellets. A lot of variation appears in scats within a species, with overlap between similarly sized species. Clockwise, from upper-left: bighorn sheep, Rocky Mountain elk, woodland caribou (dried winter scat), mountain goat, pronghorn.

Ungulate summer scats, page 274. Amorphous patties which could be confused with bear scat. Pictured here: Roosevelt elk scat.

Carnivore and omnivore scats: tubes and twists, page 200. Pictured here: typical wolf scat. Overlap exists among many species. Felines tend to be highly segmented with at least one blunt end. Mustelid scats are often twisty or ropey. The appearance of omnivore scats, such as those of bears, raccoons, foxes, and coyotes, depends on diet.

Carnivore and omnivore scats: amorphous, page 200. Often appearing as ill-defined piles, bear scats can be confused with scats of some hoofed mammals. Black, tarry, runny scats indicate a rich diet and are common in canines and other carnivores after feeding on a fresh kill. Pictured here: grizzly bear scat containing grasses and forbs.

KEY TO FEEDING SIGNS ON ANIMAL REMAINS

Large carcass entire, buried, or covered: mountain lion, *Puma concolor* (page 207), bears (page 228)

Large carcass dismembered and dispersed: bears (page 228), canines (page 212)

Part of carcass cached in shallow hole or under something: canines (page 212), weasels and their kin (mustelids) (page 244)

Headless small mammal or animal parts found on perch on tree, stump, fence post, or similar: birds of prey (page 331)

Other presentation: carnivores (page 200)

Feathers sheered and/or plucked: Carnivores (page 200)

Meat eaten from breast, feathers plucked, not sheared: birds of prey (page 331)

Black-tailed deer cached under a large tree by a mountain lion. West Cascades, Washington.

Sandhill crane carcass partially eaten by bald eagle. Canadian Rockies, Montana.

KEY TO FEEDING SIGNS ON PLANTS

Ripped or roughly cut vegetation: hoofed mammals (page 274)

Cleanly sheared vegetation: lagomorphs (page 107), rodents (page 121)

Gnawed seeds and cones: rodents (page 121)

Roosevelt elk feeding sign on plant stalk. Northwest Coast, Washington.

Beaked hazelnut (*Corylus cornuta*) gnawed open by a small rodent, probably a deer mouse. Puget Trough, Washington.

Mountain beaver feeding signs on the branches of a vine maple (*Acer circinatum*) show the 45-degree angled cut created by rodents and lagomorphs.

KEY TO SIGNS ON TREES AND TREE BARK

Rubs, Scrapes, and Scratches

Knee height to more than head height, bark shredded at top and/or bottom of rub, smaller branches broken by antlers, hairs stuck to rubbed surfaces: hoofed mammals (page 274)

Waist height to more than head height, rectangular chunks of bark missing and/or opposing marks from biting the tree, parallel marks from multiple claws on a paw, wavy hairs present: bears (page 228)

Bark removed in individual flakes: woodpeckers (page 328)

Claw marks, parallel scratch marks across bark: bears (page 228), felines (page 200), raccoons (page 240), porcupines (page 190), rodents (page 121)

Antler rub of a mule deer buck. Columbia Plateau, Oregon.

Excavations in Standing or Fallen Trees

Rotten logs and stumps ripped apart: skunks (page 268), bears (page 228)

Other excavations in fallen or standing trees, alive or dead: woodpeckers (page 328)

Incisor Marks and Cambium Feeding

Large incisor marks, scraping up the tree, possibly at slight angle, waist to head height: hoofed mammals (page 274)

Large incisor marks, feeding mainly at base of conifers; outer bark peeled off in large chunks, possibly accompanied by bite marks from canine teeth or large claw marks: bears (page 228)

Tiny to medium-sized incisor marks, small branches sectioned with 45-degree cuts at the tips: rodents (page 121), lagomorphs (page 107)

Large incisor marks, substantial wood removed in chips: beavers (page 127)

Excavations made by a woodpecker in a rotten tree trunk. Puget Trough, Washington.

Beaver sign. West Cascades, Washington.

KEY TO BURROWS, HOLES, DIGS, AND MOUNDS

Burrows and holes are openings in the ground associated with a continuing tunnel. A throw mound of soil may be found at the mouth of a burrow or hole.

Small holes (up to 3 in., or 8 cm): spiders (page 348), shrews and moles (page 96), voles (page 171), squirrels and chipmunks (page 137), desert-adapted mice and rats (page 184)

Foraging dig of a badger. Columbia Plateau, Oregon.

Medium holes (up to 7 in., or 18 cm): mountain beavers (page 193), pygmy rabbits (page 113)

Large holes (greater than 7 in., or 18 cm): marmots (page 132), badgers (page 266), canines (page 212)

A **dig** is a shallow to deep excavation that is not associated with a continuing tunnel or burrow.

Small, shallow, triangular digs: desert-adapted mice and rats (page 184), squirrels and their kin (page 132), skunks (page 268)

Medium digs, triangular or otherwise: canines (page 212), skunks (page 268), wolverines (page 262), large semiaquatic rodents (page 122)

Large digs: badgers (page 266), grizzly bears (page 236)

Large rectangular chunks of sod or soil flipped over: grizzly bears (page 236)

Holes at or below water level: semiaquatic rodents (page 122)

A **mound** is a pile of soil with no obvious burrow entrance or associated excavated depression.

Plugged entrance hole to one side of fan-shaped mound: pocket gophers (page 196)

Plugged entrance hole in the middle of evenly domed mound: moles (page 100)

Belding's ground squirrel burrow. Northern Great Basin, Oregon.

Townsend's mole mound. Puget Trough, Washington.

PART TWO
Mammals

A grizzly bear feeds on an elk carcass. Canadian Rockies, Montana.

Mammalia is a highly successful class of animals that has existed for about 200 million years. Mammals are warm-blooded vertebrates that are usually covered with hair over some or all parts of the body. All females produce milk in mammary glands to nourish their young, and in our region all species produce live young. Mammals have evolved in many ways to exploit many ecological niches. Track size and shape, foot shape and physiology, and dentition vary widely and in many species have become specialized for particular activities or for survival in extreme environmental conditions. Although all mammals share the same basic survival requirements, the details of how these requirements are met for different species vary widely.

Habitat Requirements

Habitat is vital for a mammal's long-term survival. Some species, such as coyotes, are habitat generalists. They can tolerate and persist successfully in many habitats, from coastal rain forests to arid deserts, and their behavior and social structure can change to adapt to local conditions. Other species are specialists associated with a particular habitat and may have highly specialized physical and physiological adaptations and behaviors. The Ord's kangaroo rat (*Dipodomys ordii*) is an excellent example of a specialist in our region. It requires loose, sandy soils in a highly arid landscape. It excavates burrows that protect it from predation, temperature extremes, and dehydration. Its fur-lined

Pronghorns require habitats of vast arid grasslands or shrub-steppes with unobstructed views. Northern Great Basin, Oregon.

external cheek pouches allow it to conserve water while transporting food to caches. It never needs to drink, as it can meet its physiological needs for water entirely through its metabolism of food.

The area in which an individual mammal lives is called its home range. This includes all the terrain it uses throughout the year to find food and shelter and to mate and rear young. The size of a home range depends on the species and can be less than 50 square feet (4.5 m^2) for a vole, to more than 350 square miles (907 km^2) for a grizzly bear. A species often uses different parts of its home range at different times of the year. Many elk spend the summer at higher elevations and move to lower elevations in the winter. Home ranges for different individuals within a species may overlap entirely with other members of the species, as is the case with elk that often travel in herds. For other species, limited overlap exists between individuals. A common pattern in many of our native carnivores is for males and females to overlap in their use of the landscape, with less overlap between members of the same sex. Often a male's home range overlaps with that of several females. This allows for mating to occur while limiting the amount of competition between individuals for potentially scarce resources.

Individuals within a species may compete for the resources available on a distinct piece of land—in particular, for food, adequate denning sites, or access to members of the opposite sex for mating. These individuals may establish and defend a core piece of land, called a territory, which is usually only a part of an animal's home range, and other members of the same species are often aggressively deterred from entering it through scent- and visual marking behaviors or physical interactions. These behaviors often increase during the breeding season, when territorial behavior can increase between members of the same sex, while other behaviors occur to attract the opposite sex.

Reproduction

Most mammals mate only when the females are in estrus, a series of physiological changes that advertise the females' fertility to males and prepare them to produce young. In most species, mating occurs at a time of year that allows the young to be reared when food resources are abundant, initially for the nursing mother and subsequently for the juveniles. In our region, many mammals give birth to young in the spring or early summer because food resources are most abundant in the summer and fall.

In smaller species, such as rodents, rabbits, and hares, multiple litters are produced annually if conditions allow. Although for some of these species, the cycle of estrus, mating, giving birth, and rearing of young can occur repeatedly throughout the year, most carry out reproductive activities starting in the late winter or early spring and ending in the fall. Young born in earlier litters are often better able to survive during their first winter than young born in later litters, in part because they have more time to build up energy reserves.

Estrus is induced in females in several ways, depending on the species. In wolves, the increasing amount of daylight after the winter solstice is believed to induce estrus. Pheromones produced by male deer in the fall induce estrus in females. In other species, such as members of the weasel family (Mustelidae), estrus occurs shortly after a female delivers a litter.

For the American marten and other mustelid species, mating occurs during the summer when males cover larger portions of their home range, thus giving them greater access to females. After copulation and fertilization, the zygote does not become implanted in the uterine lining until the end of the following winter, allowing for young to be born in the spring when food resources are expanding and environmental conditions are less harsh. This is called delayed implantation; the ova stops developing shortly after fertilization, in some cases for months.

Young are born either altricial or precocial. Altricial young are born without hair, and thermoregulation (regulation of body temperature) is undeveloped, as is locomotion, vision, and hearing. Mammals such as deer mice that bear altricial young are often small and have a high reproductive rate, giving birth to multiple large litters each year. Precocial young are born with fur, can walk shortly after birth, and thermoregulate well. Precocial young are often produced by large animals that have lower reproductive rates (because of the great energy expenditure required to produce young). All hoofed mammals in our region produce precocial young.

Although reproduction is a necessity for all species, how various mammals have evolved to maximize the effectiveness of reproduction varies. Small mammals tend to produce more young than large animals, and, in general, herbivorous species produce more young than do carnivores. Some species, such as rabbits, are fecund, with multiple large litters each year. A female might produce four litters in a year, with seven or more young in each litter. The mortality rate, however, can be as high as 90% within the first year of life.

Larger animals produce fewer young but

A bighorn ram sniffs a ewe to determine her readiness to breed. Eastern Cascades, Washington.

invest a great deal in their survival. In these species, the reproductive rate is low. A female bear gives birth to young at most only every other year, because it takes this long to rear her young to adulthood and for the female to rebuild the energy reserves required to produce and nurse cubs. In these species, young are often fiercely protected from threats, and the survival rate under stable conditions is higher than that of more fecund species.

Mating patterns also vary widely among species and can be divided into four categories. Monogamy involves a pair bond between one male and one female, sometimes for several breeding seasons or for life, and other times for only one breeding season. Often both parents, and sometimes a previous season's offspring, help care for young. In polygamy, males mate with several females, while females usually mate with only one male per breeding season. Females alone care for young. In polyandry, the female mates with several males, and males assist in rearing young. Promiscuity involves both sexes mating with several partners during a single breeding period. Either sex may provide parental care, but this is usually the female's responsibility.

Sexual Dimorphism and Behavioral Differences

Sexual dimorphism refers to differences in form between males and females of a species. It is apparent in various bird species, such as the common mallard duck: the male is brightly colored and the female is drab.

In mammals, sexual dimorphism is often less pronounced than in many bird species. In fact, it may be impossible to ascertain the sex of a species without carefully examining a collected specimen. Mammal groups that exhibit significant sexual dimorphism include most members of the weasel (Mustelidae) and deer (Cervidae) families. In mustelids, males are significantly larger and heavier than females. In cervids, males are larger overall and grow a set of antlers, while females are smaller with no antlers. (The woodland caribou is an exception; females grow antlers, but they are significantly smaller than those of males.) Sexual dimorphism may also be noticeable in the tracks and signs an animal leaves behind.

The fact that male birds sing elaborate songs is not an example of sexual dimorphism but a difference in behavior between sexes. Like birds, male and female mammal species engage in distinct behaviors that are often associated with breeding. These behaviors can include distinct scent- and visual-marking behaviors, travel patterns, and interactions with other members of their species. Tracks and signs found in the field can make these behaviors clear. Morphological and behavior features that you can use to distinguish males and females are discussed in relevant group and species accounts.

Two black-tailed deer spar during the breeding season. Klamath Mountains, California.

Food Resources

An animal's diet is a driving force in its evolution, distribution, and interaction with other animals. Mammals comprise three broad categories: herbivores, carnivores, and omnivores.

Herbivores such as deer and rabbits consume plant matter exclusively. Among herbivores, animals that specialize in feeding on seeds are called granivores. Some hoofed mammals are primarily grazers (feeding on grasses and forbs) and others are browsers (feeding on the buds and stems of woody plants). An herbivore's teeth are specialized for cutting and grinding vegetation.

Carnivores, such as mountain lions and short-tailed weasels, consume animal tissue either exclusively or in large proportion to the rest of their diet. The term "carnivore" can be misleading, because several members of the order Carnivora are not actually carnivores but omnivores. Species such as moles, shrews, and bats that feed on insects and other invertebrates are insectivores.

Omnivores have flexible diets that can consist of plant matter, fungi, and animal tissue. Their diets are often seasonal and can vary from one location to the next. Raccoons, bears, and coyotes are all omnivorous and exemplify the adaptability, intelligence, and opportunistic nature of many members of this group.

The relationship between predators and their prey is a driving force in the evolution and behavior of both. Carnivores usually occur in lower numbers than prey species on which they depend, and their population size can fluctuate depending on the abundance of prey. Several of our hoofed mammals travel in herds, a social adaptation that decreases the risks of predation to individuals. Where snowfall is heavy, snowshoe hares turn white in winter, making them more difficult to detect by predators. A large animal killed by a wolf or mountain lion is often scavenged by smaller carnivores after the initial predator is finished with the carcass, and both can be chased from their kills by bears.

Yellow-bellied marmots are herbivores. Many rodents use their dexterous front feet to manipulate food as they feed. Columbia Plateau, Oregon.

A gray wolf feeds on an elk carcass. Canadian Rockies, Montana.

Predators also provide key ecological functions on a broader level. By affecting the behavior of prey species, which must change their use patterns on a landscape to decrease the risk of being killed, predators can indirectly affect plant communities on which their prey forages. These changes can in turn affect the availability of food and shelter for many other wildlife species. This ecological relationship, called a trophic cascade, has been documented in areas where wolves have reestablished populations in the western United States (Ripple and Beschta 2004 and 2005).

Similar animals that overlap in habitat often develop specific feeding strategies that decrease direct competition and exploit diverse food resources. Douglas's squirrels, for example, share their forest environment with northern flying squirrels. While the diurnal Douglas's squirrel's diet consists largely of conifer seeds, the nocturnal northern flying squirrel focuses on the underground fruiting bodies of a few types of fungi. Both species also consume small amounts of the primary food source of the other, but the differentiation in their diets allows the two to coexist with miminal competition. Where two closely related species directly compete for food resources, one is dominant and excludes the other from prime habitat. In some carnivores, such as weasels, food resource partitioning between sexes has been documented, with larger males focusing on larger prey than that of females. In these species, males' home ranges overlap the most with females and the least with other males (the reciprocal is also true for females), and the resource partitioning would theoretically reduce competition between males and females occupying the same location.

Social Organization

Like the variety of mating relationships that mammal species have developed, many types of organization exist in broader social structures. Some animals are solitary, often using elaborate scent- and visual-marking behaviors to avoid direct interactions with others of their species, except for breeding behaviors and interactions between mothers and juveniles.

Animals that live in groups can be divided into several categories. Colonial animals, such as hoary marmots, have well-established home areas, often with burrow systems. Colonies can have defined social structures with dominant animals of each sex, or they can be loose associations of the same species that densely occupy a high-quality habitat.

Hoofed animals may form herds. Herd animals often make substantial movements across the landscape, and their social structure changes throughout the year. Males may band together during the summer and then disperse during the breeding season to associate with groups of females. In a herd, group members may be genetically related.

Some animals such as coyotes spend their time in family or extended family groups. The group's size varies depending on habitat conditions, food availability, and competition from other predators. Young may stay with the parents for one or more years before dispersing, or they may continue to travel with their parents as adults, which is more common for female than male offspring. The wolf pack is another example of an extended family group with a highly structured social order.

POUCHED MAMMALS Order Didelphimorphia

Didelphimorphia is one of several orders commonly referred to as marsupials or pouched mammals, a reference to the flap of skin that covers the mammae, where newborns proceed directly after birth to nurse and continue development. Only one family, Didelphidae, and one species, *Didelphis virginiana,* exists in our region. Its appearance, reproductive biology, and tracks are distinctive.

VIRGINIA OPOSSUM

Didelphis virginiana

Introduced from the eastern United States during the first part of the 20th century as escaped pets and novelty animals, opossums have become common throughout rural, suburban, and urban landscapes in the western portion of our region. They are adept climbers with prehensile tails but are relatively slow moving on the ground and are often killed by cars. Although occasionally taken by predators, they are not a preferred food for any of our region's carnivores. When attacked they may enter a catatonic state and emit foul-smelling secretions from their anal glands, an act likely to deter the aggressor. They are active year-round (less so during cold weather) and are nocturnal.

SIZE Length: 35–94 cm. Weight: 0.3–6.4 kg (avg. 3.7 kg males, 2.0 kg females). Males are larger. **FIELD MARKS** Pointy nose, naked black ears with white tips, unfurred and scaly tail. **REPRODUCTION** Breeding can occur in any month. Following a 14-day gestation period, females give birth to 5–14 young, usually twice a year. At birth opossums are kidney-bean sized, and immediately travel to their mother's pouch where they latch onto a teat and stay attached, nursing, for about 2 months. They can disperse as early as 4 months of age. Individuals are solitary except for breeding behavior. **DIET** Omnivorous scavengers. Most meat consumed is carrion, including other opossums. They predate on ground-nesting birds when available. Near human habitation, they pilfer pet food left outside and raid unsecured chicken coops. **HABITAT &**

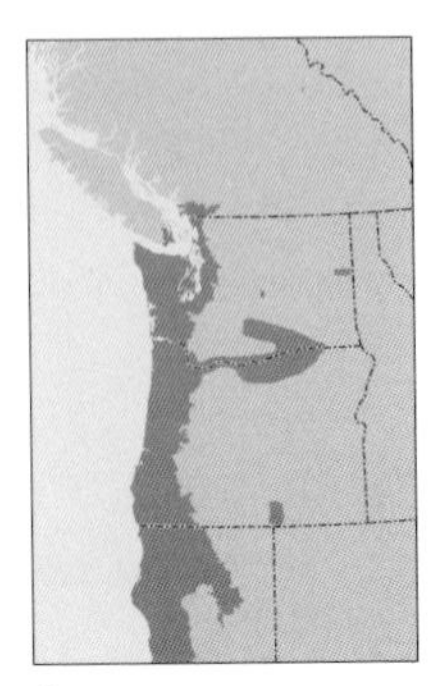

Opossum range

DISTRIBUTION Most abundant in rural, semi-wild, and suburban settings with a strong affinity for water. Most common west of the Cascades. Home ranges are 12–194 acres (0.05–0.79 km^2), often elongated, perhaps along stream courses. Male home ranges are considerably larger than those of females. **SIMILAR SPECIES** Raccoon.

TRACKS

Front and hind feet are distinctly dissimilar and smaller on average than those of a raccoon. In both the front and hind tracks, the palm pads are often deeper and at first more distinct than the toes (page 64). **FRONT** 1–1 11/16 in. (2.5–4.2 cm) L × 1–2 7/16 in. (3.7–6.1 cm) W. Five fingerlike toes connect to palm pad. Small, sharp claws often register in tracks. Toes radiate from the palm relatively evenly around the foot and are more spread than in the front foot of a raccoon. The palm pad is C-shaped overall and made of four semifused pads. Two additional heel pads rarely register. **HIND** 1 7/16–2 3/8 in. (3.5–5.5 cm) L × 1¼–2¼ in. (3.2–5.7 cm). Resembles a human hand with an opposable toe 1, which is clawless and more robust than other toes that are thin and fingerlike. Toe 1 registers at about 90 degrees (or greater) to rest of the toes, which are all roughly parallel to one another. The palm comprises several distinct pads that register deeply. **TRACK PATTERNS** Walks and trots, creating an indirect register track pattern with the hind foot slightly to the outside of the front. Front is usually cradled in the space between toes 1 and 2 of the hind. Track pattern is often confused with that of a raccoon but is distinctly different, as each set of two tracks is made by a front and hind foot from the same side of the body (typical of most walking mammals). **WALK & TROT** Stride: 9 7/16–21¼ in. (24–54 cm). Trail width: 2 3/16–6 in. (5.5–15.2 cm).

SIGNS

SCAT I have never definitively identified opossum scats in the field. **BURROWS** Use burrows and cavities under fallen logs or in standing trees for natal dens and rest sites. Underground burrows are usually refurbished holes made by other animals. Entrances have a distinctive and unpleasant odor.

Right front (top) and hind tracks of an opossum. Puget Trough, Washington.

Opossum understep walk

SHREWS AND MOLES Order Scoricimorpha

Soricimorpha has two families in our region: Soricidae (shrews) and Talpidae (moles and shrew-moles). All members of Soricimorpha feed primarily on invertebrates. Some species consume minor amounts of plant material. All have poor eyesight but an excellent sense of touch, aided by a sensitive snout.

Only one genus of shrew (*Sorex*) exists in the Pacific Northwest, with 11 species that are mostly impossible to distinguish without paying careful attention to the dentition of a specimen (something not usually possible in the field with a live animal).

Habitat and Adaptations

Shrews are mainly terrestrial but may climb into the branches of shrubs while foraging. Several species are well adapted to movement in water, where they do most of their foraging.

Moles have several special adaptations to their subterranean lifestyle. Their eyes, covered with a thin membrane of skin, can only sense differences between light and dark. Their ears are simple holes without any external features. Their fur can flatten easily in either direction, allowing them to move forward or backward in their tunnels. They have sensitive hairs on their tails and snouts, which are used for feeling their environment and helping them to navigate in both directions. Their narrow hips let them turn around in tight spaces when necessary. Mole tracks are found only occasionally, but signs of their subterranean lifestyle are common.

Moles produce tunnel systems called encampments that are used for travel, hunting, and refuge. Surface tunnels leave a raised ridge along the ground and are not constructed for permanent use. Shallow and deep tunnels are permanent, with shallow tunnels most intensively used. Deep tunnels are used most often during periods of intense cold or drought and can be up to 6 ft. (2 m) deep. For both shallow and deep tunnels, moles must expel some of the excavated dirt to the surface, producing conspicuous mole hills. Measurements for discriminating between the encampments of our two most common species are from Sheehan and Galindo-Leal (1997).

SHREWS

Sorex spp.

We are blessed with an abundance of shrews, especially west of the Cascades. The smallest mammals in our region, they can move with amazing quickness, which allows them to hunt and kill dangerous invertebrates such as centipedes, wasps, and scorpions. All have high metabolisms and can consume far more than their body weight in food in a single day. Most species construct networks of trails through forest litter that they continuously patrol, searching for invertebrates to eat. They also use the runs and trails of other small mammal species such as voles. Two semiaquatic species take to water readily and do much of their foraging in and adjacent to bodies of water.

SIZE Length: 7.5–18.0 cm. Weight: 2–21 g (less variation in individual species with most of smaller size). **FIELD MARKS** Small and dark with a tapered, pointed, furred

Unidentified Northwest shrew. Photo by Tom and Pat Lesson.

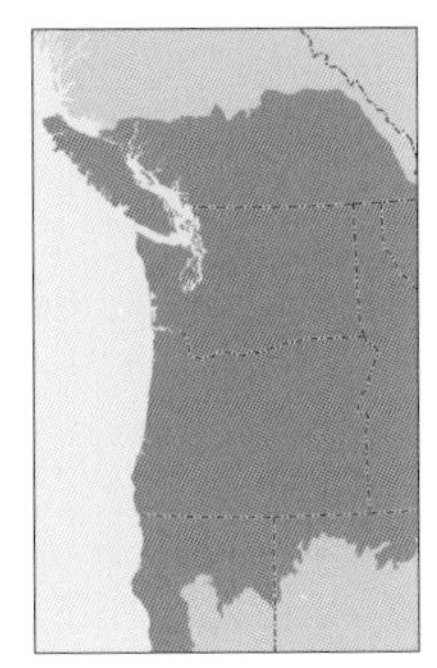

Vagrant shrew range

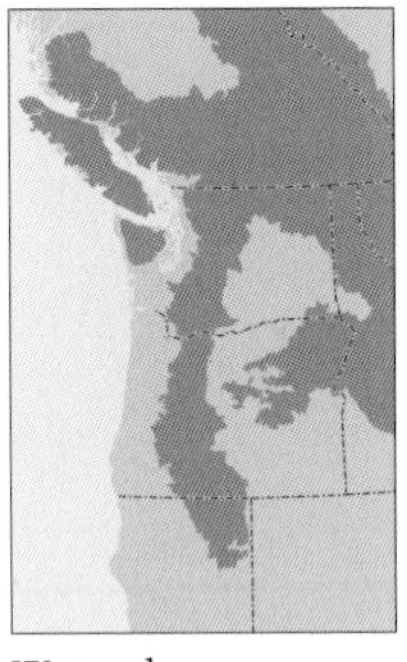

Water shrew range

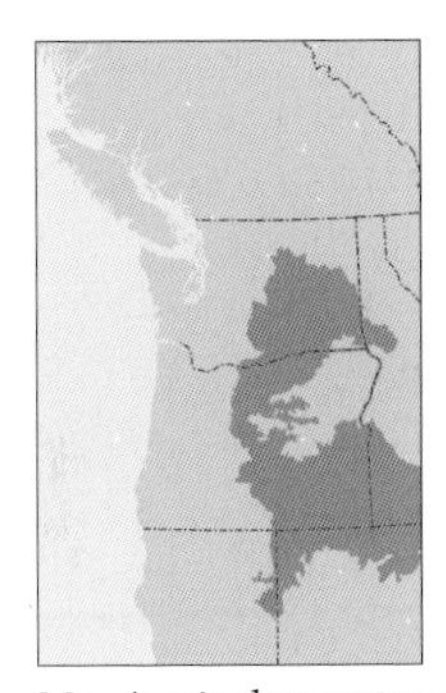

Merriam's shrew range

snout and a tail about as long as its body. Ears are larger than those of voles. **REPRODUCTION** Breeding occurs from late winter to early fall with a peak for most species in the spring. Some species can have more than one litter of 3–4 young per year. **DIET** Insectivorous. Diets vary between species but all include invertebrates. Some species eat small amounts of plant material and occasionally small vertebrates such as tree frogs. **HABITAT & DISTRIBUTION** Habitat and geographic location can help you determine which species you might encounter in the field. *Sorex vagrans* (vagrant shrew), found in moist environments, is the most widespread species in the region. *Sorex trowbridgii* (Trowbridge's shrew) is slightly larger and ranges from the east slopes of the Cascades west to the coast and inhabits slightly drier habitats in conifer forests. *Sorex palustris* (water shrew) lives in and around cold mountain streams, lakes, and wetlands in the Cascades and in other mountain ranges; *S. bendirii* (marsh shrew) occupies similar habitats at lower elevations west of the Cascades. In the Columbia Plateau ecoregion, *S. merriami* (Merriam's shrew) is the only shrew that occupies the arid, waterless landscapes, including juniper forests and sagebrush steppe. West of the Cascades, several species have limited distributions and specific but overlapping habitat preferences, making inferences about identification almost impossible. **SIMILAR SPECIES** Voles, deer mice, western harvest mice.

TRACKS

Tiny and easily overlooked, the shrew's foot

structure is one of the most primitive of all mammals. Plantigrade, with the palm comprising several distinct pads (number varies between species). Claws are sharp and usually register in clear tracks. Two diagnostic clues can help you differentiate shrew tracks from those of small rodents: the front and hind feet have five fully developed toes and both feet are similar in size. Mice and voles have four fully developed toes on their front feet, and front and hind feet are distinctly different in form and often size (page 59). **FRONT** 3⁄16–1⁄4 in. (0.4–0.7 cm) L × 3⁄16–3⁄8 in. (0.5–0.9 cm) W. Toes splay, forming an arc, with toe 3 the longest. Palm has four or more distinct pads; heel has an additional two or more proximal pads. Sometimes toe 1 does not register, leaving a track that could be confused with the front foot of a small vole. Semiaquatic species tracks have a fringe of hair, though I have never positively identified this in tracks in the field. **HIND** 3⁄16–5⁄16 in. (0.5–0.8 cm) L × 3⁄16–7⁄16 in. (0.5–1.1 cm) W. Toes 2–5 often register parallel to each other and are about the same length (similar to the hind feet in many small rodents). As in the front feet, palm has four or more distinct pads with two or more proximal pads that rarely register in tracks. **TRACK PATTERNS** Trails are often found in snow. Walks, trots, lopes, and bounds—more variations in gait than with deer mice, which leave a consistent bounding pattern. Trails are narrower than those of voles, which also walk, trot, and bound. In deep snow, shrews often leave a trough from their entire body and an intermittent tail drag. Trails may wander briefly on the

Right front and hind feet of a shrew in fine dust. Puget Trough, Washington.
Photo by Brian McConnell.

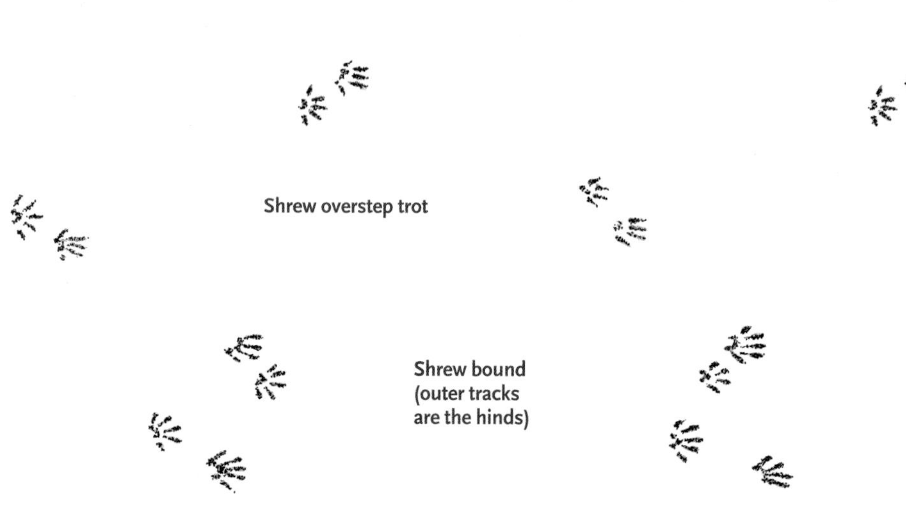

snow's surface, ending at a tree well or other access point to the subnivean zone, where they spend most of their time. **BOUND** Stride: 2¾–5½ in. (7.0–13.3 cm). Trail width: 13/16–1 9/16 in. (2.1–3.9 cm). Group length: 3/8–1 9/16 in. (1.0–3.9 cm). **WALK & TROT** Stride: 1¾–4 in. (4.5–10.1 cm). Trail width: 3/8–1 3/8 in. (1.0–3.5 cm).

This shrew changed its gait from a lope to a bound, as shown by the parallel hind feet in the set of tracks at the top. Northwest Coast, Washington.

SIGNS

SCAT To 1/16 in. (0.2 cm) D. Tiny, pointy, and dark. Filled with insect remains. Found along trails. **TRAILS** Networks of trails through forest litter, with trails less than 1 in. (2.5 cm) wide. Vegetation is usually not clipped as in vole tunnels. Shrews commonly use other species' trail and tunnel systems. **SPECIMENS** Shrews are often killed by domestic cats and other predators and not consumed, making dead individuals a common sign.

AMERICAN SHREW-MOLE

Neurotrichus gibbsii

The shrew-mole is aptly named, as it has characteristics of both shrews and moles. Its foot and limb structure are molelike, but it is more terrestrial and is similar in size and behavior to a shrew. Rather than digging deep tunnels, as is the case with moles, shrew-moles produce shallow networks of tunnels and trails in loose forest loam and forest debris. Tunnels may descend to a depth of about 12 in. (30 cm). They occasionally climb in shrubs. They are blind and hunt mainly through their sense of touch, aided by their long and delicately furred snout. Somewhat social, shrew-moles often travel in loose bands of up to a dozen individuals, though the relationships within or structure of these groups is unknown.

SIZE Length: 9.2–11.4 cm. Weight: 9–11 g. **FIELD MARKS** Shrewlike in appearance and size except for a broad, hairy tail (constricted at the base), molelike forefeet with stout claws. **REPRODUCTION** Most breeding occurs in the spring but can extend into October. Litters of 1–4 young. **DIET** Earthworms and other invertebrates. **HABITAT & DISTRIBUTION**

American shrew-mole. Photo by Alan D. St. John.

Shrew-mole range

Northwest Coast to the eastern slope of the Cascades. Occupies habitats characterized by dense ground cover and usually moist conditions in forests, marsh edges, and wet meadows. Most abundant in moist deciduous hardwood forests west of the Cascades with loose topsoil. **SIMILAR SPECIES** Shrews, moles.

TRACKS

Tracks are rarely found. Shrew-mole feet are tiny, like those of shrews, but they closely resemble moles in shape. **FRONT** Feet are round with large, blunt claws extending from each of the five toes, as with moles. Toes are slightly more developed than in moles. In tracks, often only the claws register in an arc, as in moles. **HIND** Similar to moles, five toes with claws all oriented forward. Smaller than front feet. **TRACK PATTERNS** Track pattern similar to moles.

SIGNS

TUNNELS & TRAILS Look for shallow, interconnected tunnel systems by carefully inspecting the moss and debris layer on the forest floor, especially around rotting stumps and logs in westside forests. Uncovered trails may also cross open spots in the forest. Similar networks are created and used by shrews, with whom the shrew-mole overlaps in habitat and feeding behavior. **SPECIMENS** As with other members of this order, shrew-moles may be killed but not consumed by predators, especially domestic animals. Exploration of the surrounding area may reveal its tunnels and trails.

MOLES

Coast mole, *Scapanus orarius*
Townsend's mole, *S. townsendii*
Broad-footed mole, *S. latimanus*

The coast mole, the most widely distributed mole in our region, is a forest dweller that produces surface tunnels along the forest floor. It is also active above ground around fallen logs.

The Townsend's mole is the largest mole in our region, found in deep and moist soils, open fields, and lawns. They are almost completely fossorial and highly territorial. Fights between individuals can end in death. Townsend's moles are often considered pests because of the mole hills they create in lawns, golf courses, and agricultural fields. Despite this, their tunneling activity contributes to important ecological functions such as soil aeration, soil level mixing, soil permeability improvement, and consumption of large numbers of invertebrates also considered pests by humans. Moles are most active during the wetter months of the year.

Townsend's mole. Photo by Progressive Animal Welfare Society (PAWS), Lynnwood, Washington.

SIZE Coast mole—Length: 13.3–19.0 mm. Weight: 61–91 g. Townsend's mole—Length: 18.3–23.7 cm. Weight: 50–171 g. Males are larger. **FIELD MARKS** Long, naked snout. Large outward-turned front feet with large claws; short legs; and a short, nearly naked tail. The coast mole is smaller than the Townsend's mole. The broad-footed mole is intermediate in size, lighter than the coast mole, and has a slightly furred tail. **REPRODUCTION** Breeding occurs in the late winter or early spring, and a single litter of 2–4 young are produced in March or April. During the breeding season, males construct long tunnels connecting their encampment with those of neighboring females. **DIET** Omnivorous. Earthworms are major prey, but moles also eat insect larvae, beetles, centipedes, and plant matter such as grass roots. **HABITAT & DISTRIBUTION** Coast mole—Adaptable to a variety of habitats. Signs are most abundant west of the Cascades but also found in ponderosa pine forests and sagebrush shrub-steppe farther east. Common in conifer forests and river floodplains in areas with loose soils, such as sand or forest loam. Occasionally found in meadows and fields, where it overlaps with Townsend's moles. Broad-footed mole—Slightly larger; occurs in southwestern Oregon and northern California, occupying slightly more arid habitats than the coast mole. Townsend's mole—Found west of the Cascades, where they are most abundant in fields, meadows, and suburban lawns but also occupy open forests and river floodplains. **SIMILAR SPECIES** Northern pocket gophers, toads, turtles. **CONSERVATION ISSUES** In their limited range in British Columbia, Townsend's moles are considered threatened by habitat destruction from agriculture and urban development.

Coast mole range

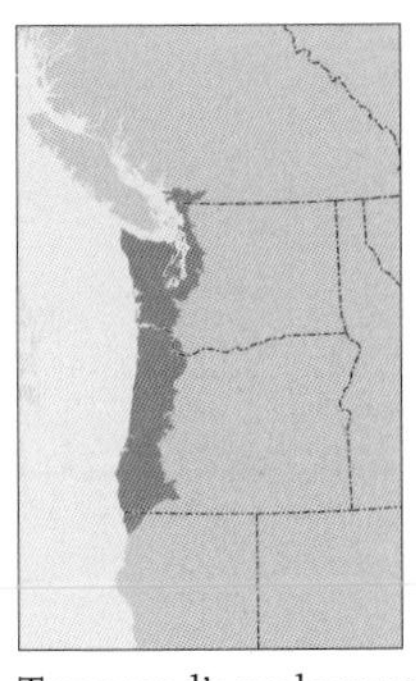

Townsend's mole range

TRACKS
Front feet and limbs are adapted for digging; broad feet turn out, with stout claws used for digging and moving soil. Tracks of Townsend's and coast moles are indistinguishable, with habitat being the best clue to help you distinguish between them. Townsend's mole track sizes tend to be toward the larger size of the range, while coast moles tend to be smaller (page 59). **FRONT** 7/16–1¼ in. (1.1–3.1 cm) L × 3/8–13/16 in. (0.9–2.0 cm). Five squat toes with large, broad claws and a large, naked palm. Foot wider than long, but usually only claws register, creating an arc that curves out away from the trail axis. **HIND** 3/16–13/16 in. (0.5–2.0 cm) L × 7/16–11/16 in. (1.1–1.7 cm). Smaller than front. Five toes with large claws (often the only parts to register clearly in tracks) that face forward, parallel to the axis of travel. Tracks longer than wide, tapering significantly in the heel. **TRACK PATTERNS** Because of their highly specialized bodies, moles' gaits and track patterns are unique but can be confused with those of a turtle or toad. Track pattern shows hind foot from one side registering close to front foot from

Left front (above) and hind tracks of a coast mole. West Cascades, Washington.

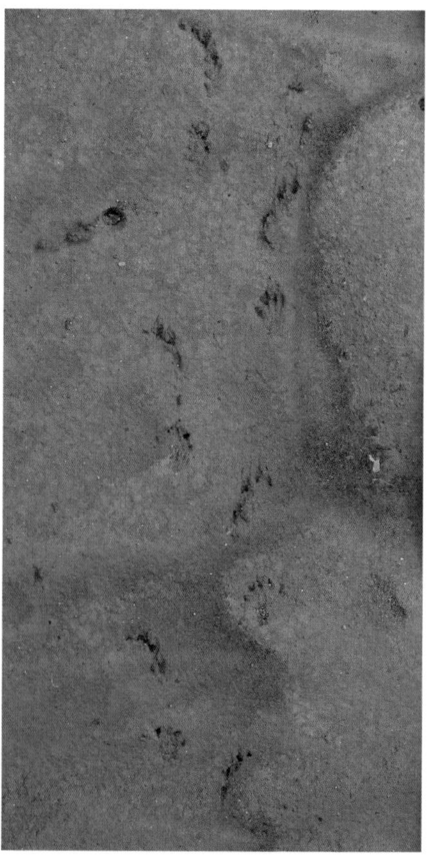

Paired tracks on alternating sides of the trail show the mole's front and hind feet, with the hind foot understepping the front. Puget Trough, Washington. Photo by Liz Snair.

the other side of the body. Sometimes creates an understep pattern. **UNDERSTEP WALK** Stride: 4½–7⁵⁄16 in. (11.5–18.5 cm). Trail width: 1¹¹⁄16–2¹¹⁄16 in. (4.3–6.8 cm).

SIGNS

SCAT When well-formed, tubes are about 3⁄16 in. (0.5 cm) D. I have never found mole scats in our region. Reported to vary in shape, depending on diet. **MOLE HILLS** Coast mole—Average diameter: 11¹¹⁄16 in. (29.7 cm). Townsend's mole—Average diameter: 17¼ in. (44 cm). Mounds are relatively symmetrical. Soil pushed straight up and out from vertical shaft in the middle of the mound. The shaft is plugged after the soil is expelled, but with careful excavation you can locate it. Pocket gophers also leave mounds, but they tend to push soil to the side, leaving a fan-shaped mound radiating from the hole. In the spring, pregnant females may construct large mounds under which they build natal nests. Townsend's mole mounds are larger and more abundant than those of the coast mole and more conspicuous in their preferred habitat of open fields. Coast mole mounds are more often connected by a network of surface tunnel ridges. Encampments larger than listed here are likely those of a Townsend's mole: mound height 5.9 in. (15 cm), width 15.7 in. (40 cm); shallow tunnel diameter 1.8 in. (4.5 cm). **SURFACE TUNNELS** Both species create raised tunnel ridges associated with forag-

Townsend's mole encampments are common in grassy fields throughout the Puget Trough ecoregion.

Common mole track pattern

The excavation tunnel has not been filled in this mole hill; the entrance in the center of the mound is diagnostic for moles. Puget Trough, Washington.

ing activity and dispersal. In sandy soils and forest loam, coast moles produce many raised ridges, and the ground may appear cracked. Excavate to reveal the tunnel just below the surface. In winter, surface tunnels may be dug at the snow–soil interface, leaving only the bottom sections intact when the snow melts. **SPECIMENS** Dead moles are often found, as they may be killed but not consumed by wild and domestic predators. In low-lying areas, they can be drowned by flooding.

Mole surface tunnel in the sandy soils of a river floodplain. Puget Trough, Washington.

BATS Order Chiroptera

This diverse order of mammals is second only to Rodentia in terms of the number of species worldwide. Our region has more than 15 species of bats, most of which are vesper bats in the family Vespertilionidae. Common species include the little brown bat (*Myotis lucifugus*), the big brown bat (*Eptesicus fuscus*), the pallid bat (*Antrozous pallidus*), and various myotis bats (*Myotis* spp.). Bats are the only mammals in our region that truly fly (flying squirrels glide), and they are often seen on warm summer nights darting back and forth, hunting insects on the wing. They find insects using echolocation, which works like radar. All our bats are nocturnal and hibernate in the winter. Rabies is occasionally, though infrequently, found in bats, but this poses little threat to us humans unless we are handling or bitten by a bat.

SIZE Length: 7–15 cm. Weight: 4–15 g. **FIELD MARKS** Brown to gray. Forelimbs have large flight membrane. **REPRODUCTION** Many species breed in the fall as they gather for winter hibernation. Egg fertilization is delayed until spring. Often a female gives birth to a single young each spring; young are raised in nursery colonies of breeding females and their young. **DIET** Insectivores, with most species preferring flying insects such as moths. Can eat up to half its body weight in insects in a single night. The pallid bat will take invertebrates off the ground or from trees. **HABITAT & DISTRIBUTION** During active months, found throughout our region. Roosts in trees, under building eaves and in attics, in caves, and under bridges. Winter hibernation sites are more protected and stable than simple day roosts, as most species require high humidity, stable temperatures above freezing, and lack of disturbance for hibernation. Old mine shafts and deep caves are often used, and occasionally buildings. For some species, such as Townsend's big-eared bats (*Corynorhinus townsendii*), spring nursery sites can be similar to hibernacula (winter shelter). Other species, including little and big brown bats, commonly use warmer locations such as attics or rock crevices, where temperatures can fluctuate. **CONSERVATION ISSUES** During the 20th century, factors such as hab-

An unidentified bat species hangs from its roost under a bridge. Klamath Mountains, California.

itat destruction and pesticide use reduced bat populations in many areas. The current status of many bat species in our region is unknown, and several are likely threatened with regional extirpation.

SIGNS

SCAT Tiny to small, depending on the species. Pointed at the tip. Insect remains may be detectable if the scat is broken apart.
ROOSTS Bats return to the same roost again and again. Along with scats on the ground around these locations, you may see white urine stains.

Large bat scats. Klamath Mountains, northern California.

Urine stains appear under a bridge where bats frequently roost. Klamath Mountains, northern California.

HARES, RABBITS, AND PIKAS Order Lagomorpha

Two families in this order can be found in our region: Leporidae (rabbits and hares) and Ochotonidae (pikas). All are ground dwellers and herbivores. Females are slightly larger than males.

Reproduction and Social Habits

All members of Lagomorpha in our region are solitary and do not readily share core territory with other members of the species. All have a high reproductive rate, birthing multiple litters in a single year, and a high mortality rate for young (up to 90% in the first year of life). During the peak of breeding season, a female may be both pregnant and nursing, as estrus often occurs immediately after it gives birth. Young disperse at about one month of age. For most if not all species, mothers attend to nursing young only once a day.

Tracks and Signs

The feet of all rabbits and hares are completely furred, with no exposed metacarpal pads; fur causes the foot and heel to register indistinctly in tracks. Toes usually register the deepest on both front and hind feet, and sometimes only claws register in harder substrates. Pikas share many characteristics with rabbits and hares but have hairless toe pads.

Front. Five toes with toe 1 dropped significantly lower and to the inside of the track (often only registering as a claw).

Hind. Four toes, oriented slightly asymmetrically.

Track patterns. A bounding gait is usually employed, often with the front feet placed one in front of the other, followed by the paired hind feet in the pattern. Animals may walk while feeding.

Feeding signs. Lagomorphs' sharp incisors create clean cuts when feeding on vegetation. They do not manipulate vegetation with their forelimbs as do rodents, a factor that can sometimes help to distinguish feeding signs from these two orders.

Scat. All lagomorphs are coprophagous, producing two distinct fecal pellets. The first time food passes through the digestive track, it is expelled in soft, moist droppings that are reingested and not often observed in the field. This allows more complete digestion of plant material. The second type of scat, which is common to find for all species, are spherical, dry pellets. The specific size and shape depends on the species.

Rest sites. Many species use shallow depressions under brush, called forms, as rest spots.

AMERICAN PIKA

Ochotona princeps

Pikas are the mountain specialists of the Lagomorpha order with specific habitat associations driven in part by an intolerance to warm temperatures. They are diurnal (though they often retire during the hottest portion of the day) and are often seen or heard by people traveling through their habitat. They emit sharp trills to announce territoriality and breeding conditions but also to warn of predators. Though they are not colonial (they are intolerant of other pikas in their core territory), home ranges are small and adjacent to one another in good habitat. A pika often perches on a prominent rock in its territory, surveying for danger and incursions from other pikas. A short, sharp, loud whistle announces an oncoming predator and the animal disappears into the rocks along with all the other pikas in the vicinity. Learning to recognize this alarm call has helped me spot a variety of predators in the mountains, including coyotes, hawks, and weasels (the most effective predator of pikas).

SIZE Length: 16.2–21.6 cm. Weight: 121–176 g. **FIELD MARKS** Short, rounded ears; no tail. Gray-brown in color with a lighter underside. **REPRODUCTION** Breeds first about a month before the snowpack melts completely, and the first litter is born after snowpack melts. This allows the mother to benefit from spring vegetation for feeding while nursing. Litter size 3–6 young. Female breeds a second time soon after parturition, with second litter in midsummer. **DIET** Herbivore: mainly grasses and forbs. Meadow areas adjacent to talus fields are grazed heavily along the rocky edge. Larger plants from farther out in the meadow are collected in summer and stored in hay mounds for winter consumption. **HABITAT & DISTRIBUTION** Talus fields with adjacent meadows for feeding at mid to high elevations in all mountains of the Northwest. Some pikas inhabit rocky habitat in stream gorges at lower elevations in western Oregon, where proximity to water and abundance of shade creates a cooler microclimate that mimics that of higher elevations. Home ranges are 0.2–0.5 acres (860–2100 m^2). **SIMILAR SPECIES** Bushy-tailed woodrat, ground squirrels. **CONSERVATION ISSUES** Because of the pika's intolerance of warm temperatures, strong association with a limited habitat type, and

Pika

Pika range

limited ability to disperse among quality habitats through marginal terrain, global climate changes may negatively affect this species.

TRACKS

Front and hind feet are densely furred (page 61). **FRONT** 9/16–13/16 in. (1.5–2.0 cm) L × 1/2–5/8 in. (1.3–1.6 cm) W. Five toes. Toe 1 is greatly reduced, low on the inside of the track, with no exposed pad but a fully developed claw. Toes 2–5 have exposed pads that register in tracks. The remainder of the foot is furred and no distinct palm pad registers. **HIND** 13/16–1 1/16 in. (2.1–2.6 cm) L × 5/8–11/16 in. (1.6–1.8 cm) W. As in rabbits and hares, the hind foot is more symmetrical than the front. Each of the four toes has a naked pad and a claw. The heel is longer and more slender than the front foot but rarely registers. **TRACK PATTERNS** Bounds, with one of the front tracks often obscured by an overlying hind track. At faster speeds, hinds fall beyond the fronts, leaving a pattern more typical of other lagomorphs. **BOUND** Stride: 13 9/16–15.5 in. (34.5–39.4 cm). Trail width: 2 1/16–3 in. (5.3–7.6 cm). Group length: 2 1/2–3 in. (6.3–8.9 cm).

Three front and two hind tracks of a pika. The upper four tracks are in a characteristic bounding pattern for this species. The two hind tracks are parallel to each other and the fronts, between the hinds, are at an angle. West Cascades, Washington.

SIGNS

SCAT 1/16–3/16 in. (0.2–0.4 cm) D. Similar in shape to other lagomorph scat but smaller and round. Large deposits can be found

Typical pika scat. East Cascades, Washington.

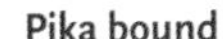

Pika bound

This brown vegetation has been clipped and brought to this location by a pika. West Cascades, Washington.

Hay mounds are often tucked into crevices in talus fields, where it is also common to find scats. West Cascades, Washington.

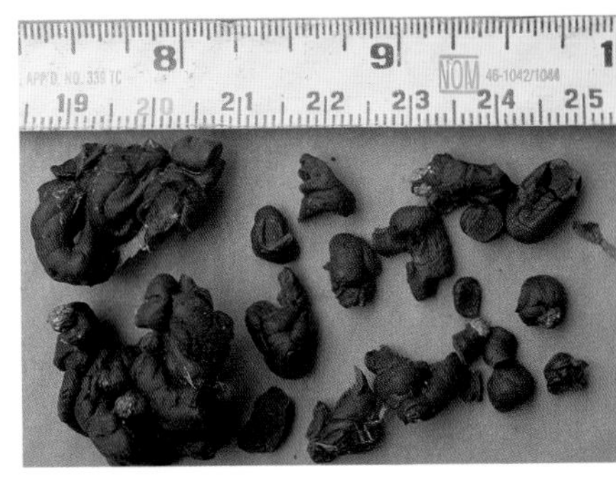

This cecal scat was found intermixed with clipped vegetation in a hay mound. East Cascades, Washington. Photo by David Moskowitz and Paul Houghtaling.

under and around the talus fields they inhabit, especially at well-used entrances to their abodes under the rocks. The softer cecal (first-round) scats are sometimes stored in hay mounds to supplement the winter diet. **FEEDING SIGNS** The most recognizable signs of a pika are the large piles of vegetation (hay mounds) they accumulate in the talus fields they inhabit. Pikas collect and store plants all summer long as a supply of emergency or supplemental food for the winter. Pikas continue to forage throughout the winter in the subnivean zone. The edges between occupied talus fields and meadows often appear mowed as pikas heavily feed close to the protection of the rocks.

COTTONTAIL RABBITS

Eastern cottontail, *Sylvilagus floridanus*
Nuttal's cottontail, *S. nuttallii*
Brush rabbit, *S. bachmani*

Our three cottontail species are similar in appearance, behavior, and track and sign patterns but can be distinguished by size

Nuttal's cottontail. Photo by Michael Francis.

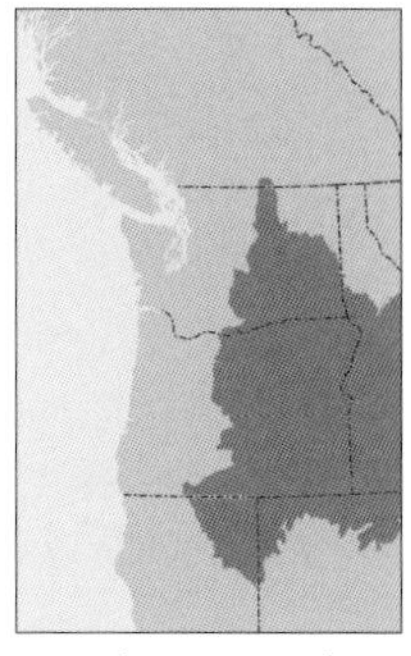

Nuttal's cottontail range

and distribution. All are herbivores with a crepuscular (twilight) activity pattern. They favor brushy habitats and rarely venture far from cover. All are solitary except during courtship and mating activities.

Nuttal's cottontails show several unique behavioral adaptations to life in their arid habitat, including climbing and feeding on western juniper trees, likely for the moisture content of the browse. They also tend to breed and disperse at night, perhaps to avoid the hotter and drier daytime conditions.

Brush rabbits are significantly smaller than our other cottontails, and this is reflected in their tracks and scats. Eastern cottontails were introduced extensively to western Washington, Oregon, and southern British Columbia and in limited locales in the eastern portions of our region. Their range overlaps in part with brush rabbits with whom they occasionally interbreed.

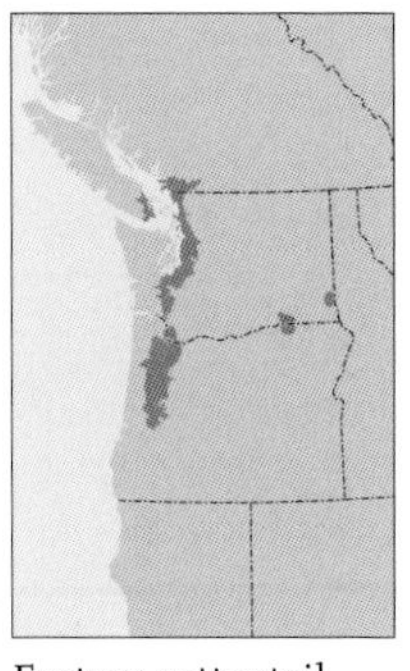

Eastern cottontail range

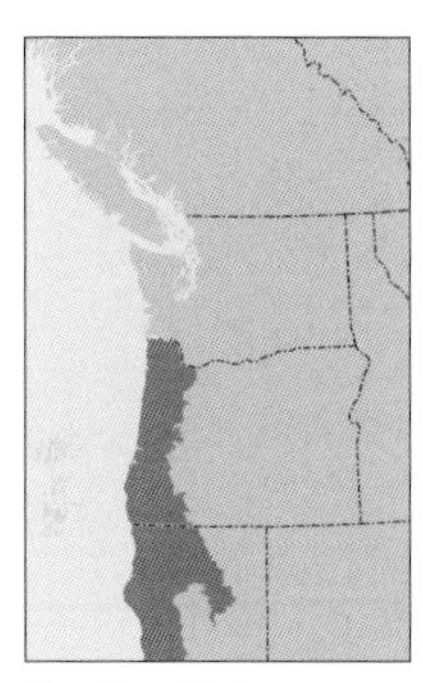

Brush rabbit range

SIZE Brush rabbit—Length: 30.3–36.9 cm. Weight: 0.511–0.917 kg. Eastern cottontail—Length: 39.5–47.7 cm. Weight: 0.801–1.533 kg. Nuttal's cottontail— Length:33.8–39.0 cm. Weight: 0.628–0.871 kg. **FIELD MARKS** Rounded ears and gray-brown pelage. Nuttal's and eastern cottontails have small, fluffy white tails. The brush rabbit's tail is similar in color to its coat and smaller than that of the eastern cottontail. **REPRODUCTION** Breeding begins in February and continues into August (shorter for Nuttal's). Courtship behavior includes leaping over each other, urinating, boxing, and male chasing female. Females may have 3–4 litters a year, each with 3–6 young, which are born altricial but develop quickly and are independent within a month, when the mother may birth another litter. Females from the first litter may breed by summer's end. **DIET** Herbivorous, with grasses and forbs as the primary food source in spring and summer. Buds and cambium of woody plants, conifer needles, and occasionally berries become more important food in late summer, fall, and winter. **HABITAT & DISTRIBUTION** Brushy locations, rarely venturing far into the open. Brush rabbit range has increased in western Oregon because of anthropogenic changes—brushy clearcuts, agricultural fence rows, and stabilized coastal dunes. The Nuttal's range is mainly east of the Cascades from low desert to ponderosa pine and Douglas-fir forests in mountains. Brush rabbits' home ranges conform to their brush and edge habitat with a radius of 33–258 ft. (9–72 m) from a core use area. **SIMILAR SPECIES** Snowshoe hare, jackrabbits, pygmy rabbit, ground squirrels. **CONSERVATION ISSUES** Concerns about the eastern cottontail pushing out brush rabbits in western Oregon have not been substantiated. These species have likely benefited from many modern changes to our region's environment.

TRACKS

Typical lagomorph shape. Feet are completely furred, often creating tracks with little detail, and sometimes only the claws register. Both front and hind often come to a point in the front of the track. Fronts are smaller than hinds but the difference is not as pronounced as in snowshoe hares. Most

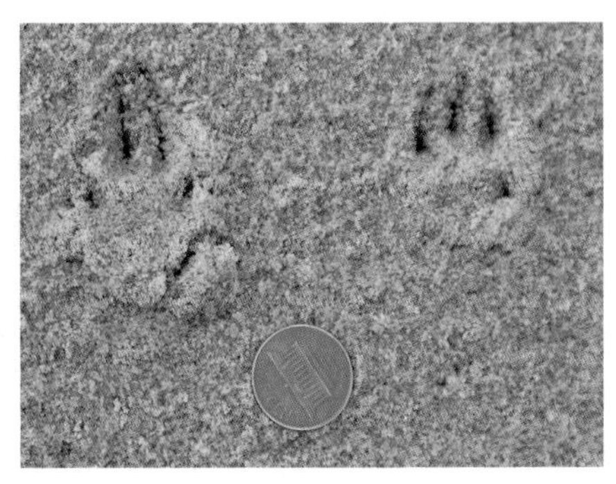

Two hind feet of a brush rabbit. Claws are often the most distinctive part of rabbit tracks. Northwest Coast, Oregon.

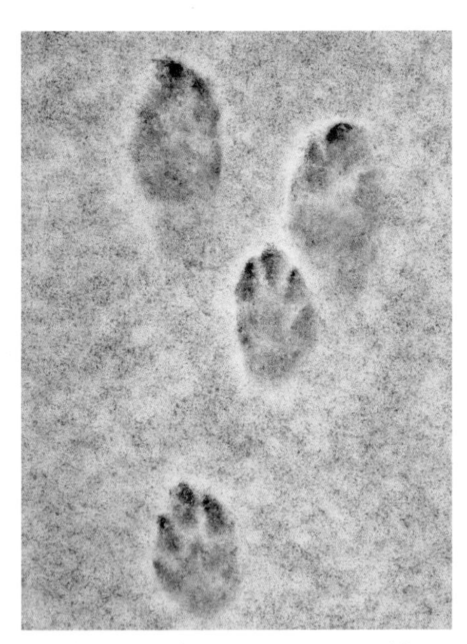

Nuttal's cottontail track pattern is typical for this genus. From bottom to top: left front, right front, right hind, left hind. Columbia Plateau, Washington.

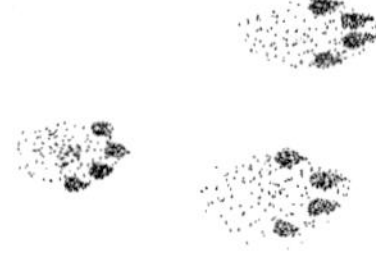

Nuttal's cottontail bound

tracks are smaller than those of jackrabbits and snowshoe hares (page 65). **FRONT** 1¹⁄₁₆–2 1⁄16 in. (1.8–5.2 cm) L × ½–1 7⁄16 in. (1.3–3.7 cm) W. Highly asymmetrical, with five toes, with toe 1 low, generally registering only as a claw if at all. The remaining four give the track a characteristic J-shape. **HIND** 1 1⁄16–2 5⁄8 in. (2.7–6.7 cm) L × 9⁄16–2 1⁄16 in. (1.5–5.2 cm) W. Slightly asymmetrical, four toes can splay widely in deep substrate. More likely to appear pointed at the tip of the track than in snowshoe hare tracks. **TRACK PATTERNS** Characteristic pattern is an angled bound. Tracks rarely appear far from cover. **BOUND** Stride: 12 3⁄16–65 3⁄8 in. (31–166 cm). Trail width: 2¾–4 15⁄16 in. (7.0–12.5 cm). Group length: 5 5⁄16–15 5⁄16 in. (13.5–38.9 cm).

SIGNS

SCAT 1⁄8–¼ in. (0.3–0.7 cm) D × ¼–½ in. (0.6–1.3 cm) L. Flattened disks of compressed vegetation are often found collected in their brushy habitat. All three species' scats appear similar, but brush rabbit pellets are usually slightly smaller than those of the eastern cottontail. **FEEDING SIGNS** Abundant feeding on grasses, forbs, and small stems of woody shrubs can resemble the work of rodents, but cottontails are more likely to remove some wood along with cambium when feeding on bark. In desert areas, Nuttal's cottontails feed on sagebrush, leaving feeding signs similar to those of the pygmy rabbit. Many scats are often scattered around well-used feeding locations.

Cambium feeding by a Nuttal's cottontail on a willow shrub (*Salix* spp.) shows that, along with the bark, some wood has been removed. Scats are often found around such feeding sites. Northern Great Basin, Oregon.

Two forms of scats from Nuttal's cottontail; scats on the right are more typical. Columbia Plateau, Washington.

PYGMY RABBIT

Brachylagus idahoensis

The smallest member of this order in our region, pygmy rabbits are also the only lagomorph native to our region to excavate and use burrows extensively. Their hind legs are shorter than those of other rabbits. Active any time of day, they rest in forms under sagebrush or in burrows and occasionally climb sagebrush to feed on the foliage.

SIZE Length: 22.0–30.1 cm. Weight: 0.26–0.46 kg. **FIELD MARKS** Small, with a tiny gray tail and short, rounded ears. **REPRODUCTION** Shorter breeding season than other lagomorphs, February to May. Up to three lit-

Pygmy rabbit. Photo by Keith Lazelle.

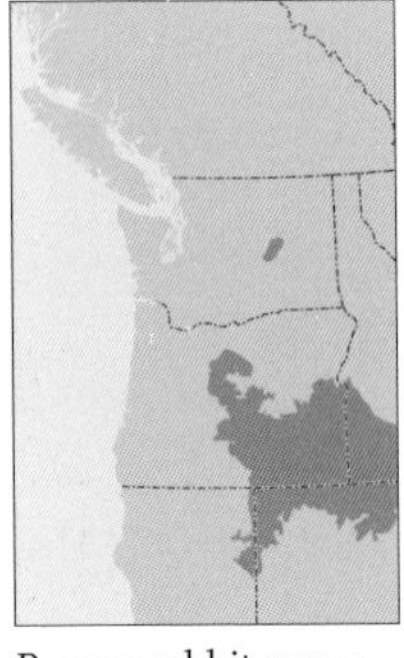

Pygmy rabbit range

ters a year with up to six young per litter. **DIET** Sagebrush (*Artemisia tridentata*) makes up about 60% of their diet (and as much as 99% of their winter diet). Grasses and forbs account for the rest. **HABITAT & DISTRIBUTION** Sagebrush-obligate species, usually occurring in locations with extensive stands of tall, mature sagebrush, though they also occupy sites that include other shrub cover. Studies show that 40% brush cover of an area is optimal, and they do not occupy areas with less than 20%. Also require loose soil for burrow excavation. Most individual activity occurs within 100 ft. (30 m) of the primary burrow. **SIMILAR SPECIES** Cottontail rabbits, ground squirrels. **CONSERVATION ISSUES** The disjunct population of this species in southeastern Washington is endangered and the focus of a captive-breeding program. Population has also declined in Oregon and elsewhere in its range and is being considered for protection under federal endangered species laws. Decline is a result of conversion of habitat to agriculture and the fragmentation of remaining habitat beyond the species' dispersal abilities.

TRACKS

Typical rabbit shape and track pattern. Smaller than tracks of the Nuttal's cottontail.

SIGNS

SCAT 1/8–3/16 in. (0.3–0.5 cm) D × 3/16–1/4 in. (0.4–0.7 cm) L. Similar in shape to other rabbits but smaller and often slightly more spherical. **BURROWS** Constructed at the base of sagebrush shrubs and into the banks of dry streams in areas with deep soils and minimal rocks. A throw mound of soil may be visible at the entrance. In actively used burrows, large amounts of scats appear in the runway leading into the burrow. Average diameter is 4 1/8 in. (10.4 cm), often wider than tall. Badgers and coyotes excavate these

Pygmy rabbit scats. Northern Great Basin, Orego

Pygmy rabbit burrow in typical habitat. Northern Great Basin, Oregon.

Burrow entrance at the base of a sagebrush shrub. The sunken trough is filled with scats. Northern Great Basin, Oregon.

Typical feeding sign in sagebrush; around actively used burrows, sagebrush often shows extensive cropping by pygmy rabbits. Columbia Plateau, Oregon.

burrows, creating larger entrance holes and throw mounds. These holes are often reinhabited by pygmy rabbits (not included in the above average size). **FEEDING SIGNS** Clipped sagebrush branches at ground level and higher in the shrub, close to burrow. Similar to Nuttal's cottontail signs.

SNOWSHOE HARE

Lepus americanus

Snowshoe hare tracks are some of the most common tracks to find in snow in most of our region's mountains. Several characteristics show its amazing adaptations to deep snow and long winters: exceptionally large feet in comparison to its weight and a change in fur color to pure white in winter. This nocturnal species is an important prey for mountain carnivores, including Canada lynx and American marten.

SIZE Length: 36.3–52.0 cm. Weight: 0.9–2.2 kg. **FIELD MARKS** Relatively short ears and large hind feet. Brown in summer, occasionally with white splotches. White in winter (except where little or no snow occurs). **REPRODUCTION** Breeds April to October. Up to three litters in a year with 3–4 young per litter. Young are precocial. Young disperse at about one month of age. Regionally they do not experience the population cycles that occur in the boreal populations of this species. **DIET** In winter, conifer needles and the bark on twig tips. Otherwise grass and forbs, and buds on shrubs such as *Vaccinium* spp. **HABITAT & DISTRIBUTION** Most productive habitat is dense conifer forest with low branches and edge habitats. Ranges from sea level to treeline on the west side of the Cascades and in mountains throughout the region. Absent in lower elevations and arid environments east of the Cascades. **SIMILAR SPECIES** Cottontail rabbits, white-tailed jackrabbit, canines.

TRACKS

Typical lagomorph shape. The feet are completely furred and large for the animal's size (page 65). **FRONT** 1 5⁄16–2 9⁄16 (3.3–6.5 cm) L × 1 1⁄8–1 15⁄16 in. (2.9–4.9 cm) W. Highly asymmetrical. Five toes, with toe 1 low and often registering only as a claw. The remaining four give the track a characteristic J-shape.

Snowshoe hare range

HIND 2 3⁄8–4 3⁄4 in. (6.1–12.1 cm) L × 1 7⁄16–3 1⁄8 in (3.6–8.0 cm) W. Slightly asymmetrical, with four large toes. Hinds distinctly larger than fronts (greater difference than in cottontails). Track appearance can depend on substrate and toe spread, often with a rounded appearance at the front of the track. Furred heel usually registers in track. **TRACK PATTERNS** Angled bound similar to that of cottontail rabbits. **BOUND** Stride: 16–68 1⁄2 in. (40.6–174.0 cm). Trail width: 4 15⁄16–7 7⁄16 in. (12.5–18.9 cm). Group length: 8 1⁄2–18 1⁄2 in. (21.6–47.0 cm).

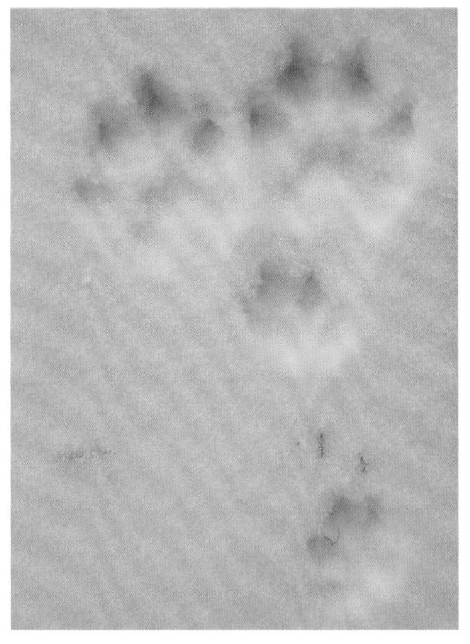

Typical track pattern of a snowshoe hare. Left front (lowest) and right front tracks are significantly smaller than the paired hind tracks. North Cascades, Washington.

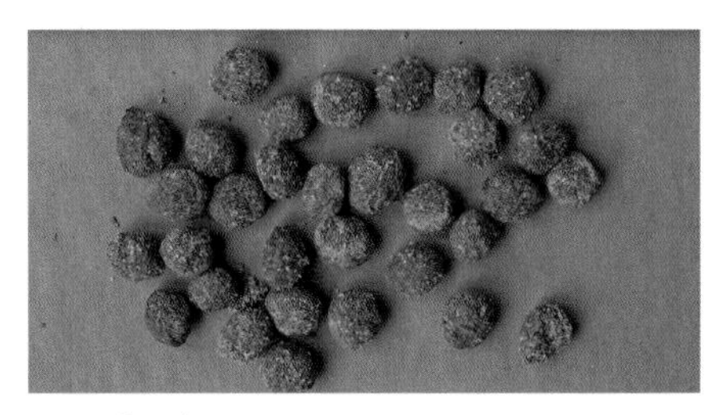

Snowshoe hare scats. West Cascades, Washington.

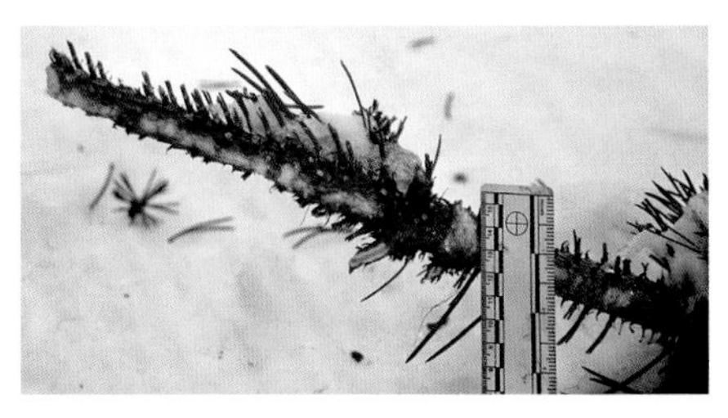

A snowshoe hare fed on the needles and bark of this lodgepole pine branch that had fallen to the forest floor during a winter storm. North Cascades, Washington.

Several herbivore species favor bark from burls, such as this one, eaten by a snowshoe hare on a lodgepole pine when deep snow provided access. Canadian Rockies, Montana.

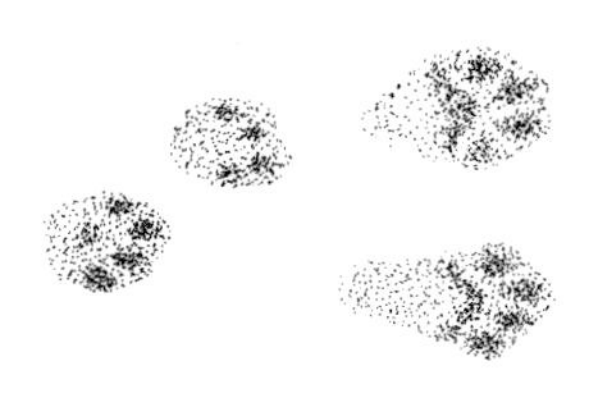

Snowshoe hare bound

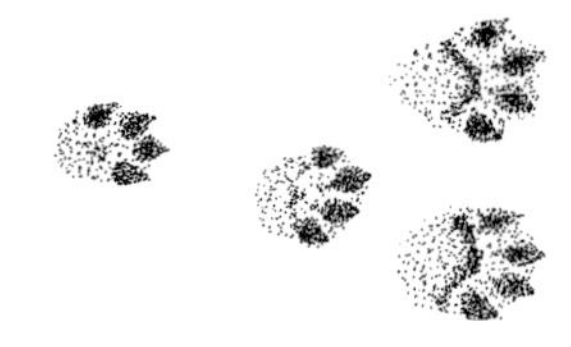

SIGNS

SCAT 1⁄4–3⁄8 in. (0.6–1.0 cm) D × 5⁄16–7⁄16 in. (0.8–1.1 cm) L. Overlaps in size and shape with that of cottontail rabbit, though generally larger. **FEEDING SIGNS** Sharp, often angled cuts on tips of conifer and shrub branches. Cambium feeding on branches and trunk is common. They do not climb, but in summer their browse signs can appear quite high because the animals stood atop deep winter snowpack to feed.

Black-tailed jackrabbit. Photo by Richard Day/Daybreak Imagery.

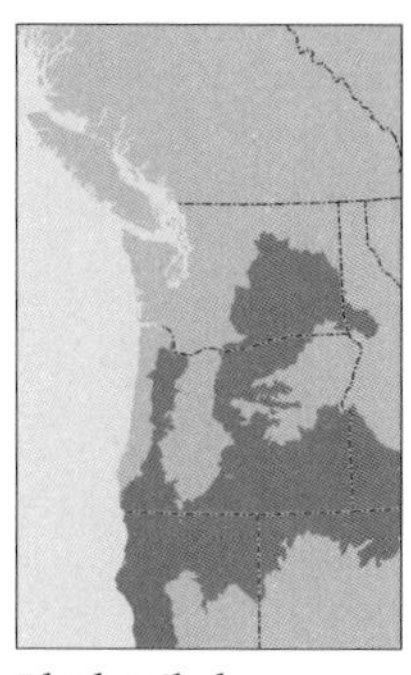

Black-tailed jackrabbit range

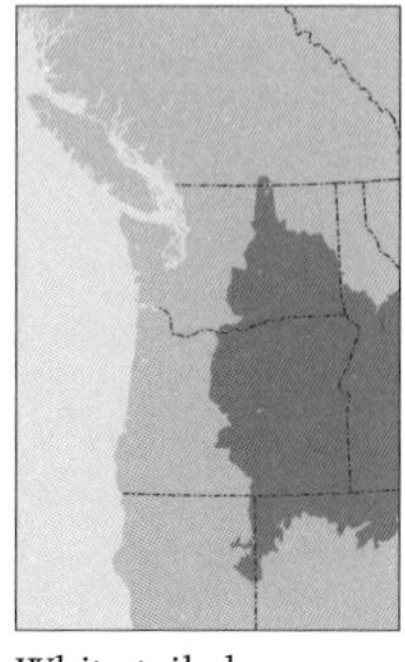

White-tailed jackrabbit range

JACKRABBITS

Black-tailed jackrabbit, *Lepus californicus*
White-tailed jackrabbit, *L. townsendii*

Jackrabbits are strong runners, so they can use habitat far from cover, unlike cottontail rabbits. They are nocturnal, resting in forms in brush by day and feeding in open grasslands by night. Well adapted to life in an arid environment, their diet provides required moisture when free water is unavailable. They have excellent hearing (their large ears also act to shed heat during hot weather) and a 360-degree field of vision, because their eyes are on the sides of their head. White-tailed jackrabbits are the largest lagomorphs in our area.

SIZE Black-tailed jackrabbit—Length: 46.3–66.0 cm. Weight: 1.3–3.6 kg. White-tailed jackrabbit—Length: 56.5–65.5 cm. Weight: 2.6–4.3 kg. **FIELD MARKS** Both species are substantially larger than cottontails, with larger black-tipped ears and long hind legs. Black-tailed jackrabbits have a black tail. White-tailed jackrabbits have a white tail, and their hair turns white in winter (hair appears mottled because it is white at the root and tip but brown in the middle of each hair). **REPRODUCTION** Black-tailed jackrabbits can breed year-round, with most young born from late winter to midsummer. Breeding season for white-tails lasts from early April to July, with 1–4 litters per year and 3–6 young in each litter. Young are precocial. Population size appears to be cyclical. Mortality rate for young is high. **DIET** Herbivores. White-tailed jackrabbits prefer grasses and forbs to shrubs when readily available. **HABITAT & DISTRIBUTION** Black-tailed jackrabbits inhabit arid and lower elevation environments east of the Cascades, plus interior valleys of western and southwest Oregon and coastal dunes on the northern California coast. White-tailed jackrabbits inhabit arid grasslands and higher elevation open habitats. **SIMILAR SPECIES** Cottontail rabbits, snowshoe hare. **CONSERVATION ISSUES** The black-tailed jackrabbit population in western Oregon has dropped from historic levels and may be in danger of disappearing from this part of their range. Populations in eastern Washington have also been reduced and may receive protection under state endangered species regulations. White-tailed jackrabbits are associated with mature arid grasslands in our region, which are rare, scattered, and threatened regionally from agricultural development and fire suppression. Population and abundance is unclear in much of our region. The species has likely been completely extirpated from British Columbia.

TRACKS

Tracks are similar in shape but larger overall than cottontail tracks. Hinds are larger than fronts. A pointy hind foot appearance is less likely than in cottontails. White-tail jackrabbit tracks are larger on average than those of black-tailed, but some overlap occurs in the size ranges of tracks and track patterns. **FRONT** Black-tailed jackrab-

Black-tailed jackrabbit tracks, from bottom to top: right front, left front, right hind, left hind. Columbia Plateau, Oregon.

All four feet of a black-tailed jackrabbit, with only the claws registering. Northern California coast.

Black-tailed jackrabbit gallop

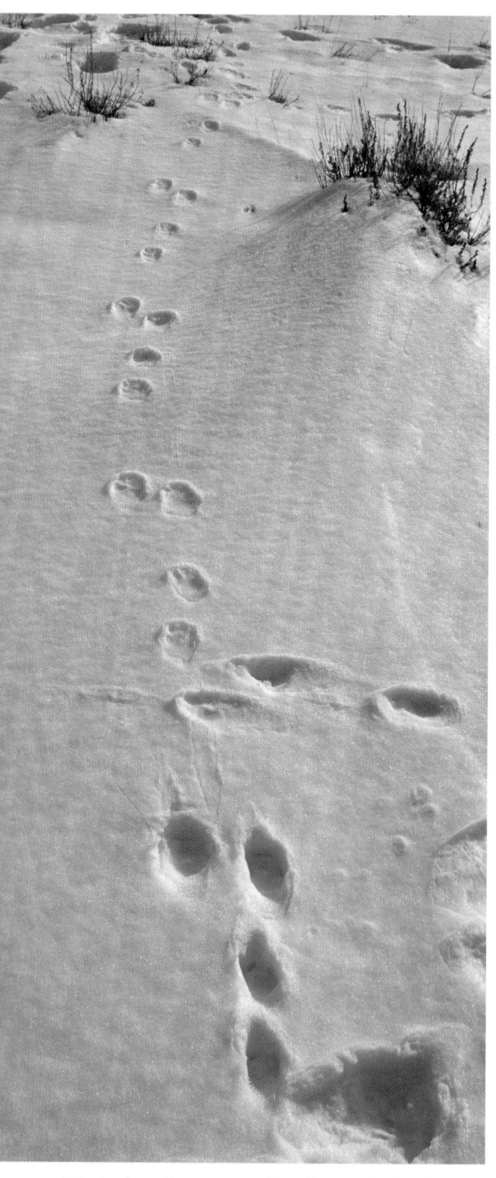

Typical track pattern of a white-tailed jackrabbit, traveling away from the viewer. Tracks and trail width are wider than typically encountered in black-tailed jackrabbits.

bit—1 3/8–2 3/16 in. (3.5–5.6 cm) L × 1 1/8–1 7/16 in. (2.8–3.7 cm) W. White-tailed jackrabbit—2 3/16–3 3/8 in. (5.5–8.5 cm) L × 1 9/16–2 3/16 in. (4.0–5.5 cm) W. **HIND** Black-tailed jackrabbit—1 3/4–4 1/8 in. (4.5–10.5 cm) L × 1 1/4–2 1/8 in. (3.1–5.4 cm) W. White-tailed jackrabbit—3 1/8–4 3/4 in. (8.0–12.0 cm) L × 1 15/16–2 3/4 in. (4.9–7.0 cm) W. **TRACK PATTERNS** Distinguished from cottontails because groups are usually more elongated with hind tracks often registering offset (rather than even as in cottontails). Tracks are commonly found in large openings, far from protective brush. **BOUND/GALLOP** Black-tailed jackrabbit—Stride: 18 1/8–72 5/8 in. (46.0–184.5 cm). Trail width: 3 3/8–5 11/16 in. (8.5–14.5 cm). Group length: 7 1/16–35 5/8 in. (18.0–90.5 cm). White-tailed jackrabbit—Stride: 29 15/16–51 9/16+ in. (76–131+ cm). Trail width: 4 15/16–6 7/8 in. (12.5–17.5 cm). Group length: 16 1/8–26 3/4 in. (41–68 cm).

SIGNS

SCAT 3/16–5/16 in. (0.5–0.8 cm) D × 3/8–1/2 in. (0.9–1.3 cm) L. Similar in shape to cottontail scats but larger. Indistinguishable between the two species. Often found in feeding areas that may be farther from cover than is common for cottontails.

Black-tailed jackrabbit scats. Columbia Plateau, Oregon.

RODENTS Order Rodentia

Rodents, often referred to as gnawing mammals, are defined in part by their unique dentition—large, continuously growing incisors; no canine teeth; and a significant space between their incisors and cheek teeth, which in some species are also continuously growing. More species of rodents exist in the world than any other order of terrestrial mammals. The Northwest is home to ten families and more than 60 species. Rodents can be terrestrial, semiaquatic, fossorial, or arboreal. They are herbivorous or omnivorous, but diet varies widely in this diverse order.

Tracks and Signs

Rodents are plantigrade. Because of the variety of ecological niches and lifestyles of the order's many members, a great deal of variation from, and exceptions to, the general descriptions of tracks and signs presented here exists.

Front. Four toes register in tracks in a 1-2-1 pattern with toes 2 and 5 oriented outward and toes 3 and 4 aligned and facing forward. A remnant toe 1 or its claw may register low on the inside of the track. Palm comprises three unfused, partially fused, or totally fused metacarpal pads. Typically two proximal pads. Claws generally register.

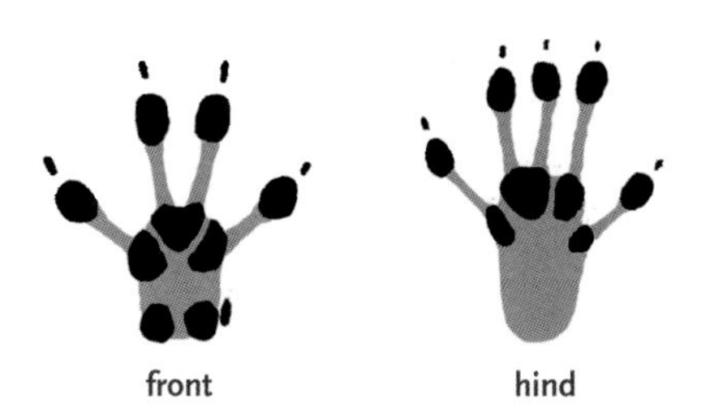

Hind. Often larger than front. Five toes register in a 1-3-1 pattern, similar to the front foot, with outer toes splaying and the central toes pointing forward. Claws often register. The palm shape varies but usually comprises four distinct metacarpal pads. One or more proximal pads or a furred heel may register.

Track patterns. Walking and bounding are common. Each group described has distinctive track patterns.

Scat. Varies in size and shape from species to species, but all produce numerous oblong pellets that reflect the maker's size. May be deposited in large collections (latrines), along trails, in feeding areas, or in and around bedding sites.

Gnawing signs. All rodents have two pairs of continuously growing, large incisors used to collect food and chew through wood and vegetation to build trails, nests, and lodges. Rodents remove bark from trees or shrubs and clip grasses and forbs at a sharp angle. Old bones and antlers in the field often show signs of rodents' gnawing for calcium. Gnaw marks' relative size and chewed item's location suggest which species created the marks.

Rodent Groups

Rodent groups and species are arranged by appearance, habitat, and similarity in tracks and signs to aid in field identification. While this organization often reflects their taxonomic classification, in several cases, members of one family are grouped with members of other families that share similar characteristics. Use the introduction to each group to narrow down your search for what sort of rodent you have detected in the field.

LARGE SEMIAQUATIC RODENTS

Families Castoridae, Myocastoridae, and Cricetidae

The Pacific Northwest is home to three species of semiaquatic rodents—muskrat, nutria, and beaver. Each species is adapted to life in the water and travels on land briefly for foraging or dispersal, with swimming being the most comfortable mode of travel. Trails on land generally lead away from and back to water or appear between two water sources, and most signs are found close to water.

Tracks. This group shows significant variance from the typical rodent structure. Hind feet, much larger than the fronts for all three species, have webbing (beaver and nutria) and stiff hairs that increase the foot's surface area (muskrats) for swimming.

Track patterns. All three species usually walk, leaving an alternating track pattern. Tail drag sometimes registers. Large hind tracks often partially or completely obscure front tracks.

Lodges and burrows. Dwellings are constructed of plant material in water or are burrowed into the banks along ponds or streams with underwater entrances.

MUSKRAT

Ondatra zibethicus

Taxonomically speaking, muskrats are exceptionally large voles. Their tracks and signs are found in or along the edges of

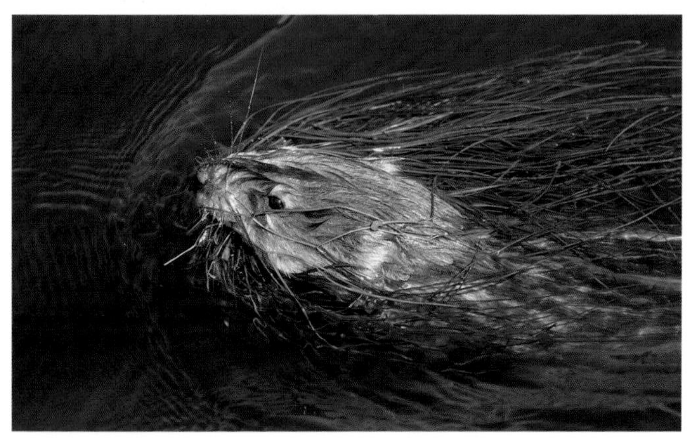

Muskrat. Photo by Larry Jon Friesen, sbnature.net.

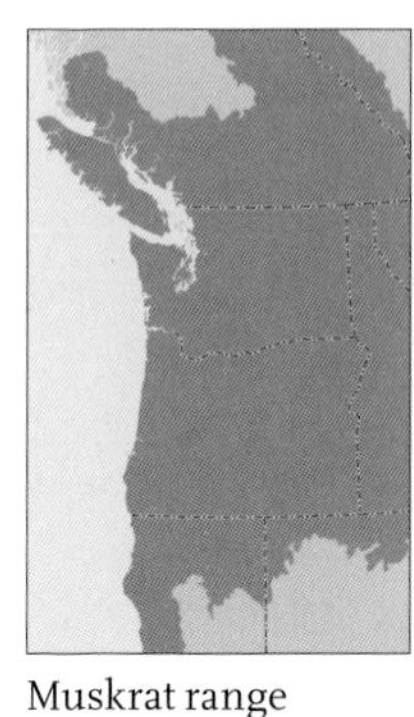

Muskrat range

marshes, slow-moving rivers, lakes, or bodies of brackish water. Adaptations to their aquatic lifestyle include trapped air in their watertight fur, which increases insulation and buoyancy; a flattened, rudderlike tail; lips that close inside incisors, allowing the animal to gnaw on items underwater; and specially adapted hind feet for swimming. Especially during the breeding season, they release musk from a pair of glands at the base of their tails on trails and scats and by lodge or burrow locations.

SIZE Length: 41–62 cm. Weight: 0.7–1.8 kg. **FIELD MARKS** Medium-sized, brown, with lighter underside. Vertically flattened tail. **REPRODUCTION** Males and females are highly territorial during the breeding season. They are usually monogamous and produce up to three litters a year with 6–7 young per litter, following a monthlong gestation period. The first litter usually occurs around April. Most young disperse in the fall. **DIET** Aquatic plants, including stems, leaves, and roots. Also fish, crayfish, and mollusks. Food is often brought to a feeding station for consumption. **HABITAT & DISTRIBUTION** Aquatic environments throughout our region. **SIMILAR SPECIES** Nutria, mountain beaver, beaver, water vole.

TRACKS

Soles of the feet are hairless. Toes are long, thin, and ribbed, with no bulbous pads at the tips (page 63). **FRONT** 1⅛–1⁷⁄₁₆ in. (2.8–3.7 cm) L × ⅞–1⅜ in. (2.2–3.5 cm) W. Four long, slender toes. Long, blunt claws, used for digging, register in tracks, sometimes more prominently than everything else. Toe 1 is greatly reduced, but the claw often regis-

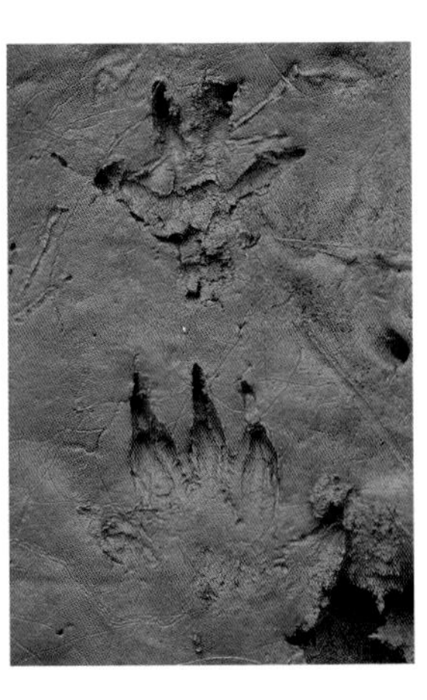

Left front (above) and hind tracks in mud. Puget Trough, Washington.

Left front (below) and hind. Only the central three toes of the hind foot have registered in firmer mud. Columbia Plateau, Washington.

Muskrat walk

A muskrat dug up cattail rhizomes (*Typha latifolia*) at the edge of a pond. Beavers leave similar signs. Columbia Plateau, Washington.

Muskrat lodge. Photo by Chris Maser.

Muskrat scats. Puget Trough, Washington.

ters along the palm pad's inside edge. Three subpads in the palm and two proximal pads form a box shape. Toes almost always connected to the palm pads in the track. **HIND** 1¼–1 15⁄16 in. (3.1–4.9 cm) L × 1 9⁄16–2 in. (3.9–5.0 cm) W. Larger than the front foot. Toes are not webbed. Stiff, bristly hairs around toes increase the foot's surface area and aid in swimming. Hairs are often visible in tracks, creating a slightly shallower "shelf" around each toe in deeper substrates. All five toes radiate out from half-moon–shaped palm. Claws are as long or longer than those of front and usually register. Heel, with a single proximal pad, rarely registers. **TRACK PATTERNS** Walks with an indirect register or overstep pattern. The long, hairless tail can drag through the center of the trail. **WALK** Stride: 7 1⁄16–14 3⁄16 in. (18.0–35.6 cm). Trail width: 2 15⁄16–5 5⁄16 in. (7.5–13.5 cm).

SIGNS

SCAT 3⁄16–5⁄16 in. (0.5–0.8 cm) D × 5⁄16–11⁄16 in. (0.8–1.7 cm) L. Pellets with rounded ends often left in a loose grouping, occasionally clumped together, along the edge of or in water. **FEEDING SIGNS** Look for cut vegetation floating along the water's edge where muskrats live. Like beavers, muskrats dig up cattail and other wetland plant roots, leaving shallow depressions in the mud. **LODGES & BURROWS** Depending on the location, muskrats will reside in lodges constructed in the water or in burrows dug into a bank. Lodges made of cattails or other vegetation are accessed via underwater entrances. (Beavers use wood and large branches, not just soft vegetation, for lodges.) Burrows are dug above the waterline with underwater entrances. When water levels drop in the late summer or fall, entrances can become exposed. Entrances are about 6–8 in. (15–20 cm) in diameter but can become larger with use. **FEEDING STATIONS** Muskrats often create feeding platforms, floating mats of aquatic vegetation that collect debris from feeding activity. You may find middens along the water's edge, where a muskrat returns to feed.

NUTRIA

Myocastor coypus

Nutria are larger than muskrats and smaller than beavers. This invasive species, native to South America, was introduced to western Oregon in the 1930s for fur production. They are active throughout the year but may be adversely affected by very cold temperatures. Their hearing is good but their vision is poor. Nutria form social groups of more than a dozen animals, with a dominant male and female. Nutria can be a confusing common name because it means otter in Spanish.

SIZE Length: 86–106 cm. Weight: 6.7–9.0 kg. **FIELD MARKS** Long, hairless, round tail and large, dark orange incisors. Exceptionally buoyant in water: when swimming, its entire back is out of the water. **REPRODUCTION** Can breed year-round. Courtship includes chases, vocalizations, and wrestling. Females may breed with more than one male during each estrus and can have two litters per year, with 3–6 young. Young are precocial, can swim shortly after birth, and disperse at about 6 weeks (documented to more than 30 miles). They can breed at 8 months or younger. Home ranges of about 10 acres (0.04 km^2, larger for males than

Nutria

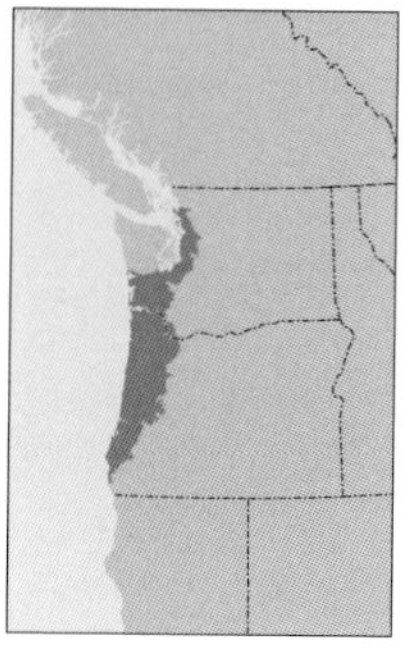

Nutria range

females). **DIET** Variety of aquatic vegetation (leaves and roots), grass adjacent to water, tree bark, and mollusks. **HABITAT & DISTRIBUTION** Lakes, rivers, and marshes in western Oregon and Washington, southwest British Columbia, and in the Columbia and Snake River systems. Most abundant west of the Cascades, where their range appears to be expanding. **SIMILAR SPECIES** Muskrat, beaver. **CONSERVATION ISSUES** In agricultural areas, nutria are often considered a pest. They likely adversely affect muskrat populations through direct competition. Alterations to aquatic plant populations may adversely affect native amphibians.

TRACKS

The front is smaller than the hind, though the difference is less than in beavers. Tracks are larger than those of muskrats (page 67). **FRONT** 1¾–2¼ in. (4.4–5.7 cm) L × 1⁹⁄₁₆–2³⁄₁₆ in. (4.0–5.6 cm) W. Toes 2–5 are long and thin with long claws. Toe 1 is greatly reduced, but the claw often registers on the inside next to the palm. Palm comprises three partially fused metacarpal pads. Two proximal pads usually register. **HIND** 2⁹⁄₁₆–4⁷⁄₁₆ in. (6.5–11.2 cm) L × 2⁹⁄₁₆–2¹⁵⁄₁₆ in. (6.5–7.5 cm) W. Five toes are long and thin. Long claws usually register. Toes 1–4 are webbed, attached below the slightly bulbous pads at the tips. Palm may show distinct pads or be totally indistinct. Key differences from beaver: Toe 5 is free of webbing, toes often radiate evenly with toe 3 distinctly longer than all others. The heel is proportion-

Left hind track with the heel registering clearly; no webbing shows between toes 4 and 5 (two left toes). Willamette Valley, Oregon.

All four feet of a nutria in a typical overstep walk pattern. The heel of the hind tracks has not registered, with the central three toes registering most strongly. Puget Trough, Washington.

Typical nutria pellets. Puget Trough, Washington.

An exceptionally large bank burrow entrance of a nutria at the edge of a marsh is exposed by low water levels. Willamette Valley, Oregon.

ally narrower and registers less consistently. **TRACK PATTERNS** Walks. Hinds may overstep or directly register. **WALK** Stride: 11 5/8–16 1/8 in. (29.5–41.0 cm). Trail width: 4 3/8–7 11/16 in. (11.1–19.5 cm).

SIGNS

SCAT 5/16–1/2 in. (0.8–1.3 cm) D × 1 3/8–1 9/16 in. (3.5–4.0 cm) L. Coprophagous. Pellets appear grooved linearly. Occasionally produce amorphous scats that could be confused with summer deer scats. **BANK BURROWS** Dug into sloped banks of waterways and marshes and often partly out of water. Diameters average about 8.25 in. (21 cm).

BEAVER

Castor canadensis

The beaver is a keystone species and ecologically one of the most important species of mammals in North America and the Pacific Northwest. Beavers make significant modifications to their environment, creating or increasing the size of wetlands and riparian habitat, vital for numerous species of plants and wildlife. They create intricate dams to impound water, which creates a refuge from predators and increases the abundance of preferred forage plants.

SIZE Length: 100–120 cm. Weight: 16–30 kg. **FIELD MARKS** Our largest rodent. Large, flat tail; brown pelage; orange incisors. Often only the head is visible while swimming. **REPRODUCTION** Colonial: parents plus young of the year and previous year. Only the parent female in the group breeds. Breeding occurs around January with 3–5 young born around May. Animals are territorial and intruders are treated aggressively.

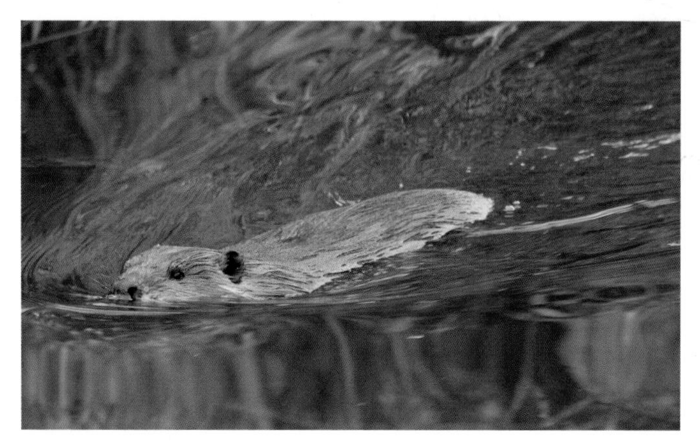

Beaver

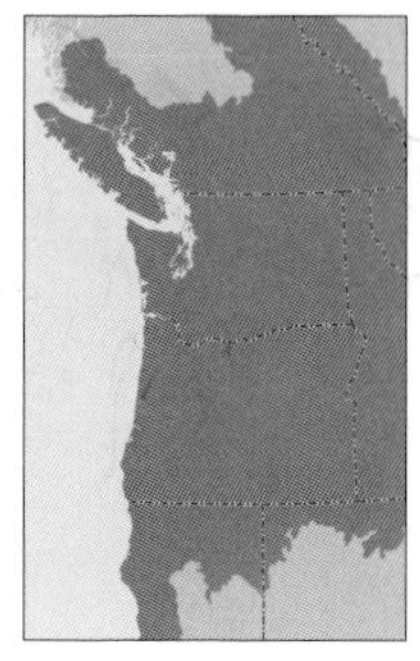

Beaver range

DIET Strictly herbivorous. Main food source is the cambium from trees and saplings that they fell. Preferred species include willows, red alder (*Alnus rubra*), and cottonwoods. Also leaves and roots of aquatic plants such as cattails. **HABITAT & DISTRIBUTION** Streams, marshes, ponds, and rivers throughout our region, including urban areas. More abundant west of the Cascades. **SIMILAR SPECIES** Nutria, muskrat. **CONSERVATION ISSUES** Extensive beaver trapping throughout North America caused massive declines in this species. The first regional protection measures were enacted in Oregon in 1899. Beaver population throughout the region has been increasing since then, with numerous transplants of animals to aid in repopulation. In some mountain areas, populations have yet to return. The presence of beavers correlates with increased ecological productivity and diversity for songbirds, salmon, and many other species.

TRACKS

FRONT 1¾–3 7/16 in. (4.5–8.8 cm) L × 2 1/16–2 11/16 in. (5.3–6.8 cm) W. Toe 1 is fully developed though it rarely registers clearly. Large, blunt claws usually register. Toes 3 and 4 curve slightly inward. Toes connect to the palm pad in tracks. Two large proximal pads give heel a boxy appearance. **HIND** 4 1/8–7 7/8 in. (10.5–20.0 cm) L × 2 5/16–5 1/8 in. (5.9–13.0 cm) W. Large for the animal's size and more than twice the size of the fronts. Toes are completely webbed, which often shows in tracks. Heel may be rounded or blunt. Often only toes 3–5 register or register more deeply than the rest of the foot. Toes 3 and 4 often appear similar in length in tracks (page 69). **TRACK PATTERNS** Walks with hind feet registering behind, next to, or on top of fronts. Hind tracks often angle in toward center of trail. Large, flat tail often drags and partially obscures tracks. Common to find trail accompanied by drag marks of branches being hauled to the water. **WALK** Stride: 11 7/16–20½ in. (29–52 cm). Trail width: 5 1/8–8½ in. (13.0–21.6 cm).

SIGNS

SCAT ¾–1½ in. (1.9–3.8 cm) D × 1¾–2¼ in. (4.5–5.7 cm) L. Usually deposited in the water and found in water or on land after floods. Scats are oval or egg shaped and contain mostly small chips of wood, resembling sawdust. **SCENT MOUNDS** Beavers leave mounds of mud along the edges of water they frequent, constructed to mark family territory. After mounding up the mud, beavers deposit castorum, a strong smelling secretion, from their anal glands. **CAMBIUM FEEDING** While felled wood is used in the construction of dams and lodges, beavers drop trees primarily to access the cambium (inner bark) for food. Beavers have the largest incisors of any rodents in our region and leave correspondingly large impressions on trunks and branches they have chewed through or debarked. After cutting down a tree, the beaver removes branches and hauls them into the water. Look for stripped branches in and around the water's edge where beavers are active. **DIGS** Beavers use their strong, blunt claws to dig up the roots of cattails and other plants growing in and around water. These signs are similar to digs made by muskrats, and with no tracks or other accessory signs, it can be difficult to distinguish digs made by these two species. **DAMS & LODGES** Beavers construct dams and lodges in many (but not all) areas where they live, using cut branches, logs, and mud. Dams made in streams and around the edges of wetlands are constructed to increase the depth and area covered by water. Lodges are built in a body of water and act as shelter and refuge. Lodges range from about 6 ft. (2 m) to more than

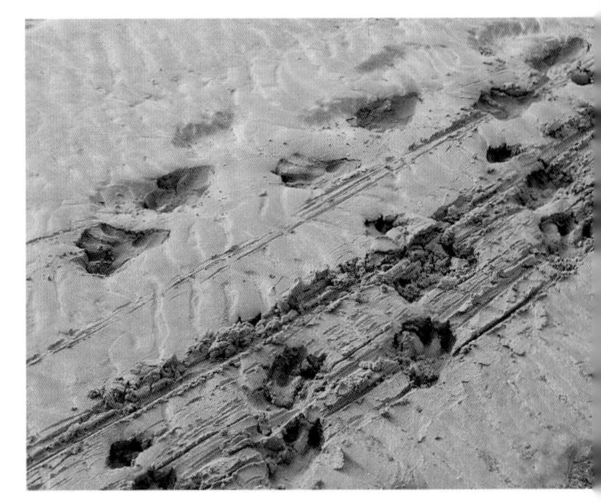

Two beaver trails: in the lower trail, the beaver dragged tree branches, indicated by the linear marks across the trail. The beaver turned its body to the side on which branches were being dragged, and the front feet registered to that side of the hind feet in the tracks. Columbia Plateau, Washington.

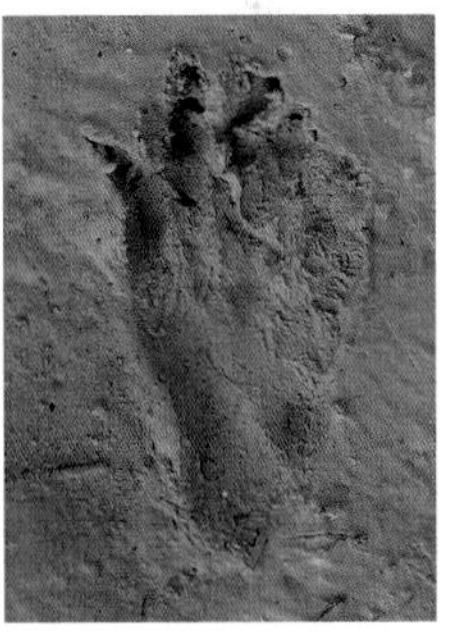

Left hind with front partially showing beyond the hind toes. Puget Trough, Washington.

Left front track. Columbia Plateau, Washington.

Beaver walk

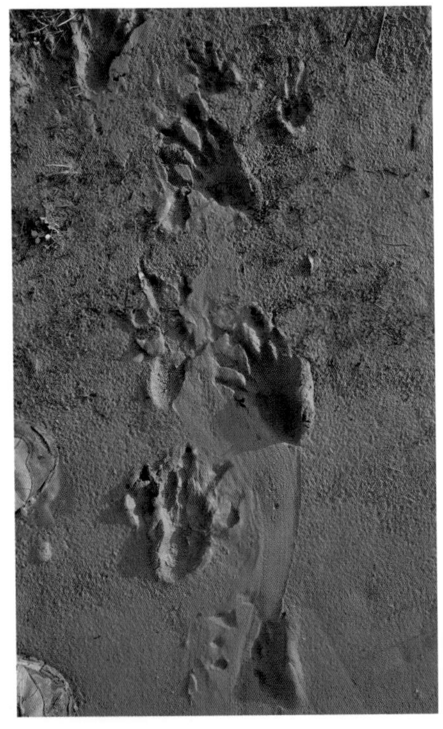

Typical walking pattern. Puget Trough, Washington.

Beaver scats. Puget Trough, Washington.

Willow branches were cut and debarked by a beaver. Puget Trough, Washington.

Large scent mound on the edge of a slow-moving stream. Columbia Plateau, Washington.

This black cottonwood was girdled by a beaver. Puget Trough, Washington. Photo by Tobias Maloy.

30 ft. (10 m) across and have underwater entrances, with the living space built above the waterline. **BANK BURROWS** Instead of constructing a lodge, some beavers excavate burrows into the banks along streams and ponds. Entrances are underwater and the living platform is excavated above the waterline. You can sometimes find burrow entrances when water levels are lower than normal. You may encounter what appear to be small sink holes along the bank, which result from old, collapsed burrows.

Behind this exceptionally large beaver lodge, traffic zooms by on a highway bridge across Seattle's Lake Washington.

A typical beaver dam constructed in an existing marsh. Puget Trough, Washington.

SQUIRRELS AND THEIR KIN

Family Sciuridae

With seven genera and more than 25 species in our region, the Sciuridae family includes marmots, ground squirrels, tree squirrels, and chipmunks. Representatives from each group are covered in individual species accounts. All except for one species, the northern flying squirrel, are diurnal.

Tracks. Front and hind feet are of typical rodent structure. Marmot and ground squirrel feet can register somewhat pigeon-toed or with toes curving inward toward the centerline of travel (most prominent on front tracks). The front feet of ground squirrels are more asymmetrical than those of other squirrels.

Track patterns. Common gaits are walking or bounding. When bounding, tree squirrels usually register their front feet side by side, while marmots and ground squirrels register their front feet at an angle (as in lagomorphs).

HOARY MARMOT

Marmota caligata

Marmots are the largest members of the Sciuridae family in our region. They are semifossorial and live in colonies. Their shrill whistles, used to alert other marmots of approaching danger, are well known to alpine travelers. Different predators elicit unique calls depending on the level of threat they pose to the marmots. Hoary marmots are active for only about 4 months during the brief mountain summer, when they must consume enough food to put on fat for a long winter hibernation. Two closely related species, the Vancouver Island marmot (*Marmota vancouverensis*) and the Olympic marmot (*M. olympus*), have limited ranges west of the Cascades and likely developed as unique species through geographic isolation in the mountains of Vancouver Island and the Olympic Peninsula.

SIZE Length: 62.5–85.0 cm. Weight: 5–7 kg. Males are larger. **FIELD MARKS** Grizzled gray pelage, white between eyes with dark snout. Fur relatively long. Olympic marmots are yellowish or brown. **REPRODUCTION** Females are fertile only every other year to build up enough energy reserves for litter production and nursing. In a colony, only one dominant male usually breeds with one or two females (with alternating yearly cycles of estrus). Young stay with the mother until at least their second summer and may subsequently become integrated into the colony. **DIET** Herbivorous, mainly grasses and forbs. **HABITAT & DISTRIBUTION** Alpine and subalpine meadows. Olympic and Vancouver Island marmots occupy similar habitats in their respective ranges. No overlap occurs in the three species' range. **SIMILAR SPECIES** Yellow-bellied marmot, American marten, badger, wolverine. **CONSERVATION ISSUES** The Vancouver Island marmot is listed as endangered in Canada, with as few as 200 animals known to exist. The Olympic marmot population appears to be in decline in its already limited range.

TRACKS

Typical rodent structure. Plantigrade. Toes curve inward with bulbous toe pads. Claws usually register. Palm pads are robust (page 68). **FRONT** 2 1⁄8–3 in. (5.4–7.6 cm) L × 1 5⁄8–2 9⁄16 in. (4.1–6.5 cm) W. Toe 1 is greatly

Hoary marmot

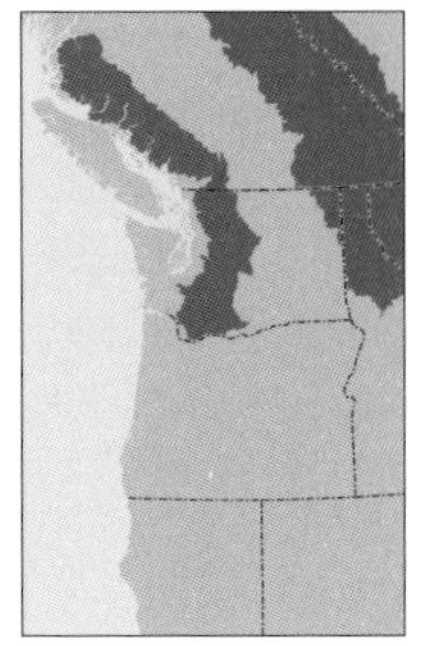

Hoary marmot range

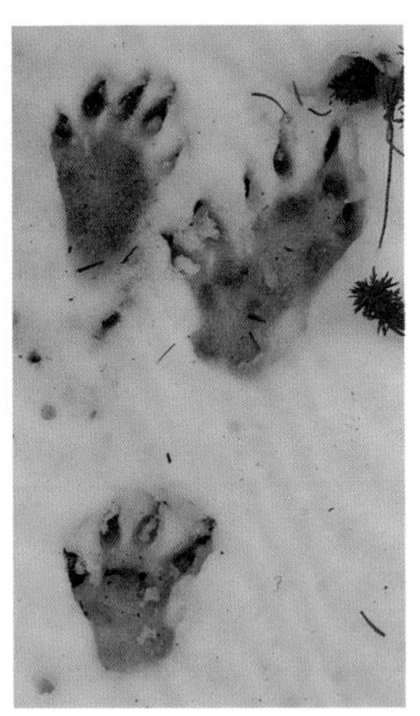

All four feet of a bounding hoary marmot. Hind tracks are outermost at the top of the frame. North Cascades, Washington.

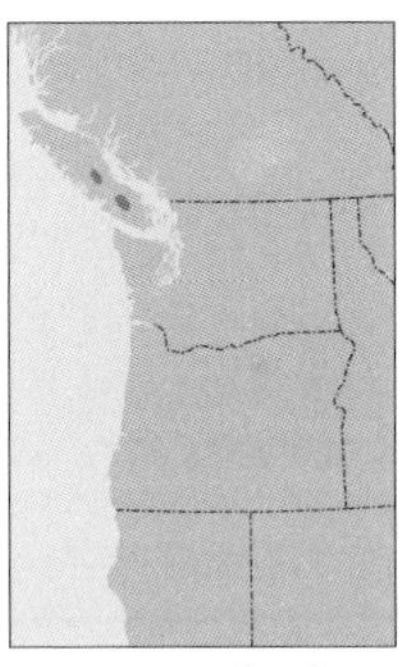

Vancouver Island marmot range

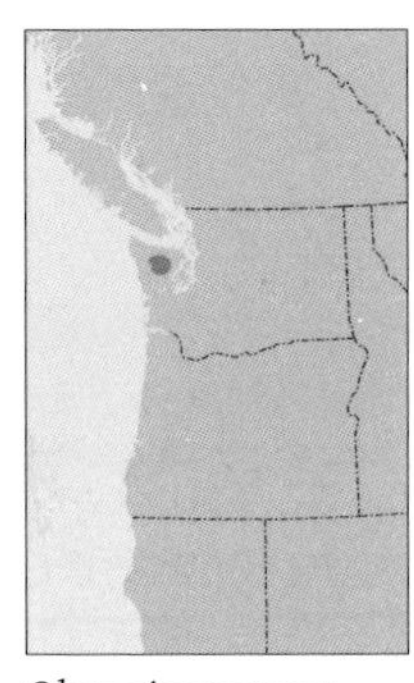

Olympic marmot range

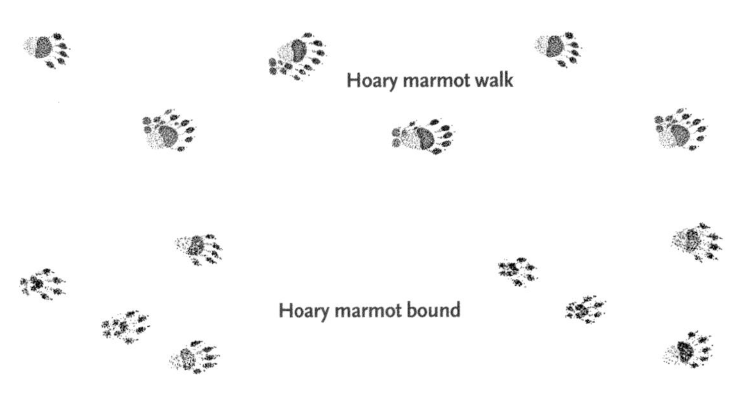

Hoary marmot tracks, right hind over right front. North Cascades, Washington.

Hoary marmot scat. North Cascades, Washington.

Tracks of a hoary marmot bounding away (at right) and toward the viewer. North Cascades, Washington.

In spring, hoary marmots become active before the deep snowpack melts. Conical access holes in the snow are often surrounded by dirt and debris carried out on the marmot's feet during their repeated trips in and out of burrows. Here the snowpack was still about 8 ft. (2.4 m) deep in mid-June. North Cascades, Washington.

Burrow and throw mound in a meadow just above treeline. North Cascades, Washington.

reduced with a nail that may register in clear tracks. Two proximal pads usually register, giving heel a square shape. The roundness of proximal pads on the actual foot is diagnostic for this species, but in tracks, I have not found the shape to be consistently distinctive from those of yellow-bellied marmots. **HIND** 2½–3½ in. (6.3–8.9 cm) L × 1¾–2 11⁄16 in. (4.4–6.9 cm) W. Long and rounded heel usually registers. Slightly larger than front. **WALK** Stride: 13 3⁄8–18 1⁄8 in. (34–46 cm). Trail width: 5 5⁄16–7 in. (13.5–17.8 cm). **ANGLED BOUND** Stride: 16½–32 in. (42.0–81.3 cm). Trail width: 5 1⁄8–8¼ in. (13–21 cm). Group length: 6 11⁄16–18 in. (17.0–45.7 cm).

SIGNS

SCAT ¼–11⁄16 in. (0.7–1.8 cm) D × 2 3⁄8–3¼ in. (6.0–8.2 cm) L. Varies depending on diet: oval pellets, sometimes loosely connected; tubular cords with pointed end(s); or amorphous and runny. Latrines created around burrows and lookouts. Can be confused with coyote or other carnivore scats because of the shape, though close inspection will reveal contents comprising entirely of vegetation. **BURROWS** In alpine meadows, often dug into the base of a large boulder or rock pile that may act as a lookout for the animals. Well-worn trails often radiate from a burrow entrance or connect one entrance to another. Burrow diameters average 7 in. (18 cm).

YELLOW-BELLIED MARMOT

Marmota flaviventris

Marmots living in arid regions come out of hibernation much earlier than those in higher elevation montane sites and enter dormancy earlier as well to avoid the hottest and driest parts of the year. Yellow-bellied marmots are often found in small colonies. Loud whistles are used to alert one another to danger.

SIZE Length: 47–67 cm. Weight: 1.6–5.2 kg. Males are larger. **FIELD MARKS** Yellow underside and neck with small band of white across the nose; otherwise gray. Stocky appearance; about the size of a house cat. **REPRODUCTION** Breed soon after the end of hibernation, around late February at low elevations, later at higher elevations. A dominant male establishes and defends his territory, mating with females within it and chasing away other males. Birth occurs one month later. Young emerge from the natal den a month after this, around May in lower elevations. Litters average 3–8 young. **DIET** Mainly herbivorous and occasionally insects and larvae. **HABITAT & DISTRIBUTION** Eastern slope of the Cascades and farther east in both montane and arid environments. Often associated with rock outcroppings or abandoned human structures that provide protection. Habitat must include access to green vegetation for forage, often irrigated agricultural fields in arid environments. **SIMILAR SPECIES** Hoary marmot, badger, Columbian ground squirrel.

TRACKS

Tracks and track patterns are similar to those of the hoary marmot but average slightly smaller. **FRONT** 1¾–2 3⁄16 in. (4.5–5.6 cm) L × 1¼–2 in. (3.1–5.1 cm) W. **HIND** 1 7⁄16–2¾ in (3.6–7.0 cm) L × 1 5⁄16–2 1⁄8 in. (3.3–5.4 cm) W.

SIGNS

SCAT 5⁄16–9⁄16 in. (0.8–1.5 cm) D × 1–1 15⁄16 in. (2.5–5.0 cm) L. Similar to that of hoary marmot. Look for large latrines on and around rock piles near burrows. **BURROWS** Often associated with rock piles, talus, or cliffs. **TERRITORIAL MARKING** During breeding season, males mark their terri-

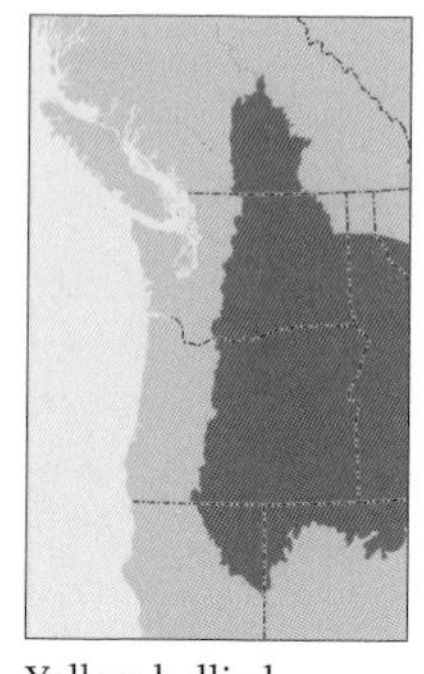

Yellow-bellied marmot range

Left front and partial left hind tracks of a yellow-bellied marmot. Columbia Plateau, Washington.

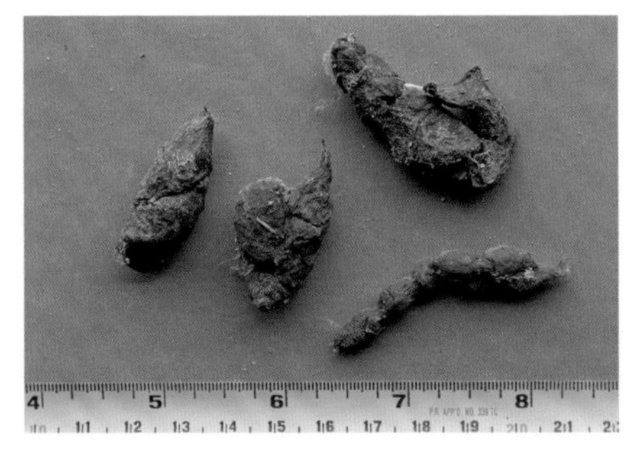

Dry scat. Columbia Plateau, Washington.

Yellow-bellied marmots gnawed on the bark of black locust trees around burrow entrances at an old homestead site along the Columbia River, Washington. Photo by Roy Ashton.

Throw mound and burrow entrance at the base of a rock pile. Columbia Plateau, Oregon.

tories with scent glands on their cheeks. I have observed them gnawing on tree bark in their area, similar to the behavior of tree squirrels.

GROUND SQUIRRELS

California ground squirrel, *Spermophilus beecheyi*
Belding's ground squirrel, *S. beldingi*
Washington ground squirrel, *S. washingtoni*
Columbian ground squirrel, *S. columbianus*

Seven species of unlined ground squirrels inhabit our region, all with similar appearance, behaviors, tracks, and signs. All are associated with arid habitats and hibernate through the winter. Several species also become dormant during midsummer to escape the most intense periods of heat and dryness, a behavior known as estivation. Ground squirrels dig burrows near rocks or shrubs or out in the open. They are colonial and diurnal. They use vocalizations to alarm of approaching hazards. Those who inhabit treed areas, such as the California ground squirrel, occasionally climb into them. In most species the male is slightly larger than the female. Four distinctive species are described.

REPRODUCTION Females are in estrus for only a single day each year, in the spring, shortly after they emerge from hibernation. Males become active one to several weeks before females and competition is often fierce for access to females. One litter is produced in the late spring, with 3–9 young. Young are active above ground about a month and a half after birth. **DIET** Omnivorous. Primarily plant matter—grasses, forbs, seeds, acorns, and other nuts—along with insects and carrion. **SIMILAR SPECIES** Western gray squirrel, red and Douglas's squirrels, chipmunks, wood rats. **CONSERVATION ISSUES** The California and Belding's ground squirrels have extensive ranges in our region and beyond. The California ground squirrel has been steadily spreading its range northward. The Washington ground squirrel has a limited range and is a candidate for both federal and Washington state endangered species lists.

Columbian ground squirrel. Canadian Rockies, British Columbia.

Two closely related species of the genus *Spermophilus*, golden-mantled ground squirrels, are discussed in a separate section because of their distinct appearance and habitat.

California Ground Squirrel

SIZE Length: 36–50 cm. Weight: 0.25–0.89 kg. **FIELD MARKS** Grizzled fur with light patch on shoulders. **HABITAT & DISTRIBUTION** Oak woodlands, agricultural and semiarid environments. In Washington, found only on the east slope of the Cascades, where its range is expanding northward.

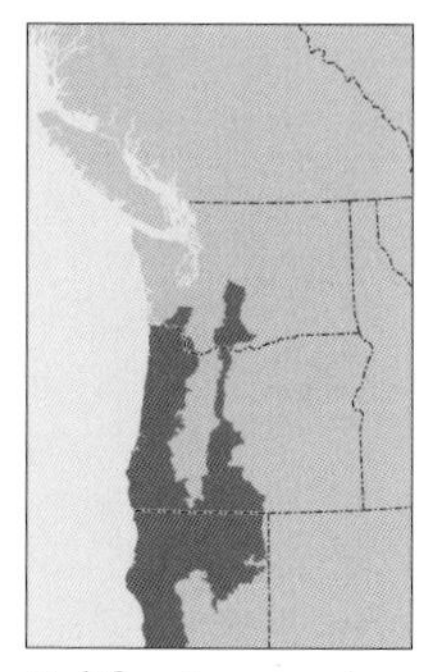

California ground squirrel range

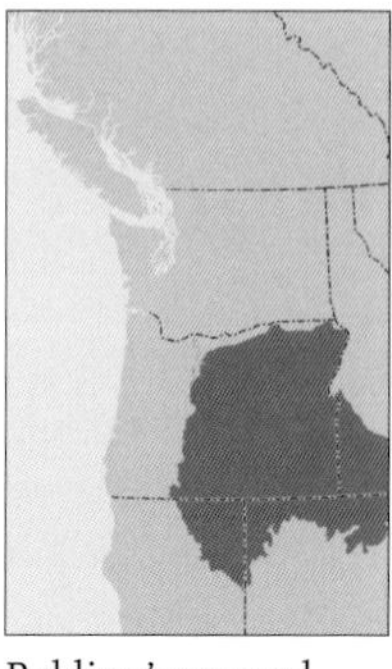

Belding's ground squirrel range

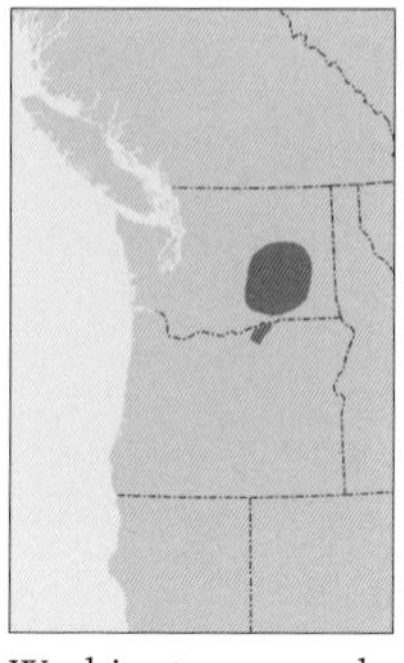

Washington ground squirrel range

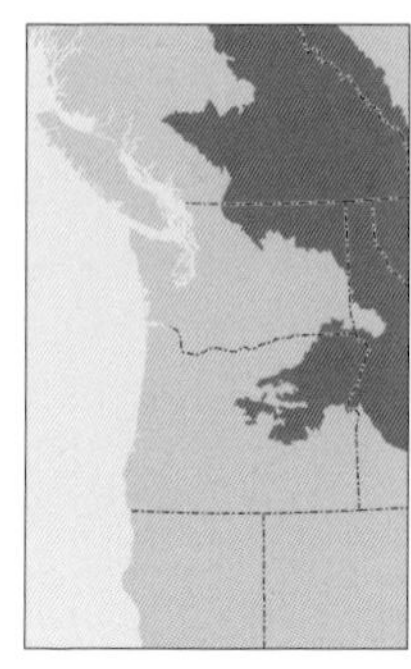

Columbian ground squirrel range

Belding's Ground Squirrel

SIZE Length: 26.5–31.5 cm. Weight: 230–400 g. **FIELD MARKS** Top of back reddish-brown, sides gray. Tail with a dark tip. **HABITAT & DISTRIBUTION** Meadows, agricultural areas, and sagebrush country close to water, eastern Oregon and south.

Washington Ground Squirrel

SIZE Length: 18.5–24.5 cm. Weight: 120–300 g. **FIELD MARKS** Small, with white flecks along gray back. Belly is light and distinctly delineated from back. **HABITAT & DISTRIBUTION** Arid landscapes in the Columbia Plateau. Numerous small disjunct populations in Washington and Oregon. Thought to be in decline because of habitat fragmentation.

Columbian Ground Squirrel

SIZE Length: 23.5–21.0 cm. Weight: 340–812 kg. Medium to large in size. **FIELD MARKS** Reddish-orange coloration on nose and head. Light shoulder patch and tail with a light tip. **HABITAT & DISTRIBUTION** Mountain meadows, fields, and rangeland. Associated with higher elevations than for other ground squirrels. Mountain ranges in the eastern portion of our region, ranging west into the North Cascades.

TRACKS

Front claws are much longer than those on the hind feet, a reliable distinction between ground squirrels and tree squirrels. Toes may register long and thin or slightly bulbous-tipped. Track measurements presented include data from Belding's, California, and Columbian ground squirrels. Tracks and trail patterns average smallest for Belding's ground squirrels and largest for California ground squirrels (as expected based on their relative sizes) but measurement ranges overlap almost completely for all three species. Smaller species would likely have smaller measurements (page

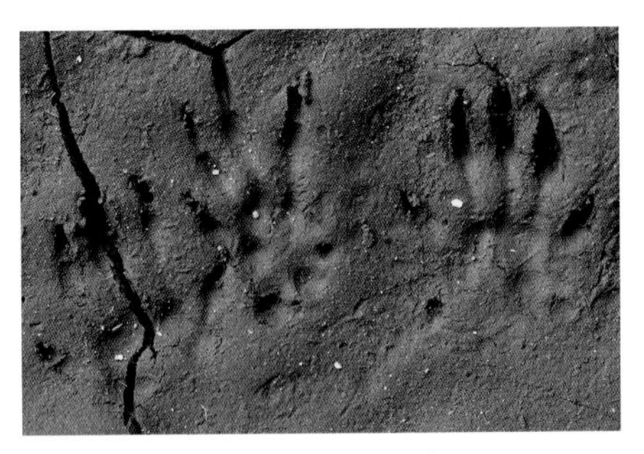

Columbian ground squirrel tracks. From left to right: left hind, left front, right hind. Middle Rockies, Idaho.

Columbian ground squirrel walk

Columbian ground squirrel bound

Typical track pattern of a bounding Columbian ground squirrel. Middle Rockies, Idaho.

Belding ground squirrel scats. Columbia Plateau, Oregon.

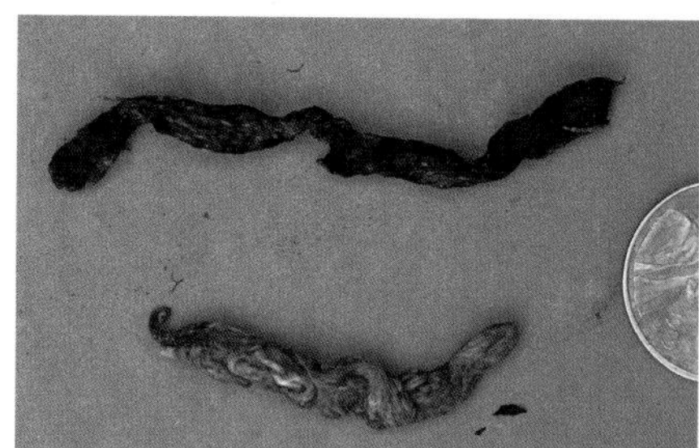

Top: Scat of a Columbian ground squirrel containing plant fibers. Bottom: Scat of a long-tailed weasel containing small mammal hair. Both were found at the entrance to an active ground squirrel burrow. Canadian Rockies, Montana.

61). **FRONT** ⅝–1½ in. (1.6–3.8 cm) L × 9⁄16–1 3⁄16 in. (1.5–3.0 cm) W. Narrower than hinds. Toes oriented in more asymmetrical fashion than in tree squirrels, with toe 3 registering most forward. Long claws. **HIND** ⅝–1 5⁄16 in. (1.6–3.3 cm) L × 11⁄16–1⅜ in. (1.7–3.5 cm) W. Typical rodent structure. Claws are distinctly shorter than those of front foot. Furred heel usually does not register in tracks. **TRACK PATTERNS** Most common gait is an angled bound. Also walks for short distances. **BOUND** Stride: 13 3⁄16–34⅝ in. (33.5–88.0 cm). Trail width: 2 5⁄16–4 5⁄16 in. (5.8–11.0 cm). Group length: 2 9⁄16–7 11⁄16 in. (6.5–19.5 cm). **WALK** Stride: 5½–9¼ in. (14.0–23.5 cm). Trail width: 1¾–2⅞ in. (4.4–7.3 cm).

SIGNS

SCAT ⅛–¼ in. (0.3–0.6 cm) D × ¼–1⅛ (0.7–2.9 cm) L. Washington ground squirrel—to 11⁄16 in. [1.7 cm]) L. Scats can be found along runs or near burrow entrances. Large variations in shape among all species. Pellets are sometimes connected in strands. **BURROWS** Found either at the base of a rock or shrub thicket when available, or in the open. May be dug into flat ground or slopes. Some entrances surrounded by a throw mound. Well developed trails often connect different burrow entrances and entrances

California ground squirrel burrow entrance at the base of a large rock with a throw mound. Klamath Mountains, Oregon.

Drag marks in dust of a scent-marking California ground squirrel. Northern Coast, California.

Throw mounds of Columbian ground squirrel burrows in a subalpine meadow. North Cascades, Washington.

Belding ground squirrel scent-marking. Columbia Plateau, Oregon.

with feeding areas. Often found in loose clusters. Average diameter: Belding's, 2¾ in. (7 cm); California and Columbian, 3¼ in. (8.2 cm); Washington, 1 15/16 in. (4.9 cm). **SCENT-MARKING** Ground squirrels have scent glands on their bellies, which they drag across the ground to mark their territory. In dusty locations, these can be identified as furry drag marks about as wide as the animal that leaves it.

GOLDEN-MANTLED GROUND SQUIRRELS

Golden-mantled ground squirrel, *Spermophilus lateralis*
Cascade golden-mantled ground squirrel, *S. saturatus*

These two closely related species occupy distinct geographic areas. Their behavior, signs, and track parameters overlap almost entirely. They are active from early spring to mid-August (later for juveniles), followed by a 5- to 8-month period of hibernation. They are solitary and diurnal and store large quantities of seeds that are consumed during brief periods of activity within winter hibernation. The white-tailed antelope squirrel (*Ammospermophilus leucurus*), which occurs in the deserts of the far southeastern corner of our region, is differentiated from the golden-mantled ground squirrel by its white tail carried over its back.

SIZE Length: 23.0–31.5 cm. Weight: 170–300 g. **FIELD MARKS** Single light stripe along back, with black on either side, more distinctly in *Spermophilus lateralis*. Larger than a chipmunk, with no eye stripe. **REPRODUCTION** Breeds shortly after emergence from hibernation in late March. Most young are born in late spring or early summer. Average litter size is 4–6. Newborns are altricial, and those born later in the summer are challenged to put on enough fat and store enough food for winter torpor. They may disperse as late as October, more than

Cascade golden-mantled ground squirrel

Golden-mantled ground squirrel range

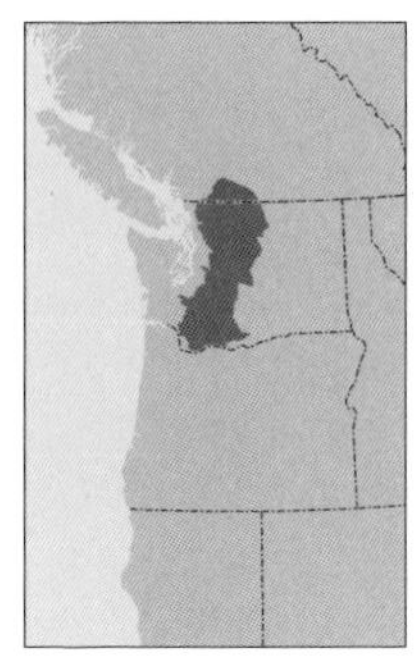

Cascade golden-mantled ground squirrel range

a month after most of the population has entered hibernation. **DIET** Omnivorous. Fungi, seeds, green vegetation and flowers, insects, bird eggs, occasionally lizards and mammal tissue (likely carrion). **HABITAT & DISTRIBUTION** Cascade Crest and east. Talus fields, meadows, moist and dry forests, rocky locations in semiarid locations. Most abundant in open timber with rocky outcroppings or rock piles. Their use of forested environments is unique among ground squirrels in our region. **SIMILAR SPECIES** Other ground squirrels, chipmunks, woodrats, pika.

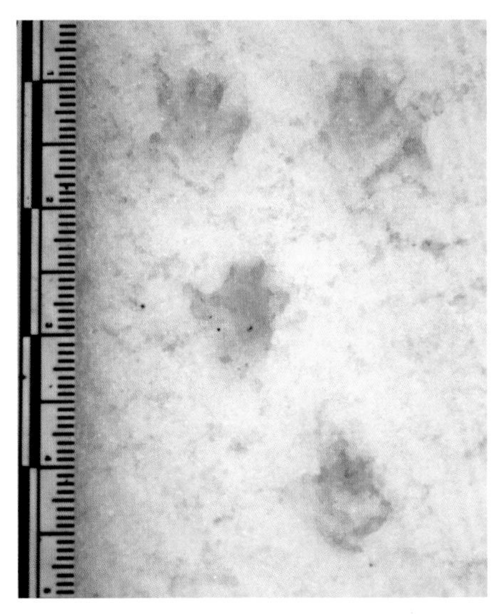

Typical track pattern of a Cascade golden-mantled ground squirrel; top two tracks are the hinds. North Cascades, Washington.

TRACKS

Typical rodent foot structure. Tracks share characteristics with larger ground squirrels and chipmunks (which are slightly smaller). **FRONT** Claws are longer than those on hinds, but the difference is not as great as in larger ground squirrels. The asymmetry in the toe arrangement is less exaggerated on the front foot than in other ground squirrels. **TRACK PATTERNS** Walks and bounds similar to other ground squirrels.

SIGNS

SCAT 3⁄16–1⁄4 in. (0.5–0.6 cm) D × 3⁄8–3⁄4 in. (1.0–1.9 cm) L. Similar to other ground squirrels. **BURROWS** At the base of rotting stumps, root wads, and in rock outcrops where these animals are found along with pikas. Entrance diameter averages 2 5⁄8 in. (6.6 cm). **FEEDING SIGNS** Like chipmunks, with whom they share habitat, they often feed at a prominent perch close to the food, leaving a small amount of refuse (seed husks, flower parts, and so on) rather than a huge collection, as in the middens of red squirrels.

Golden-mantled ground squirrel burrow at the base of a rock. Middle Rockies, Idaho.

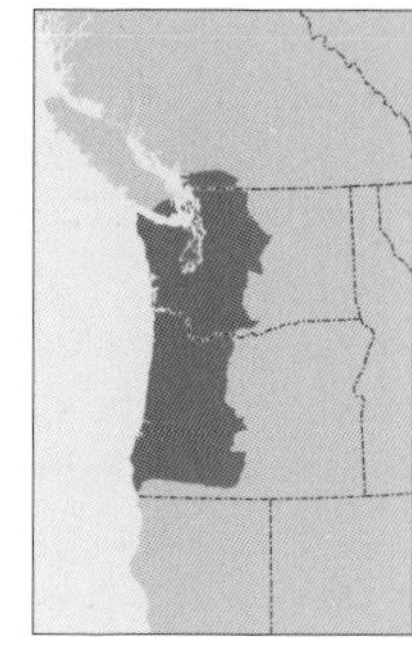

Townsend's chipmunk range

Townsend's chipmunk. Northwest Coast, Washington.

CHIPMUNKS

Townsend's chipmunk, *Tamias townsendii*
Yellow-pine chipmunk, *T. amoenus*
Least chipmunk, *T. minimus*

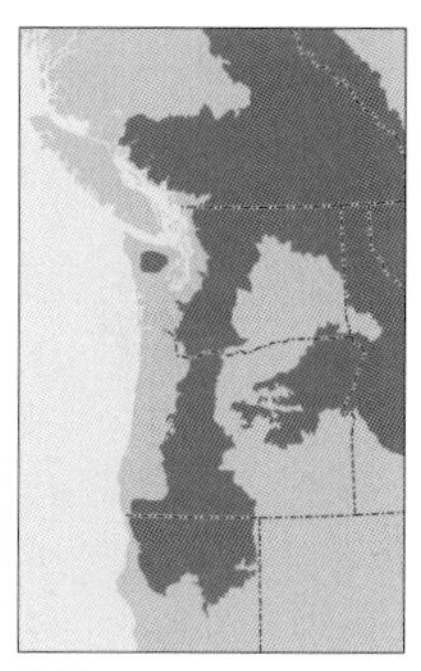

Yellow pine chipmunk range

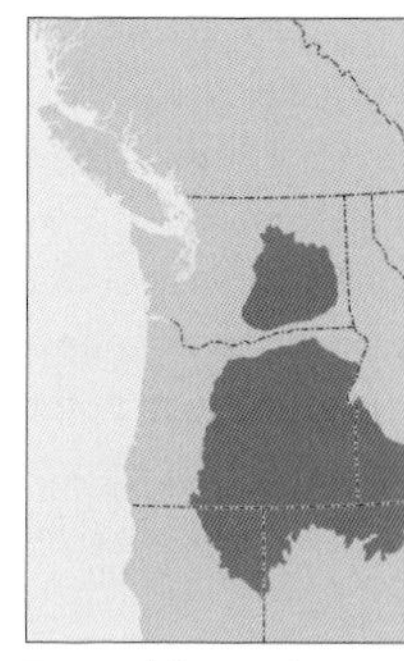

Least chipmunk range

There are eight species of chipmunks in our region. The three discussed here have the most extensive ranges in our area. The tracks, signs, and behaviors of all three are similar. Chipmunks are diurnal and mainly inactive during the winter months; however, they do not accumulate fat and truly hibernate. They rely on large quantities of stored food collected during the spring, summer, and fall. Food is gathered in large internal cheek pouches and then stored in burrows, natural crevices, or tree cavities. They forage on the ground and in shrubs and trees. Their food-gathering antics are enjoyable to watch, perhaps until your lunch turns up as part of theirs!

SIZE Females are slightly larger than males. Townsend's chipmunk—Length: 23–28 cm. Weight: 90.0–118.0 g. Yellow-pine chipmunk—Length: 18.6–23.8 cm. Weight: 36–50 g. Least chipmunk—Length: 18.5–21.6 cm. Weight 32–50 g. **FIELD MARKS** Smallest sciurids, with light and dark eye stripes and striped backs. The least chipmunk holds its tail straight up when running. **REPRODUCTION** Breeding occurs in spring, shortly after chipmunks become active. One litter per year, with 2–8 young. **DIET** Omnivorous: flowers, forbs, grasses, berries, seeds, roots, tree and shrub buds, fungi, bird eggs, and insects. Fungi often make up more than half their diet. **HABI-**

TAT & DISTRIBUTION One species or another can be found in almost every environment in our region. In many areas, two or more species' ranges overlap. In the Cascades, Townsend's and yellow-pine chipmunks overlap. On the eastern edge of the Cascades, the yellow-pine is joined by the least chipmunk, which specializes in the most arid landscapes, such as areas dominated by sagebrush. **SIMILAR SPECIES** Golden-mantled ground squirrels, red and Douglas's squirrels, woodrats, Norway and roof rats.

TRACKS

Typical rodent structure. Toes are long and thin on both front and hind feet, although slightly less distinctly than in the larger red and Douglas's squirrels. Ranges include measurements from all three species (page 59). **FRONT** 7/16–3/4 in. (1.1–1.9 cm) L × 5/16–9/16 (0.8–1.5 cm) W. Two round proximal pads usually register in tracks but are smaller than those of red and Douglas's squirrels, with the heel about the same width as the palm. **HIND** 7/16–1 1/8 in. (1.1–2.8 cm) L × 1/2–7/8 in. (1.2–2.2 cm). Larger than the fronts. Heel registers occasionally. **TRACK PATTERNS** Bounding gaits used almost exclusively; may walk for short distances while foraging. **BOUND** Stride: 3 3/8–29 1/4 in. (8.5–74.3 cm). Trail width: 1 9/16–2 11/16 in. (4.0–6.8 cm). Group length: 1 1/16–4 3/4 in. (2.7–12.1 cm).

SIGNS

SCAT 1/16–3/16 in. (0.2–0.5 cm) D × 3/16–3/8 in. (0.5–1.0 cm) L. Small, oblong pellets, sometimes pointed at one end. Often found in a small collection where the animal was feeding. **FEEDING SIGNS** Often feed in prominent locations such as the top of a stump, a large rock, or a fallen log, carrying their morsel of food to the location and consuming it while watching for danger. Feeding spots are often close to the food's original location and can be identified by a small

All four feet of a Townsend's chipmunk in a typical bound pattern. Lowest two tracks are fronts. West Cascades, Washington.

Scat of yellow-pine or least chipmunk. East Cascades, Washington.

Townsend's chipmunk bound

A gall from the branch of a lodgepole pine was cut from the tree and carried to a feeding perch on a fallen log, where the chipmunk fed on the bark. Canadian Rockies, Montana.

The top of this stump is a typical chipmunk feeding station. East Cascades, Oregon.

amount of discarded plant material and a scat or two. They rarely if ever accumulate the large middens typical of tree squirrels. Chipmunks often make shallow digs while foraging for food, possibly looking for insects or fungi. **BURROWS** 1½–2 in. (3.8–5.1 cm) D. Hard to find; often situated in thick brush or rock piles. No throw mounds at the entrance, because the chipmunk disposes of excavated soil at a different location and seals the entrance once excavation is complete, thus helping to camouflage it.

GRAY SQUIRRELS

Western gray squirrel, *Sciurus griseus*
Eastern gray squirrel, *S. carolinensis*

Gray squirrels are diurnal and active year-round. Arboreal in their habits, they usually nest in tree cavities and occasionally in stick and leaf nests (dreys). They forage extensively on the ground under and adjacent to trees. The western gray squirrel is the largest tree squirrel in our region. The introduced eastern gray squirrel is now the most common tree squirrel in many urban, suburban, and rural landscapes in our region. A second introduced tree squirrel, the fox squirrel (*Sciurus niger*), has established itself in parts of the Willamette Valley and Puget Trough.

SIZE Western gray squirrel—Length: 51–77 cm. Weight: 500–950 g. Eastern gray squirrel— Length 38.3–52.5 cm. Weight: 338–750 g. **FIELD MARKS** The larger western gray squirrel has a silver-gray back with a white underside and a large, fluffy tail. Eastern gray squirrels have narrower and shorter tails, and overall color is less gray and more brown, with hints of red. Fox squirrels have darker, reddish undersides and lack the white fringe on their tails. **REPRODUCTION** Breeding begins in mid- to late winter. Western grays bear a single litter annually between late February and May. Eastern grays produce two litters—one in late winter (January–February) and another in the summer (June–July). Litters of 1–6 young are weaned in about 3 months. Young are altricial at birth and are usually born in a tree cavity nest, though the mother may move them into a drey as they grow larger. **DIET** Somewhat omnivorous. Fruits, nuts, seeds, insects, and fungi such as truffles (*Rhizopogon* spp.). Occasionally birds' eggs and carrion. For western grays, acorns (where available) are a staple and are cached for winter along with conifer cones. **HABITAT & DISTRIBUTION** Western gray squirrel—Associated with oak woodlands, mixed broadleaf forests, riparian habitat, ponderosa pine forests, and parks adjacent to natural habitat. Coastal to eastern slope of the Cascades as far north as southern Washington, with several isolated populations farther north. Primarily lower elevations. Eastern gray squirrel—In and around deciduous trees in urban, suburban, and rural areas in the Puget Trough ecoregion and Spokane area (eastern Washington). **SIMILAR SPECIES** Douglas's and red squirrels, yellow-bellied marmots. **CONSERVATION ISSUES** Western gray squirrel populations are declining as a result of competition with eastern gray squirrels and habitat loss. Several discrete populations in Washington are listed as endangered.

TRACKS

Typical rodent structure. Western gray squirrel tracks appear large for a squirrel, substantially larger than those of Douglas's and red squirrels. Eastern gray squirrels are intermediate in size. Tips of toes often appear bulbous and not connected to the robust palm pad in tracks. Sharp claws register (page 63). **FRONT** Western gray squirrel—1¼–1⅞ in. (3.2–4.8 cm) L × ¾–1¾ in. (1.9–4.5 cm) W. Eastern gray squirrel—1¼–1¾ in. (3.1–4.5 cm) L × 11⁄16–1⅛ in. (1.8–2.9 cm) W. Two proximal pads register and are the same width as the palm pad; posterior appears square. The small pad

Western gray squirrel

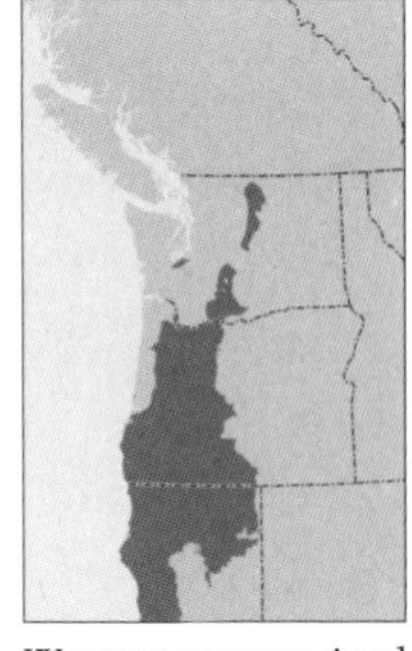

Western gray squirrel range

of the vestigial toe 1 registers occasionally in clear tracks alongside the proximal pads. **HIND** Western gray squirrel—1¼–2 1/16 in. (3.1–5.2 cm) L × 1–1¾ in. (2.6–4.4 cm) W. Eastern gray squirrel—1–2 1/16 in. (2.5–5.2 cm) L × ⅞–1⅜ in. (2.2–3.5 cm) W. Hinds are larger than the fronts, but may appear shorter if the full heel does not register (common on firm substrates). The unfurred heel has two small proximal pads, distinguishing it from red and Douglas's squirrels, but rarely registers. **TRACK PATTERNS** Similar bound pattern to other tree squirrels. Walks for short distances while foraging. **BOUND** Western gray squirrel—Stride: 15 1/16–31⅞ in. (38.2–81.0 cm). Trail width: 3⅞–5 11/16 in. (9.8–14.5 cm). Group length: 4–12 3/16 in. (10.2–31.0 cm). Eastern gray squirrel— Stride: 3 15/16–29 15/16 in. (10–76 cm). Trail width: 3½–4 15/16 in. (8.9–12.5 cm). Group length: 2 9/16–6½ in. (6.5–16.5 cm).

All four tracks of a western gray squirrel.
Photo by Jonah Evans.

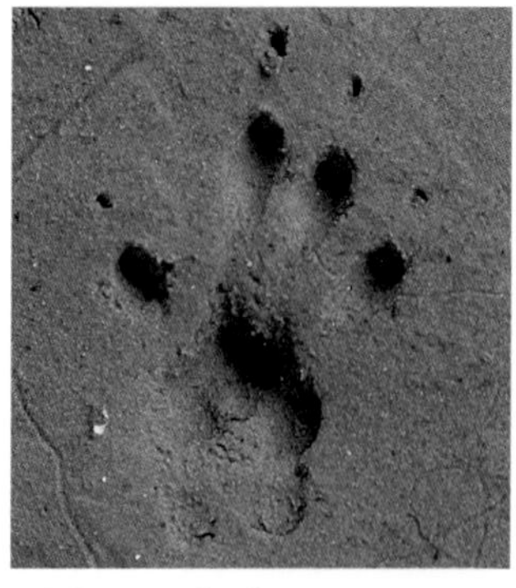

Left front track of an eastern gray squirrel. Puget Trough, Washington.

SIGNS

SCAT ⅛–¼ in. (0.3–0.7 cm) D × ¼–7/16 in. (0.7–1.0 cm) L. Uneven pellets with a pointy tip at one end. Brown to black. **FEEDING SIGNS** Nuts such as walnuts are gnawed and broken open and appear rough. Found under trees where the squirrel has been feeding. **DIGS, CACHES, & MIDDENS** Western gray squirrels produce many small digs in forest litter. Digs may be signs of foraging for underground fungi or associated with caching or retrieval of a cache. Digs are rarely more than 2 in. (5.1 cm) deep. Similar to red and Douglas's squirrels, western gray squirrels produce piles of conifer cone scales and cores beneath favorite feeding locations. **SCENT-MARKING** Gnaws tree bark at scent-marking locations and rubs the area with cheek glands, sometimes urinating there as well.

Western gray squirrel bound

Eastern gray squirrel scats. Puget Trough, Washington. Photo by David Moskowitz and Paul Houghtaling.

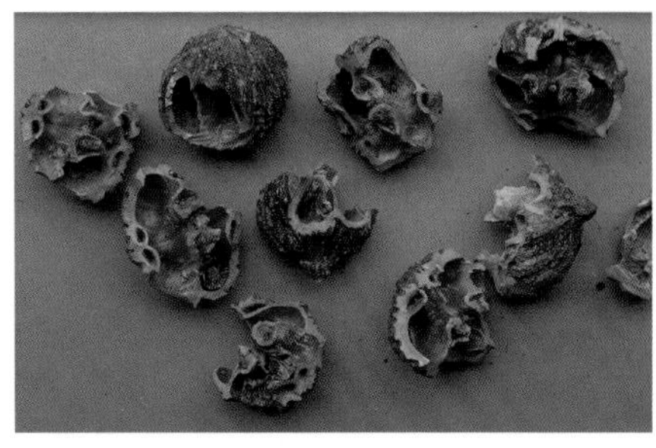

These walnut shells were broken open by an eastern gray squirrel in a suburban park. Puget Trough, Washington.

Remains of Garry oak (*Quercus garryana*) acorns eaten by a western gray squirrel (arranged by the author). Klamath Mountains, Oregon.

Western gray squirrel dig in the litter of a ponderosa pine forest. East Cascades, Oregon.

RED SQUIRRELS

Red squirrel, *Tamiasciurus hudsonicus*
Douglas's squirrel, *T. douglasii*

Red and Douglas's squirrels are the most common tree squirrels in conifer forests throughout our region. The two closely related species hybridize where their range overlaps in parts of eastern Oregon and Washington. They are active year-round, diurnal, and highly visible. Their loud chattering is often directed at predators.

SIZE Length: 27–36 cm. Weight: 140–310 g. **FIELD MARKS** Smaller than gray squirrels, with reddish-brown back and a medium-sized tail with a dark tip. Red squirrel—Underside is off-white and eye ring is white. A dark band along its flank may divide its white underside and reddish back. Douglas's squirrel—Underside and eye ring are orange or yellow. **REPRODUCTION** Most litters born in May, though some are born in late summer. One litter produced each year, with 4–6 young. **DIET** Mostly conifer seeds and fungi, plus seasonal foods such as insects, berries, and green vegetation in summer and cambium in the winter. Douglas-fir (*Pseudotsuga menziesii*) cones are a

Douglas's squirrel. Oregon Coast.

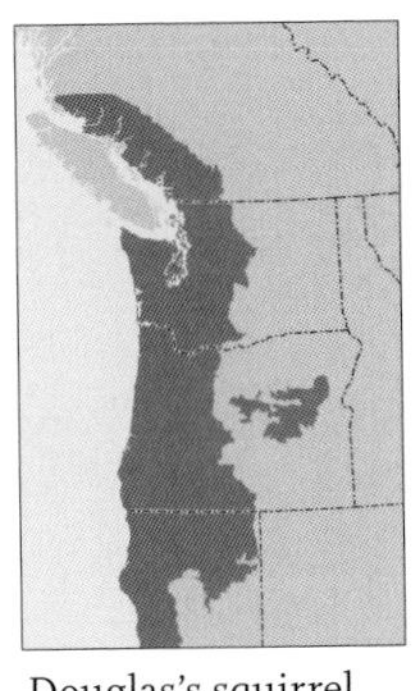

Douglas's squirrel range

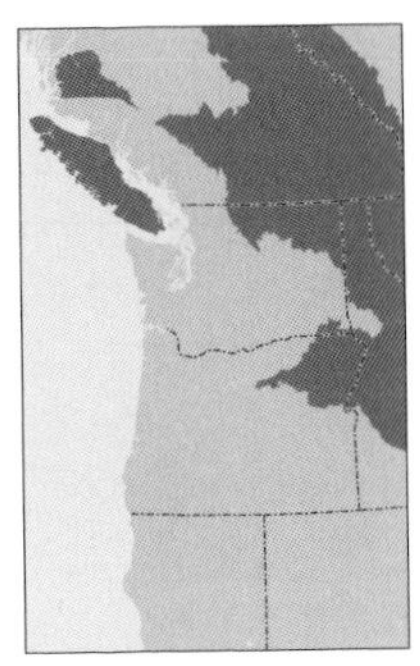

Red squirrel range

Left front track of a Douglas's squirrel. West Cascades, Washington.

Right front (below) and hind tracks of a red squirrel. Middle Rockies, Idaho.

Douglas's squirrel bound in snow

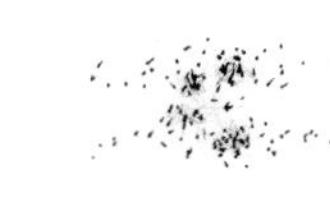

mainstay of the Douglas's squirrel's diet. Lodgepole pine cones are the red squirrel's main food source. **HABITAT & DISTRIBUTION** Conifer forests in all parts of our region. Douglas's squirrel range extends from the coast to east of the Cascades. Red squirrels are found east of the Cascades, on Vancouver Island, and coastally north of the island. **SIMILAR SPECIES** Gray squirrels, northern flying squirrel, chipmunks, ground squirrels, woodrats, Norway and roof rats, weasels.

Red squirrel bounding trail. Each group shows all four feet. The pointy drag marks in and out of each group are helpful diagnostic features of this species in snow. Okanagan, Washington.

TRACKS

Typical rodent structure. Sharp claws often register in tracks for both fronts and hinds. Toes are long and thin with slightly bulbous pads at their tips. Toes usually connect to the palm pad (page 60). **FRONT** 13/16–1 1/4 in. (2.0–3.1 cm) L × 3/8–7/8 in. (1.0–2.2 cm) W. Two large proximal pads usually register and are often wider than the palm. The vestigial toe 1 pad registers in clear tracks adjacent to the proximal pads. **HIND** 11/16–1 1/8 in. (1.8–3.0 cm) L × 9/16–1 in. (1.5–2.5 cm) W. Slightly wider than front track. Heel is completely furred. **TRACK PATTERNS** Bounds almost exclusively with the front feet landing in various positions. In deep snow, often shows consistent drag marks in and out of each track group. **BOUND** Stride: 9 1/16–50 13/16 in. (23–129 cm). Trail width: 2 9/16–3 3/4 in. (6.5–9.5 cm). Group length: 1 9/16–6 7/8 in. (4.0–17.5 cm).

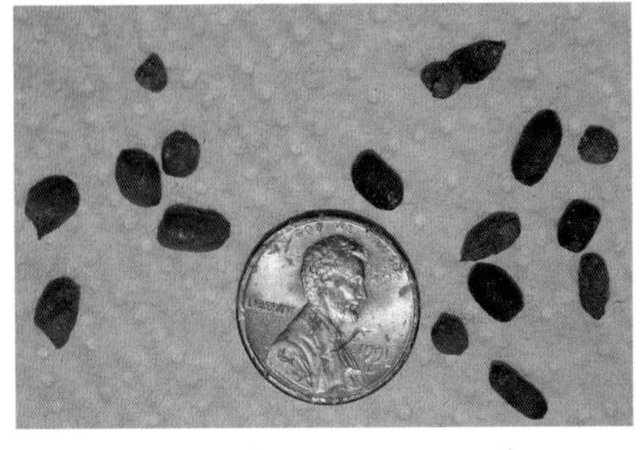

Douglas's squirrel scats. Puget Trough, Washington. Photo by Liz Snair.

SIGNS

SCAT 1/8–1/4 in. (0.3–0.6 cm) D × 3/16–7/16 (0.5–1.1 cm) L. Elongated pellets, occasionally found in feeding areas. **FEEDING SIGNS** Conifer cones are an important food source for both species, which cache cones in the late summer and fall for use through the winter. Squirrels eat the cones' seeds, methodically biting and tearing the cone scales to reach them, leaving woody cores and a pile of scales behind. Look for conifer branch tips cut at 45-degree angles and dropped to the forest floor; the animals access cones or the twigs' terminal buds from the ground. **SCENT-MARKING** Look for incisor marks on low-hanging branches of favored trees. **MIDDEN** Look for a collection of discarded bracts and cores from conifer cones. Middens come in all sizes and surround a favorite feeding perch such as a tree branch, stump, or fallen tree's root wad. Often used to store collected cones. **NESTS & BURROWS** Both species use tree cavity nests for shelter. Burrows are also used, especially in winter. Burrows are often built into the base of a fallen tree root wad and associated with a feeding midden.

These ponderosa pine cones were dismantled by a Douglas's squirrel. East Cascades, Washington.

Midden of a red squirrel at the base of a Douglas-fir. The hole in the snow shows where the squirrel accessed a food cache or its winter nest. Okanagan, Washington.

NORTHERN FLYING SQUIRREL

Glaucomys sabrinus

Flying squirrels are the most unique members of the Sciuridae family. Despite their name, they do not truly fly. A skin membrane that connects the front and hind legs and a broad, flattened tail enable this squirrel to glide long distances through the forest. It launches from high on one tree and glides to the base of another, where it scampers up to launch again. Flying squirrels are active year-round. Unlike all other squirrels in our region, they are nocturnal. During cold weather, they may congregate in a single tree hollow to conserve energy. Stick nests constructed in tree limbs are probably used more in the summer than the winter.

SIZE Length: 27.5–34.2 cm. Width 75–140 g. **FIELD MARKS** Gray with white belly. Large eyes. Gliding membrane is apparent. Rarely seen during the day, they make a unique twittering sound during their nighttime activities that is heard far more often than it is recognized. **DIET** Primarily underground fruiting bodies of mycorrhizal fungus, commonly known as truffles. Seasonally, lichens and plant material. Many truffles on which they feed have a symbiotic relationship with forest trees, with tree roots providing sug-

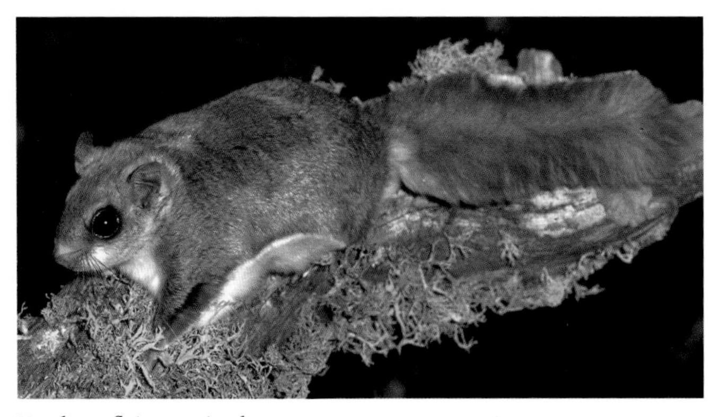

Northern flying squirrel. Photo by Dr. Lloyd Glenn Ingles, California Academy of Sciences.

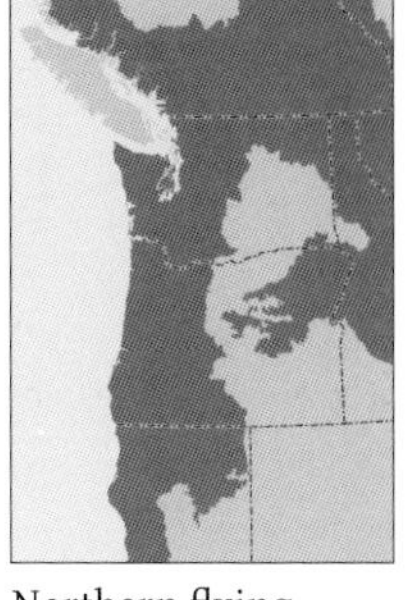

Northern flying squirrel range

ars to the fungus and the fungus enabling the tree to uptake nutrients more effectively. The squirrels' foraging behavior likely helps inoculate the tree roots, thus benefiting the trees, the truffles, and the squirrels. **REPRODUCTION** Breeding occurs in March, and a single litter of 2–4 young are born in May or June. Females raise altricial young without help from the male. **HABITAT & DISTRIBUTION** Found throughout the Northwest. In conifer forests east of the Cascades, decades of fire suppression that enabled the growth of dense stands of Douglas-fir has increased the abundance of flying squirrels. West of the Cascades, studies indicate that the squirrels reach greater densities in old-growth forests than in younger forests, possibly because of a greater abundance of truffles and more tree cavities for nests. **SIMILAR SPECIES** Red and Douglas's squirrels. **CONSERVATION ISSUES** Northern flying squirrels are an important prey for the spotted owl (*Strix occidentalis*), an endangered species strongly associated with old-growth forests. A pair of spotted owls may consume as many as 500 flying squirrels in a year. Mycorrhizal fungi die following logging of their host trees, making large clearcuts damaging to the symbiotic relationship between the fungus, trees, and squirrels.

TRACKS

Tracks are smaller on average than those of red and Douglas's squirrels, but museum specimens I inspected showed a similar amount of fur on the soles of the feet. However, tracks are usually detected in snow and appear to show less detail than is generally seen in red and Douglas's squirrel tracks, suggesting a more heavily furred foot. Toes are long, thin, and bulbous at tips. The unfurred metacarpal and toe pads register clearly in appropriate substrate (page 60). **FRONT** ½–⅞ in. (1.2–2.2 cm) L × ½–13⁄16 in. (1.3–2.0 cm) W. Four toes (toe 1 is absent). Palm has three distinct metacarpal pads, similar to other squirrels. Two proximal pads usually register in tracks. **HIND** 11⁄16–13⁄16 in. (1.7–2.0 cm) L × 11⁄16–1 5⁄16 in. (1.7–3.3 cm) W. Palm has distinct subpads, similar to other squirrels, but is slightly thinner and less arced. As with red and Douglas's squirrels, the heel is completely furred. Toes 1 and 5 often do not spread as much

The bounding trail of a northern flying squirrel traveling toward a stump. The wider and larger pair of tracks in each group are the hind feet. Canadian Rockies, Washington.

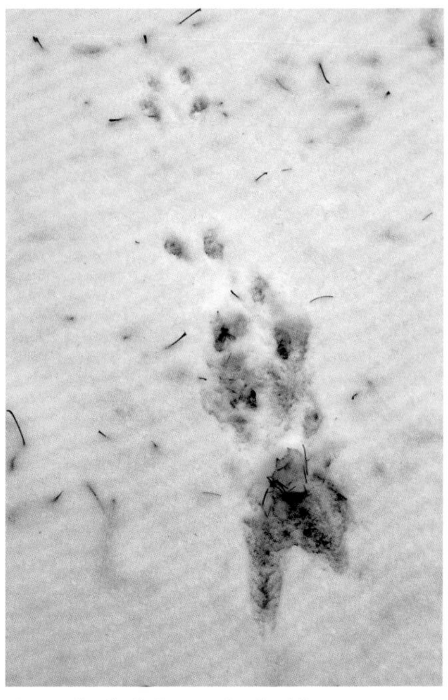

Sitzmark of a flying squirrel with a trail bounding off at the top of the frame in a dusting of snow. Canadian Rockies, Washington.

Two truffle digs in forest debris. East Cascades, Washington.

Dig with the spherical shape of the removed truffle apparent. West Cascades, Washington.

Northern flying squirrel bound

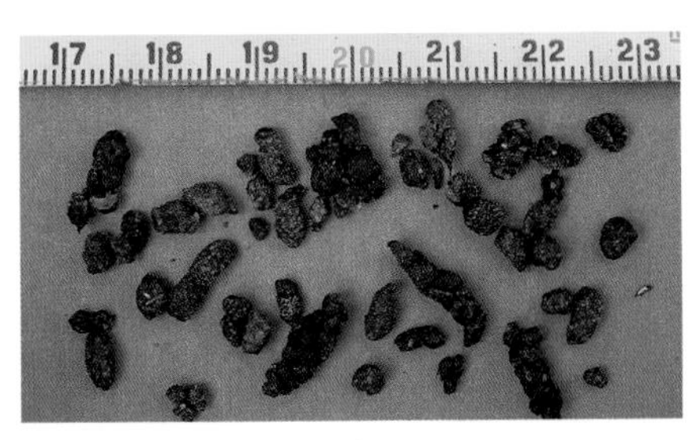

Northern flying squirrel scats
(from captive animal)

as in Douglas's and red squirrels; instead, they more closely hug the central toes in the track. **TRACK PATTERNS** Most distinctive pattern is the landing spot (sitzmark), where a gliding individual touches down in snow. Bounding pattern may show the front registering between or even ahead of the hind feet. When the hinds pass the fronts (as is common in red and Douglas's squirrels), the front feet are more widely spread, perhaps because of the flight membrane that connects front and hind limbs, giving the track pattern a boxy appearance. Occasionally the flight membrane registers in deep snow, which could be confused with feet drag marks in red and Douglas's squirrel tracks. **BOUND** Stride: 11 7/16–37 in. (29–94 cm). Trail width: 2 3/8–3 3/8 in. (6.0–8.5 cm). Group length: 2 3/8–4 13/16 in. (6.0–12.3 cm).

SIGNS

SCAT Irregular in shape and deposited at feeding locations such as tree branches. **FEEDING SIGNS** Digs made to reach truffles are the most abundant and visible signs. Digs are shallow and V-shaped with a small throw mound and a width of 1 3/8–2 9/16 in. (3.5–6.5 cm). A small, spherical space may appear at the bottom of the dig, where the fruiting body of a truffle has been removed. Often found in small congregations in the forest.

MICE AND RATS

Families Cricetidae, Muridae, and Dipodidae

This group includes deer mice, harvest mice, grasshopper mice, jumping mice, woodrats, and invasive Old World rats. Young are altricial at birth and for most species the reproductive rate is quite high, with multiple litters produced per year.

Tracks. Tracks fit the typical rodent structure (jumping mice being the notable exception).

Track patterns. All group members use bounds extensively, if not primarily, and may also walk.

Scat. Mice and rats produce numerous elongated pellets.

DEER MICE

North American deer mouse, *Peromyscus maniculatus*
Northwest deer mouse, *P. keeni*
Canyon deer mouse, *P. crinitus*
Pinyon deer mouse, *P. truei*

In North America, when people think of mice, they picture a member of this genus. The Pacific Northwest is home to four species, all with similar appearance and habits. The North American deer mouse is the most widespread and abundant species and can be found in all portions of our region, overlapping with one or more of the other species. Deer mice are nocturnal and active year-round. They forage on the ground and in trees. Deer mice found in and around human habitations are common pests for people. The introduced house mouse (*Mus musculus*) is not covered here as it has little or no range in areas away from human constructions in our region.

Northwest deer mouse. Puget Trough, Washington.

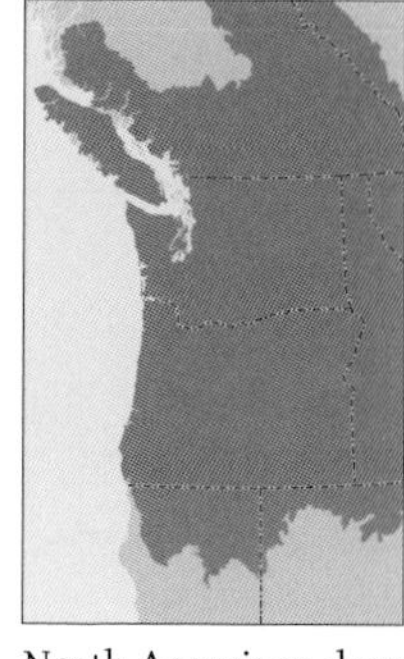

North American deer mouse range

SIZE North American deer mouse—Length: 12.0–22.5 cm. Weight: 10–30 g. Pinyon deer mice can be slightly larger; others fall within the size range of North American deer mice. **FIELD MARKS** Large, naked ears; large eyes; long, thin tails; dark backs; and light bellies. Pinyon deer mouse has exceptionally large ears. Where they overlap, the northwest deer mouse is difficult to distinguish from the North American deer mouse in the field. Tail length is used to distinguish these two species: if it is longer than 3⅞ in. (9.8 cm), it most likely is northwest deer mouse. House mice are a single color with a dark belly, which distinguishes them from deer mice. **REPRODUCTION** High reproductive rates in favorable environmental conditions. Females can breed year-round with multiple litters per year with an average of five young. Gestation is about 23 days. Young are weaned at about 3 weeks and can begin to breed at 2 months. Monogamous or polyandrous. **DIET** Omnivorous, varies based on habitat and season. Plant materials, especially seeds, but also insects and fungi. **HABITAT & DISTRIBUTION**: North American deer mouse—Nearly every habitat throughout our region. Northwest deer mouse—From the coast east as far as the east slopes of the Cascades. Canyon deer mouse—Desert areas of southeast and central Oregon, almost exclusively near cliffs and rock outcroppings. Pinyon deer mouse—Wide range of habitats, from moist forests to desert communities in the Klamath Mountains, Northern Great Basin, and Columbia Plateau ecoregions in Oregon. **SIMILAR SPECIES** Other mice, voles, shrews.

TRACKS

Deer mouse tracks are the most common small mammal tracks found in many locations. Tracks are larger on average than those of pocket mice and harvest mice. Typical rodent structure. Tips of the toes are more bulbous than those of voles and often register disconnected from the palm. The palm is a proportionally larger part of the track than in voles. Small, sharp claws may register (page 59). **FRONT** ¼–½ in. (0.6–1.2 cm) L × ¼–7⁄16 in. (0.6–1.1 cm) W. Round to oblong. Toes 2–5 are bulbous at the tips and often not connected or lightly connected to the palm in tracks. Toe 1 is greatly reduced but retains a claw that occasionally registers in clear tracks, adjacent to the medial proximal pad. Palm comprises three distinct pads, and two small, round proximal pads

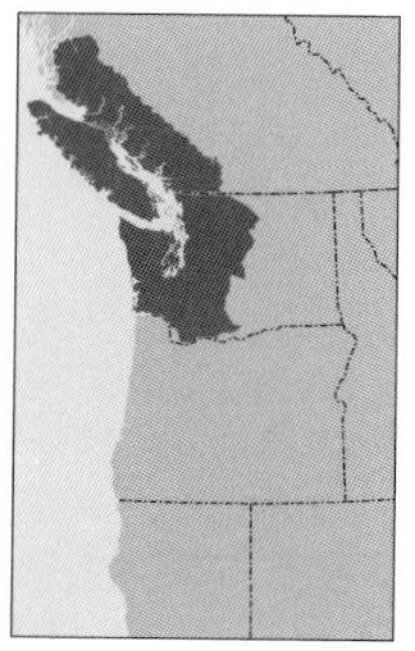

Northwestern deer mouse range

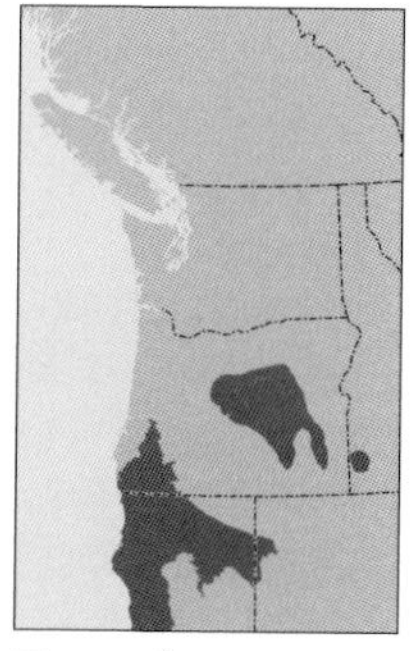

Pinyon deer mouse range

Multiple tracks of a deer mouse. The lowest two tracks are from the left and right front feet; the next three up are left and right hinds with a front in between. The toes are disconnected from the palm, and the claws are small (not registering in some tracks). Puget Trough, Washington.

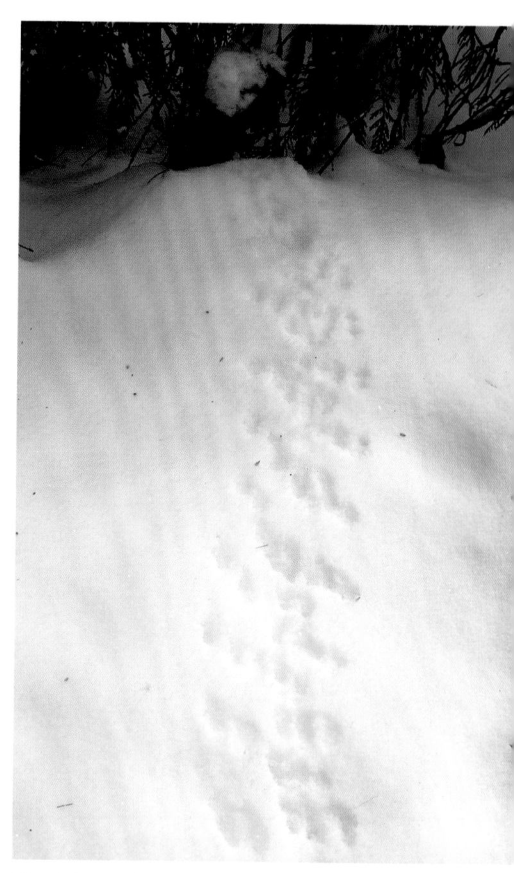

Multiple North American deer mouse trails go back and forth across an open stretch of forest, a pattern that is common for this genus in every sort of substrate and environment in our region. Canadian Rockies, Washington.

North American deer mouse bound

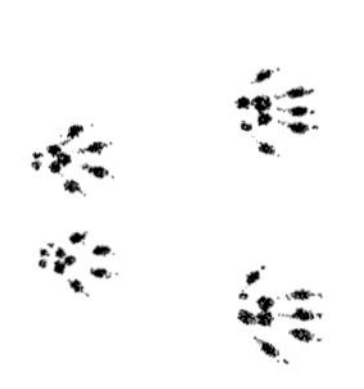

Scats of a deer mouse. Columbia Plateau, Oregon.

Deer mouse nest discovered under an abandoned board. Columbia Plateau, Oregon.

Abandoned bird's nest used by a deer mouse as a feeding station for consuming maple seeds (*Acer* spp.). Puget Trough, Washington.

Sitka spruce (*Picea sitchensis*) cones dismantled by rodents to access the seeds. The rightmost cone, discovered inside a small cavity at the base of a tree, was fed on by a deer mouse and has a characteristically smooth appearance. The three to the left, with a rougher appearance, were consumed by a Douglas's squirrel and retrieved from a midden pile. West Cascades, Washington.

are smaller relative to the overall size of the track than for chipmunks. **HIND** 1/4–9/16 in. (0.7–1.5 cm) L × 5/16–1/2 in. (0.8–1.3 cm) W. Slightly larger than fronts (the difference is less than in pocket mice). Overall shape is round. Toes 2–4 register parallel to each other. Toe 1 is smaller and lower than toe 5, which both angle away from the track midline. Palm is relatively robust and comprises three distinct pads that form a single shape in the track and a fourth, smaller pad that is often disconnected from the others on the inside of the track close to toe 1. Heel rarely registers. **TRACK PATTERNS** Most common gait is a paired bound. Overall pattern is squatter and wider than pocket mice and more consistent than harvest mice. Occasionally leave an alternating pattern of two when walking. In snow, tail drag is sometimes found, and the hind feet may land directly in the tracks of the fronts. **BOUND** Stride: 2 9⁄16–9 1⁄16 in. (6.5–23.0 cm). Trail width: 1 1⁄16–1 13⁄16 in. (2.7–4.6 cm). Group length: 7⁄8–2 1⁄4 in. (2.2–5.7 cm).

SIGNS

SCAT 1⁄16 in. (0.1 cm) D × 1⁄8–3⁄16 in. (0.3–0.4 cm) L. Deer mice scats, the bane of many a person cleaning out cupboards, are tiny pellets, inconsistent in size and shape and pointy at the tips. Found scattered and commonly in conjunction with foraging and feeding signs. **FEEDING SIGNS** Diverse, matching their varied diet. Seeds are often neatly gnawed open, rather than torn apart, as with larger rodents such as squirrels and rats. Food often carried to protected locations to be eaten, such as under logs or rocks or inside tree cavities—another good distinction from squirrels and chipmunks that feed in locations with a good view, such as up in a tree or on top of a stump. **NESTS** Constructed of grasses and other fine materials; can be located in a range of locations—under boards and logs, in tree cavities or rock crevices, or within woodrat nests.

WESTERN HARVEST MOUSE

Reithrodontomys megalotis

The western harvest mouse is one of the smallest mammals in our region. They rarely live more than a year in the wild. Individuals often move between several nests located on the ground or in shrubs. They are not territorial, and in winter they often huddle together in a shelter for warmth. They are active year-round and are nocturnal. Foraging activity occurs both on the ground and in shrubs. They climb adeptly and most food is consumed where it is found.

SIZE Length: 11.8–17.0 cm. Weight: 8–15 g. **FIELD MARKS**: Very small. Bicolored tail is longer than the body. Grayish-brown back, with darker stripe along the top, and a light-colored belly. Ears are light in color with an orange tinge on the hairs inside the ear. **REPRODUCTION** Breeds throughout the year, but reproduction slows down in winter. Multiple litters per year with 1–6 young, sometimes with only a month between parturition of one litter and the next. Apparently monogamous with overlap between home

Western harvest mouse

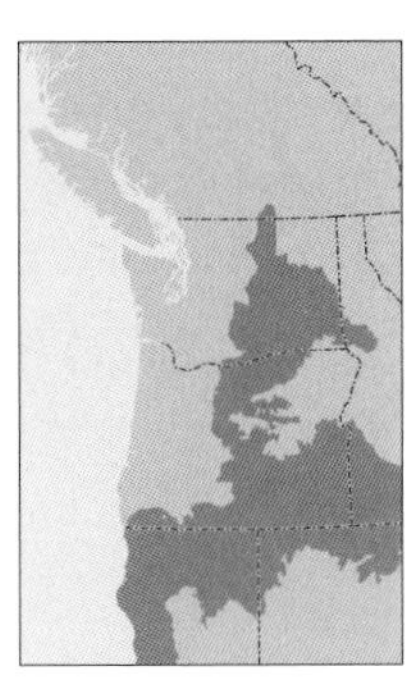

Western harvest mouse range

ranges of different individuals. **DIET** Omnivorous, mostly plant material (grasses, forbs, seeds, buds of shrubs) and insects. Prefer flower and seed heads to leaves and can hang onto shrubs or grasses with their hind legs while feeding on the tips of branches or grass flower stalks. **HABITAT & DISTRIBUTION** Arid grasslands, sagebrush, edges of streams and ponds, and marshes east of the Cascades and dry valleys in southwest Oregon. Marsh habitats seem to have abundances of these mice, where they are often associated with montane voles (*Microtus montanus*), which leave more extensive signs than the harvest mouse. Western harvest mice often use vole runways. **SIMILAR SPECIES** Deer mice, pocket mice, shrews, voles.

TRACKS

Typical rodent structure. Tracks are smaller than deer mice and similar in size to pocket

Left front and both hinds (outermost tracks) of a western harvest mouse. Columbia Plateau, Washington. Photo by Roy Ashton and David Moskowitz.

mice, but different in structure. Small, sharp claws may register in tracks. Front and hind tracks are similar in size, a good distinction from pocket mice, whose hinds are substantially larger than its fronts. Toes are long and thin with less bulbous tips than deer mice. Toes usually register as connected to the palm. In wet habitats where they are most abundant, they often leave

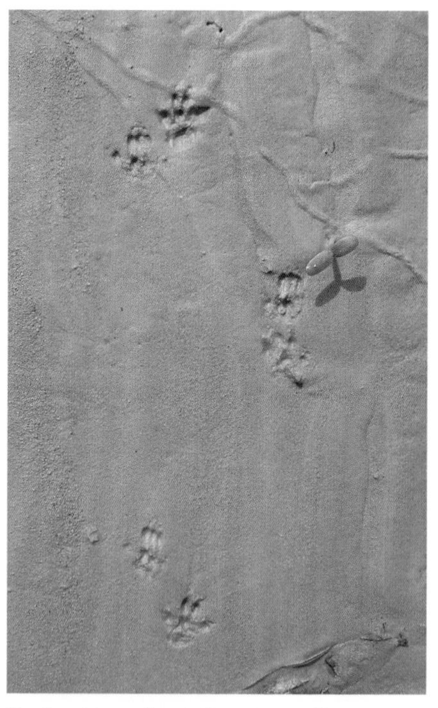

Track pattern of a trotting western harvest mouse, with the hind foot on each side of the body overstepping the front in the first two sets of tracks. Columbia Plateau, Washington. Photo by Roy Ashton and David Moskowitz.

clear individual prints (page 59). **FRONT** ¼–⅜ in. (0.6–0.9 cm) L × ¼–5⁄16 in. (0.6–0.8 cm) W. The vestigial toe 1 is clawless (unlike in deer mice). Toes are oriented in a more asymmetrical fashion than is typical in rodents with toe 3 farthest forward. Palm comprises three pads as in deer mice, but they are smaller, giving palm a more diminutive size relative to the track overall. Two proximal pads sometimes register in clear tracks. **HIND** ¼–7⁄16 in. (0.6–1.1 cm) L × ¼–⅜ in. (0.6–0.9 cm) W. Asymmetrical. Toe 1 is lower on the foot than in other mice and relatively larger, perhaps an adaptation for climbing grasses and shrubs. Toes 2–4 are long and thin and often register tightly parallel to each other. Toe 5 registers close to toe 4 rather than more splayed, as is often the case in deer mice. Overall track appearance similar to a miniature opossum hind track. **TRACK PATTERNS** Walks and bounds. Bounding patterns are less consistent than in deer mice or pocket mice. Front feet may register even with each other or offset, between the fronts or before them, rarely if ever overlapping (as in pocket mice). **BOUND** Stride: 4⅛–8 3⁄16 in. (10.5–20.8 cm). Trail width: 1–1½ in. (2.6–3.8 cm). Group length: ½–2 in. (1.2–5.1 cm).

SIGNS

NESTS Constructed of fine grasses and lined with cattail down or other fine material. Most are constructed on the ground under protective cover, such as a clump of grass or tumbleweed, but some are in shrubs. Nests average 3–5 in. (7.5–12.5 cm) in diameter and are spherical when off the ground and hemispherical on the ground. The entrance is little more than ⅜ in. (1 cm) diameter and may be closed with a grass plug when the inhabitant is at home. Nests are often occupied by a male and female. May enter burrows constructed by other rodents but do not produce their own.

Western harvest mouse nest. The mouse was home when we discovered the nest, and a grass plug filled the entrance hole. Columbia Plateau, Washington.

NORTHERN GRASSHOPPER MOUSE

Onychomys leucogaster

This fascinating mouse is highly predatory, consuming large insects and even smaller mice. In small mammal trapping studies it is consistently found in low densities in the Columbia Plateau ecoregion, usually associated with sand dune habitats. It is rarely observed otherwise. It is also known for its howls, which have reminded some of a miniature wolf!

SIZE Length: 12–19 cm. Weight: 25–50 g. **FIELD MARKS** Light sandy gray or brown with a whitish belly. Tail is short (less than half the length of the rest of its body), and ears are slightly smaller than those of deer

Northern grasshopper mouse. Photo by B. Moose Peterson/WRP.

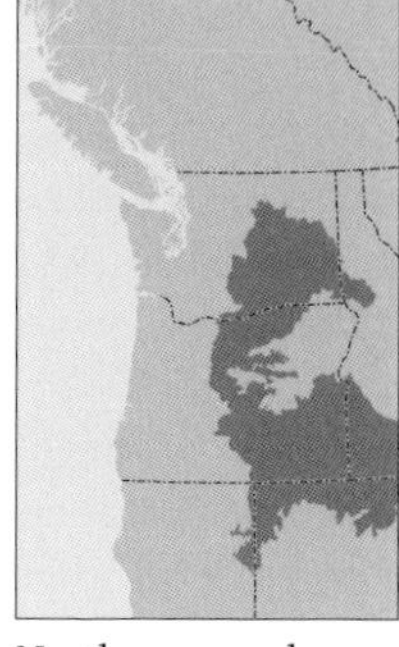

Northern grasshopper mouse range

mice. **REPRODUCTION** Can produce multiple litters a year from late winter to late summer. Thought to be monogamous, with male and female cooperating to raise young. **DIET** Large insects such as grasshoppers and beetles, but occasionally hunts and consumes other mice. Rarely consumes vegetation. **HABITAT & DISTRIBUTION** Arid and open sandy environments east of the Cascades. **SIMILAR SPECIES** Other mice.

TRACKS

I have not observed these mice in the field. Museum specimens I examined suggest that their tracks would be much like those of the similar-sized deer mice.

SIGNS

BURROWS Several types; the largest is a nest burrow, roughly 1.5 in. (3.8 cm) in diameter, with a 4–6 in. (10–15 cm) shaft traveling nearly straight down before opening up into a sleeping chamber. Smaller burrows are produced as refuges from predators and to cache food. **FEEDING SIGNS** Remains of insects would likely be detectable where they have been feeding.

JUMPING MICE

Pacific jumping mouse, *Zapus trinotatus*
Western jumping mouse, *Z. princeps*

Jumping mice are in the family Dipodidae. As their name implies, they are excellent leapers, easily jumping 3 to 6 ft. (1 to 2 m) when disturbed, before hiding in dense grass or shrubs. They are active only during the warm months of the year when they accumulate fat for a long period of hibernation. At higher elevations, the western jumping mouse may be active for as little as 2½ to 3 months per year. In the more mild climates west of the Cascades, Pacific jumping mice hibernate for about 7 months, with activity beginning around April (earlier for males than females).

SIZE Length: 21.1–25.0 cm. Weight: 18–30 g. **FIELD MARKS** Long hind legs and large feet; very long tail. When disturbed, its jumps are startlingly long for such a small animal. The ears of the western jumping mouse are tinged in white; ears of the Pacific jumping mouse are brown. **REPRODUCTION** Breeding occurs in May or June depending on the

Western jumping mouse. Canadian Rockies, British Columbia.

location. Gestation is only 18 days, with litter size of 2–7. Young are born in nests of woven grass on the ground or slightly above ground in grasses or shrubs. **DIET** Omnivorous. Grass and plant seeds are a primary source when available. In early to midsummer, when seeds and fruit are not yet abundant, they eat mostly insects. **HABITAT & DISTRIBUTION** Wet meadows and riparian habitat. Home ranges are often linear, along the edges of a waterway, marsh, or other body of water. Pacific jumping mouse—Cascades and west to the coast. Western jumping mouse—mountainous regions east of the Cascades and in a small section of the Cascades in southern Oregon and the Klamath Mountains ecoregion. Both species are present in the eastern section of the North Cascades ecoregion. **SIMILAR SPECIES** Deer mice, voles, Norway and roof rats, frogs, small perching birds.

Pacific jumping mouse range

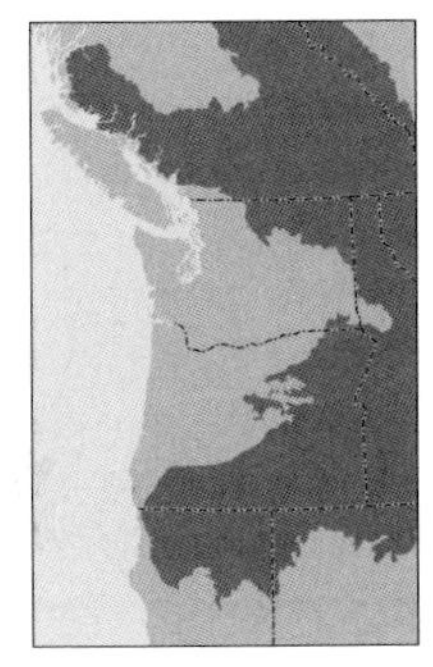
Western jumping mouse range

TRACKS

Tracks are the most commonly encountered signs and are easily confused with small bird or frog tracks. Both fronts and hinds diverge from the typical rodent structure. On all feet, toes are long and slender and often connect to the palm. Claws generally register clearly in tracks (page 59). **FRONT** 3⁄8–9⁄16 in. (1.0–1.4 cm) L × 5⁄16–1⁄2 in. (0.8–1.3 cm) W. Smaller than hinds. Track often appears canted outward from the direction of travel. (In frog and toad tracks, the fronts are turned inward.) Toes 2 and 5 often curve outward, a distinctive feature in jumping mice. The palm is relatively small, and the two small proximal pads rarely register. **HIND** 1⁄2–15⁄16 in. (1.3–2.4 cm) L × 7⁄16–11⁄16 (1.1–1.7 cm) W. Toes 2–4 are very long, join

All four feet of a Pacific jumping mouse in a typical bounding track pattern (hinds are the outermost tracks). West Cascades, Washington.

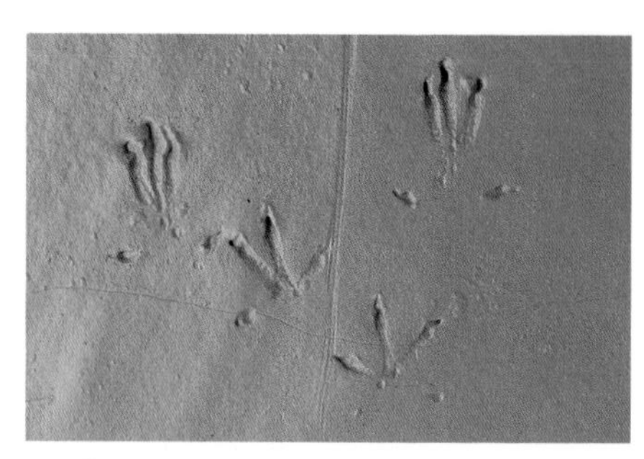

Pacific jumping mouse tracks with a slight tail drag (uncommon). West Cascades, Washington. Photo by Brian McConnell.

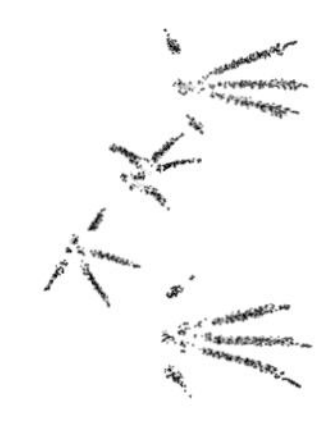

Pacific jumping mouse bound

the palm close together, and splay out from there, creating a shape similar to the front three toes of many small perching birds. Closer inspection will reveal the rest of the foot: Toe 1 is lowest, smallest, and points inward. Toe 5, also smaller than the middle three toes, points outward or curves slightly back. The palm is often less distinct than the middle toes and the heel rarely registers clearly. **TRACK PATTERNS** Primarily bounds. The fronts register either behind, parallel, or in front of the hinds. Less consistency in the pattern than in deer mice. Hinds are often more conspicuous than fronts. Strides can be long and erratic. **BOUND** Stride: 3¾–79 in. (9.5–200.0 cm). Trail width: 1⁵⁄₁₆–2½ in. (3.3–6.4 cm). Group length: ¹¹⁄₁₆–1 ¹¹⁄₁₆ in. (1.8–4.3 cm).

SIGNS

NESTS Summer nests are globular and above-ground, constructed of grasses. **FEEDING SIGNS** Feeds on grass seeds and plants in their wetland habitat. Look for collections of clipped plants with seed heads removed.

WOODRATS

Bushy-tailed woodrat, *Neotoma cinerea*
Dusky-footed woodrat, *N. fuscipes*
Desert woodrat, *N. lepida*

Neotoma is an industrious genus of rodents. The bushy-tailed woodrat is the most abundant and widespread regionally. Their penchant for collecting and storing items in their nests has earned them the name "pack rat." Once, while unpacking on a high mountain pass in the middle of the night after a long climb, my partner and I were accosted by a bushy-tailed woodrat who ran up and stole my friend's bowl practically out of his hands. The woodrat disappeared into the darkness and the bowl was never recov-

Bushy-tailed woodrat. Northwest Coast, Washington.

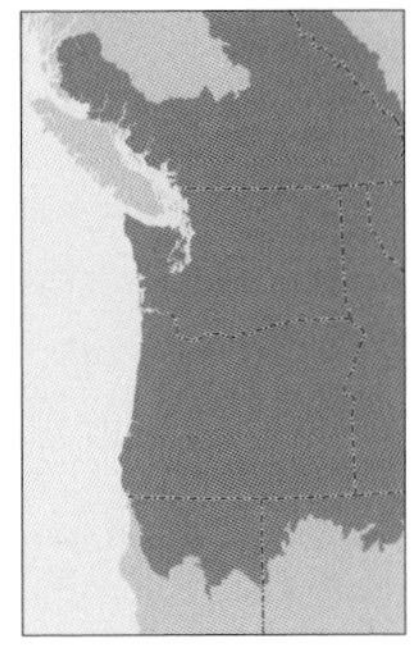

Bushy-tailed woodrat range

ered. We were not the first mountaineers to encounter this creature—a beautiful granite peak, Snafflehound Spire, in southeastern British Columbia is named after it. I have encountered bushy-tailed woodrats and their signs on numerous mountaintops in the Cascades, including one completely surrounded by glaciers! All three species are primarily nocturnal and active year-round.

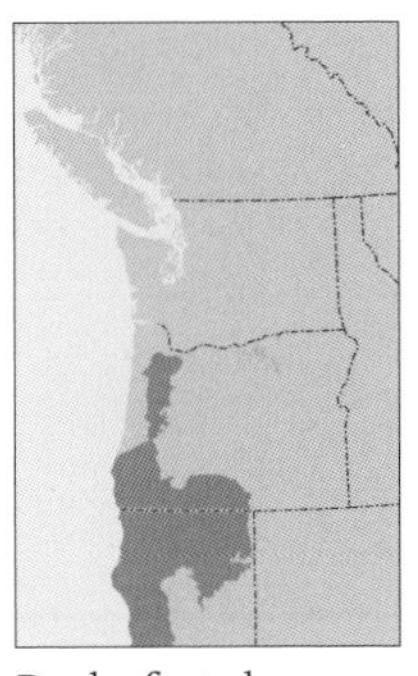

Dusky-footed woodrat range

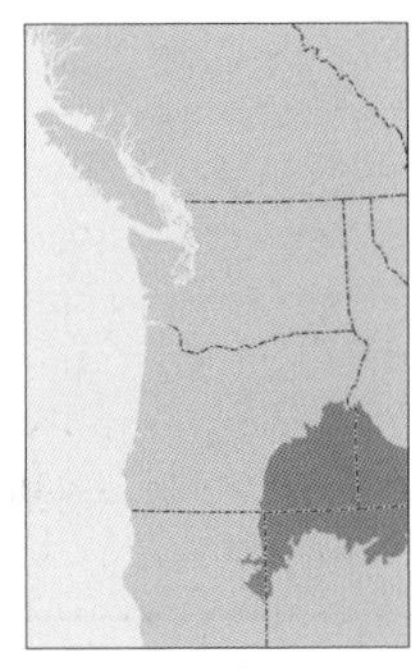

Desert woodrat range

SIZE Males are larger than females in bushy-tailed and desert woodrats. The desert woodrat is significantly smaller than the other two similar-sized species. Bushy-tailed woodrat—Length: 27.2–41.0 cm. Weight: 166–585 g. Dusky-footed woodrat—Length: 33.5–46.8 cm. Weight: 205–360 g. Desert woodrat—Length: 22.5–38.3 cm. Weight: 130–160 g. **FIELD MARKS** All three species have light bellies that distinguish them from *Rattus* species. Bushy-tailed woodrats have a heavily furred, bushy tail; the similar sized dusky-footed woodrat's tail is thinner, and the desert woodrat's tail is bicolored with a whitish tip. **REPRODUCTION** Breeding occurs in late winter to summer. Litters average 2–4 young, and a second litter may be born in early summer if conditions are favorable. Desert woodrat young may reproduce in their first year. Bushy-tailed woodrat young often do not disperse far, and related individuals tolerate one another. Bushy-tailed woodrats are likely polygamous. Dusky-footed woodrats are colonial. **DIET** Opportunistic but primarily herbivorous, consuming a variety of plant materials. In forest locations, will forage in trees. **HABITAT & DISTRIBUTION** Bushy-tailed woodrat—Throughout the region, from sea level to above treeline and from coastal rain forests to desert areas. Requires rock outcroppings, fallen logs, or other structures for nests, often nesting under bridges and in old buildings when available. Dusky-footed woodrat—Range overlaps with bushy-tailed in parts of the Klamath Mountains and in the Great Basin desert in the southern portion of the Columbia Plateau. Desert woodrat—Desert basins in the southern portion of the Columbia Plateau ecoregion. Unlike the bushy-tailed woodrat, the other two species are not limited to rock outcroppings or areas with natural cavities. **SIMILAR SPECIES** Norway and roof rats, ground squirrels.

TRACKS

Typical rodent structure: tracks appear pudgy with bulbous tips of toes that are either disconnected from the palm or connected by a shallow, often striated registration of the rest of the toe. Outer toes on front and hind feet tend not to splay as widely as those of Old World rats. Claws are small and may register in tracks. Bushy-tailed and dusky-footed woodrat tracks are similar in size (measurements include ranges for both species). Desert woodrat tracks are significantly smaller (not included in measurements) (page 60). **FRONT** 7/16–3/4 in. (1.1–1.9 cm) L × 7/16–13/16 in. (1.1–2.0 cm) W. Palm is more robust than in squirrels, chipmunks, and Old World rats and comprises three distinct pads that are visible in clear tracks. Two proximal pads often register, more narrowly than the palm width and significantly smaller than the palm pads, also good distinctions from squirrels and chipmunks. Toe 1 is vestigial and has no claw

Right-front track of either a bushy-tailed or dusky-footed woodrat. Northern Great Basin, Oregon.

Left front and hind tracks of a woodrat. Photo by Jonah Evans.

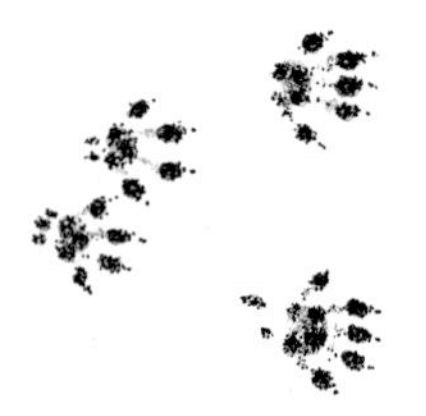

Bushy-tailed woodrat bound

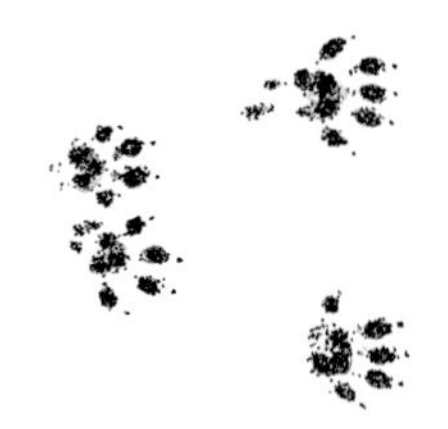

but registers in clear tracks. **HIND** 9/16–15/16 in. (1.4–2.4 cm) L × 11/16–1 in. (1.7–2.5 cm) W. Slightly larger than front track. Toes are bulbous and relatively shorter than in squirrels, chipmunks, and Old World rats. The robust palm comprises three semifused pads that form an L-shape and a smaller disconnected forth pad that sits in the opening of the L. Heel rarely registers, but when it does, it shows two offset proximal pads. **WALK** Stride: 4¼–6 11/16 in. (10.8–17.0 cm). Trail width: 2 3/16–2 7/16 in. (5.5–6.2 cm). **BOUND** Stride: 7¾–11 5/8 in. (19.7–29.5 cm). Trail width: 2 7/16–3 5/16 in. (6.2–8.5 cm). Group length: 2–2 3/8 in. (5–6 cm).

Dusky-footed woodrat scats (bushy-tailed similar). Klamath Mountains, California.

SIGNS

SCAT Bushy-tailed and dusky-footed woodrats—1/8–3/16 in. (0.3–0.5 cm) D × 3/8–½ in. (0.9–1.3 cm) L. Desert woodrat—1/16–1/8 in. (0.2–0.4 cm) D × ¼–7/16 in. (0.6–1.1 cm) L. Tubular and blunt tipped. Found in large quantities around nest sites, on trails, and in feeding areas. **NESTS** Often the most prominent sign of this animal on the landscape. Large nests are used by successive generations and also used by other small mammals, reptiles, amphibians, and insects, making them an important ecological feature in areas where woodrats exist. Nest construction is distinctly different for bushy-tailed woodrats than for the other two species. Bushy-tailed nests are often located in rock ledges, hollow logs, or tree cavities. Sticks, scats, and debris can be found flowing out of crevices and cavities in use. Human structures such as abandoned buildings and bridges are also used. Will not occupy areas lacking rock outcroppings, fallen logs, or other similar protective structures for their nests. Dusky-footed and desert woodrat nests are often constructed in locations next to a shrub, fence post, log, or

A jumble of sticks spreads from under a large boulder in a subalpine talus field—characteristic for bushy-tailed woodrat nests. Closer inspection would reveal numerous scats mixed in with the sticks. North Cascades, Washington.

Dusky-footed woodrat nest at the base of a juniper tree (*Juniperus occidentalis*). Klamath Mountains, California.

Bushy-tailed woodrat nest in the base of a hollow juniper tree, consisting of branches, juniper fronds, and scats. Columbia Plateau, Oregon.

This desert woodrat nest, constructed of sticks and cow and horse droppings, had been excavated, probably by a coyote. Northern Great Basin, Oregon.

Bushy-tailed woodrat urine deposits and a few scats in a basalt talus field. East Cascades, Washington.

other supporting structure. Nests are large, often more than 3 ft. (1 m) high and wide, and sometimes larger. Dusky-footed woodrats frequently build multiple nests in close proximity, connected by well-worn paths. They also construct nests in cavities or in rock outcroppings when available. **URINE DEPOSITS** Bushy-tailed woodrats urinate on rocks and ledges in areas that they use. Documented mostly in males, this is likely a territorial marking behavior. It leaves a conspicuous white residue on the rocks that is visible from great distances and can be confused with whitewash from a large bird's roost or nest. In some areas, similar white deposits that appear on rocks are of geological rather than biological origin. **FEEDING SIGNS** Look for debarked shrubs and saplings and juniper branches clipped to retrieve berries. Clipped branches show the typical 45-degree angle characteristic of rodents. Where available, acorns and other nuts are eaten, and husks are both gnawed and ripped open by these relatively large species.

OLD WORLD RATS

Roof rat, *Rattus rattus*
Norway rat, *R. norvegicus*

These two invasive species followed European explorers and traders all over the world. Vectors of human diseases such as plague, they often invade human habitations, gnawing through walls and destroying stored food. Both species have also naturalized in rural environments, and their tracks and signs can be found along rivers and in riparian areas in many agricultural areas.

SIZE Length: 31.5–46.0 cm. Weight: 115–485 g. Norway rats average slightly larger than roof rats. **FIELD MARKS** Dark belly distinguishes them from woodrats. Roof rat ears are larger than Norway ears. The roof rat's tail is longer than the rest of its body; the Norway rat's tail is shorter. The tail is hairless and scaly on both species. **REPRODUCTION** Both species are prolific and can breed at any time in favorable conditions,

Roof rat. Photo by Larry Jon Friesen, sbnature.net.

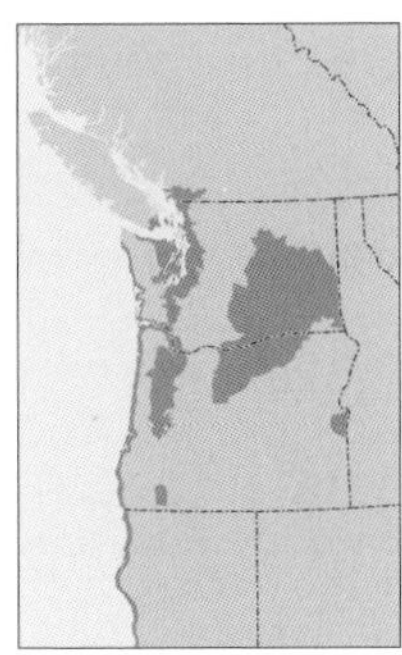

Norway rat range

producing up to 14 litters per year, with 5–6 young in each. Females can be sexually mature and reproduce at around 2 months of age. Polygamous, with dominant males defending a territory that can include multiple breeding females. **DIET** Omnivorous. Can consume large amounts of stored human food, plus carrion, including cannibalism. **HABITAT & DISTRIBUTION** Often associated with human structures in urban and rural environments, wetland and riparian habitats in agricultural areas, and along coastal rivers especially around ports. In our region, the Norway rat is more widespread than the roof rat. In addition to the shaded areas shown in the range map, Norway rats can be found in nearly every urban area in our region, large or small. **SIMILAR SPECIES** Woodrats, tree squirrels, ground squirrels, mink.

TRACKS

Typical rodent structure. The palm overall and individual pads are relatively smaller than in woodrats. Fronts are smaller than hinds. Outer toes of front and hind tracks splay more widely and toes appear less bulbous than in woodrats. These species' tracks are similar, but research suggests that they can be differentiated using a consistent substrate to capture tracks and statistical analysis. However, in field conditions, distin-

Right front (lower) and hind tracks of an Old World rat (unknown species). Puget Trough, Washington. Photo by Jason Knight.

Typical scats of a roof rat. Puget Trough, Washington.

Roof rat walk

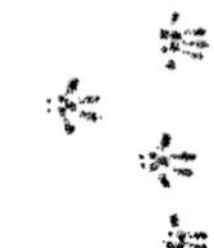

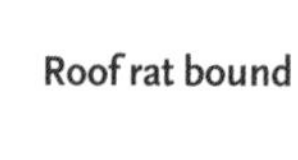

guishing the two may not be consistently possible (page 60). **FRONT** ½–13⁄16 in. (1.3–2.0 cm) L × 3⁄8–13⁄16 in. (1–2 cm) W. The two proximal pads in the heel are smaller than in squirrels and chipmunks. **HIND** ½–13⁄16 in. (1.3–2.1 cm) L × ½–1 in. (1.3–2.6 cm) W. Heel rarely registers but shows two offset proximal pads, the inner lower than the outer (similar to woodrats). **TRACK PATTERNS** Walks, trots, and bounds. Patterns of two usually show indirect registration. When bounding, the fronts are offset. **WALK & TROT** Stride: 3 1⁄8–7 5⁄16 in. (8.0–18.5 cm). Trail width: 1 3⁄8–2 9⁄16 in. (3.5–6.5 cm). **BOUND** Stride: 6 7⁄8–14 3⁄8 in. (17.5–36.5 cm). Trail width: 2–3 3⁄16 in. (5.1–8.1 cm). Group length: 1½–3 1⁄8 in. (3.8–7.9 cm).

SIGNS

SCAT 1⁄8–3⁄16 in. (0.3–0.5 cm) D × ¼–9⁄16 in. (0.7–1.4 cm) L. Rat scats are usually an unpleasant encounter in and around human habitations and structures. Pellets are less symmetrical than woodrat scats. **BURROWS** Elaborate; entrances are often concealed by boards or other obstacles.

VOLES

Subfamily Arvicolinae

Five genera of voles live in our region. Meadow voles (*Microtus* spp.) include the most species and greatest distribution. All are in the Cricetidae family. Voles' tails are shorter than their body lengths. They have small, dark, beady eyes, and their ears are partly obscured by hair. All voles in our region are active year-round. In areas with deep winter snowpack, winter activity is mainly subnivean. Most species are terrestrial but one species is semiaquatic and another arboreal.

Most voles dig shallow burrows and create tunnels through vegetation. Tree voles (*Arborimus* spp.) have a different and unique life history and subsequent set of signs. Red-backed voles (*Clethrionomys* spp.) are adapted to forest environments, while sagebrush voles (*Lemmiscus curtatus*) inhabit arid, shrubby habitats in the eastern portion of our region. *Microtus* voles are most abundant in meadows, and east of the Cascades they are strongly associated with ponds, springs, and riparian corridors. Several species undergo massive population density cycles over the course of years.

Tracks. Typical rodent structure. Unlike mice, toes are long and thin, often connecting to the palm in tracks, and claws are long and prominent.

Track patterns. Alternating track patterns from walking and trotting. Bounds occasionally.

Runways. For many species, elaborate trail systems that often weave under mats of grass are the most common signs encountered.

Nests. Above ground and below ground, constructed of fine grasses. Red tree voles construct nests exclusively in trees.

Scat. Elongated pellets, usually rounded on both ends and green when fresh. Deposited in latrines, along runways, and at feeding locations.

MEADOW VOLES

Creeping vole, *Microtus oregoni*
Townsend's vole, *M. townsendii*
Montane vole, *M. montanus*
Long-tailed vole, *M. longicaudus*

Eight species of *Microtus* inhabit our region, many with overlapping ranges, making spe-

Townsend's vole. Puget Trough, Washington.

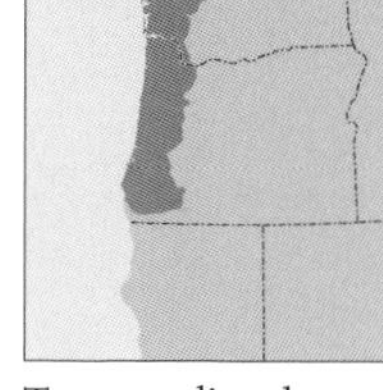

Townsend's vole range

cies identification difficult even when you actually see the animal. Habitat partitioning is common and reduces competition between species, but is often subtle and is not always a reliable clue to aid with species identification. Meadow voles can be incredibly fecund and their populations can go through rapid increases and crashes. Signs of their presence are diverse and fascinating. They are critical prey for many mammalian and avian predators and their survival depends on their high reproductive rate. The water vole is treated separately because of its distinctive size and life history.

SIZE Length: 13.0–22.5 cm. Weight: 17–83 g. Varies between species. The creeping vole (*Microtus oregoni*) is our smallest vole, and our largest (with the exception of the water vole, covered separately) is the Townsend's vole (*M. townsendii*). **FIELD MARKS** Dark gray or brown fur. Eyes are small and beady and ears are short, rounded, and obscured by hair. Stocky build and relatively short tail (less than length of the body). **REPRODUCTION** Some species may breed throughout the year if conditions are favorable, and multiple annual litters are common. In colder locations, breeding occurs from late spring to late summer, with 2–6 young in each litter. In some species, young can begin breeding at just 3 weeks of age. Voles rarely live longer than 2 years, and most don't live more than a year. **DIET** Herbivorous, with grasses, sedges, and forbs being the most important components. Also roots, bulbs, seeds, bark, and fungi. **HABITAT & DISTRIBUTION** At least one species is found in every part of our region, primarily in grasslands and meadows, but several species occupy forested habitats. In arid environments, they are strongly associated with riparian habitats and around springs. In the arid Columbia Plateau ecoregion, the montane vole is common near permanent water sources. In the Great Basin, montane and long-tailed voles are both found in wet habitats. The larger Townsend's vole inhabits lower elevations west of the Cascades in meadows, fields, and gardens. **SIMILAR SPECIES** Other voles, deer mice, jumping mice.

TRACKS

Typical rodent structure. Tracks are often similar in size to those of deer mice. Unlike deer mice, vole toes are long, thin, and

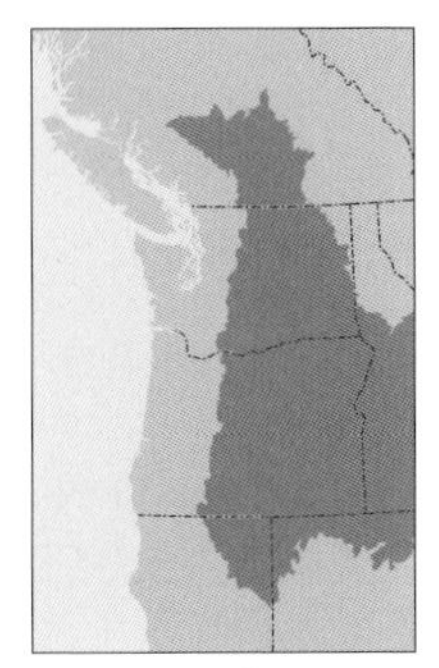

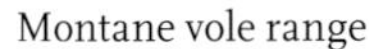

Montane vole range

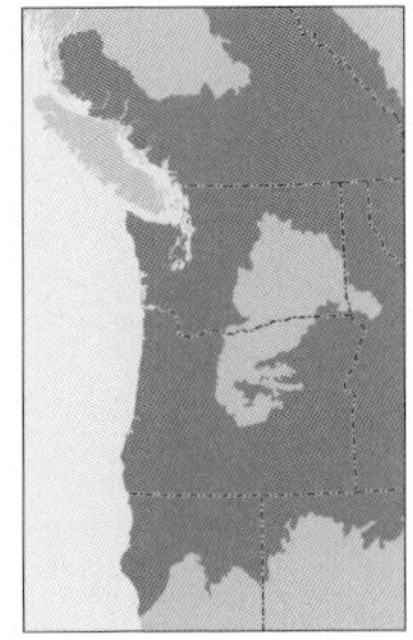

Long-tailed vole range

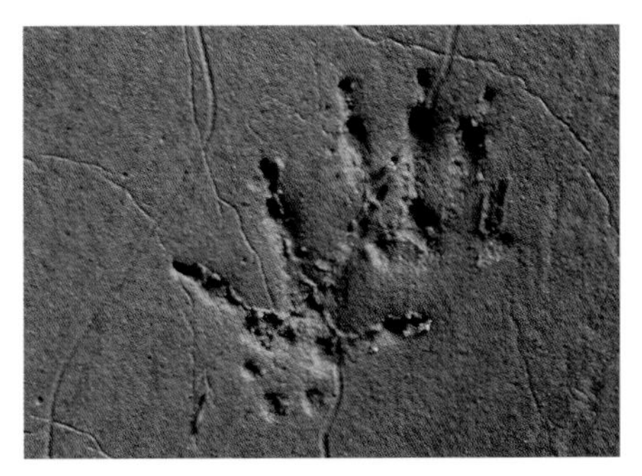

Right front (lower) and hind tracks of a Townsend's vole. Toe 1 has not registered on the hind foot. Note long, slender toes and long claws on both front and hind feet. Puget Trough, Washington.

ribbed; lack bulbous pads at the tips; and connect to the palm in tracks. Palm pads are relatively small on both front and hind feet. Claws are longer than in deer mice, often register, and are usually connected to the tips of the toes in tracks (page 59). **FRONT** ¼–½ in. (0.7–1.3 cm) L × ¼–½ in. (0.7–1.3 cm) W. Two proximal pads and vestigial toe 1 sometimes register. **HIND** ¼–9⁄16 in. (0.7–1.4 cm) L × ¼–9⁄16 in. (0.7–1.4 cm) W. Slightly larger than the front foot. Heel rarely registers. **TRACK PATTERNS** Voles commonly walk and trot and only occasionally bound. This distinguishes them from deer mice, which usually bound. Smaller vole tracks may be confused with those of shrews that commonly use these gaits. In winter, voles travel beneath the snow. **WALK & TROT** Stride: 2¼–8.5 in. (5.7–21.5 cm). Trail width: 15⁄16–1 11⁄16 in. (2.4–4.3 cm). **BOUND** Stride: 6 5⁄16–9 1⁄16 in. (16–23 cm). Trail width: 1 5⁄16–1 7⁄8 in. (3.4–4.7 cm). Group length: 9⁄16–1¼ in. (1.5–3.2 cm).

Bounding track pattern of a large Townsend's vole. Smaller deer mouse tracks are facing down and to the right for comparison. Notice the arrangement of the front feet (center two tracks) in comparison to the placement of these feet in deer mice. Many species of voles are closer in size to deer mice than this Townsend's vole. Puget Trough, Washington.

Townsend's vole trot

Aboveground nest (right) and burrow entrance (upper left) of a Townsend's vole, revealed from under a piece of debris. Puget Trough, Washington.

Montane vole feeding signs and latrine; debarked twigs lie next to the scat pile and the bark is removed from the trunk of the Douglas-fir sapling. East Cascades, Washington.

SIGNS

SCAT 1⁄32–1⁄8 in. (0.1–0.3 cm) D × 1⁄8–5⁄16 in. (0.3–0.8 cm) L. Tubular, blunt-tipped scats are found along runways, in latrines, at trail junctions, or outside burrows. Fresh scats are often green. **RUNWAYS** Microtus voles construct elaborate runway systems through grasses in fields and meadows. Runways connect burrows with feeding areas and are fastidiously maintained. Small sections of clipped blades of grasses or forbs line runways, and scat collections are often found nearby. Runways are often partly or completely obscured by vegetation that the voles do not remove to camouflage the trail. Runway widths average 1 3⁄8 in. (3.5 cm). **BURROWS & NESTS** Any time no snow is present (and year-round in areas with little or no snow), voles excavate and live in subterranean burrows. Entrance diameters range from 1 to 2 in. (2.5–5.1 cm). You can locate burrows by following runways. In the winter, voles often construct a sleeping nest in the subnivean zone above the ground. Nests are made up of spherical balls of grass and other fine material, up to about 8 in. (20 cm) in diameter. **ESKERS** During the winter, voles travel underground and in the subnivean zone. In the spring, you can often find eskers, cores of earth above the ground where a vole backfilled a subnivean tunnel with the material from subterranean excavations. These eskers resemble the work of pocket gophers, only smaller. Esker diameters are similar in size to summer runways. **FEEDING SIGNS** Clipped grass, sedges, and forbs are cut at 45-degree angles and often in short sections, 1–2 in. (2.5–5.1 cm) long. Clipped vegetation and scats can be found at feeding locations. **CAMBIUM FEEDING SIGNS** Voles debark shrubs and trees to eat the cambium, similar to activities of other rodents, with incisor marks that are smaller

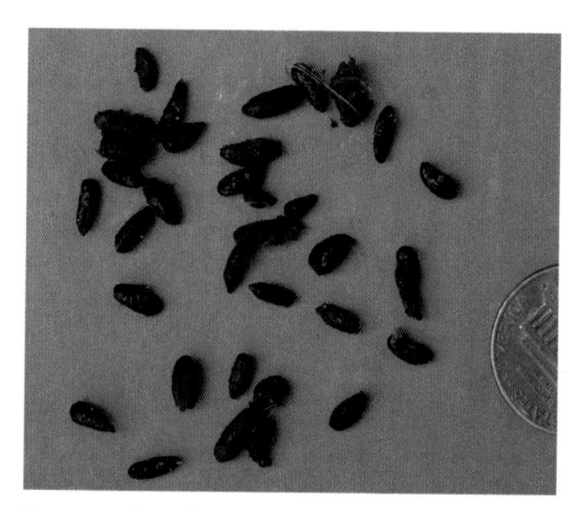

Townsend's vole scats. Puget Trough, Washington.

These meadow vole tunnels were carved through grass and exposed by lifting a piece of abandoned plywood. Similar tunnels can be found by sifting through high grass in meadows. Puget Trough, Washington.

These clipped grasses were found in a pile at the junction of Townsend's vole trails in high grass on the edge of a marsh. Puget Trough, Washington.

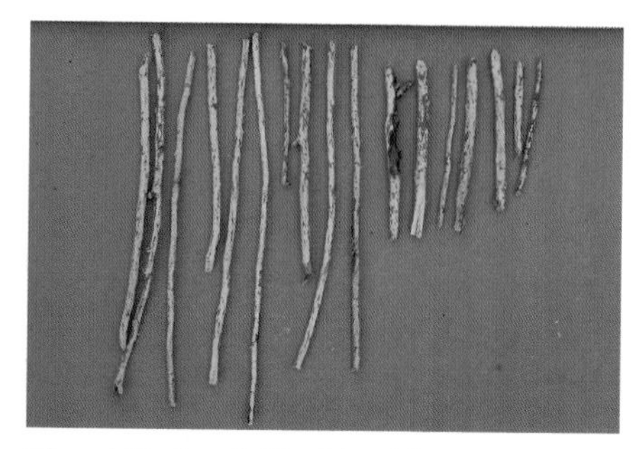

These bitterbrush (*Purshia tridentata*) twigs were clipped and debarked by a meadow vole (probably a montane vole) while the shrub was buried in snow. East Cascades, Washington.

than those of its cousins such as porcupines, mountain beavers, and beavers. I have observed this often on branches that are or were within the snowpack, perhaps providing safer access to this food source. Like grasses, twigs may be cut into small pieces and are found at feeding stations.

WATER VOLE

Microtus richardsoni

The water vole is distinguished by its large size and semiaquatic behavior, which is similar to that of a muskrat. Strong swimmers, these voles spend most of their time close to water—usually cold, clear mountain streams with extensive brush along the banks. They are active year-round and at any time of day or night.

SIZE Length: 23.4–27.4 cm. Weight: 68–150 g. Males are larger. **FIELD MARKS** Similar in appearance to other voles, but 2–3 times larger. Large hind feet. **REPRODUCTION**

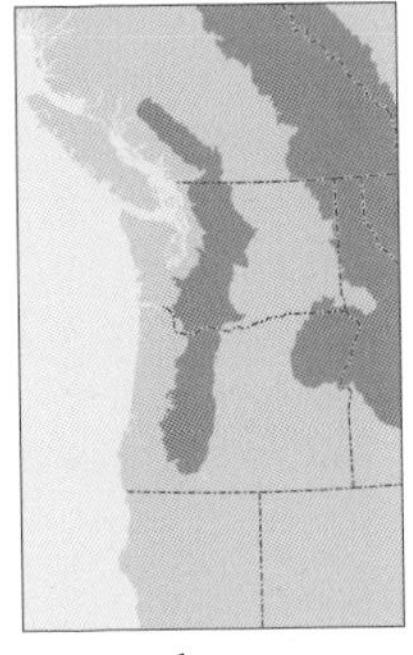

Water vole range

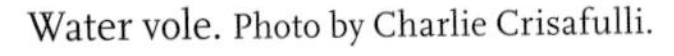

Water vole. Photo by Charlie Crisafulli.

Grasses and forbs have been cut and moved along a water vole trail emerging from the water and leading to a streamside burrow on a slow-moving subalpine stream. North Cascades, Washington.

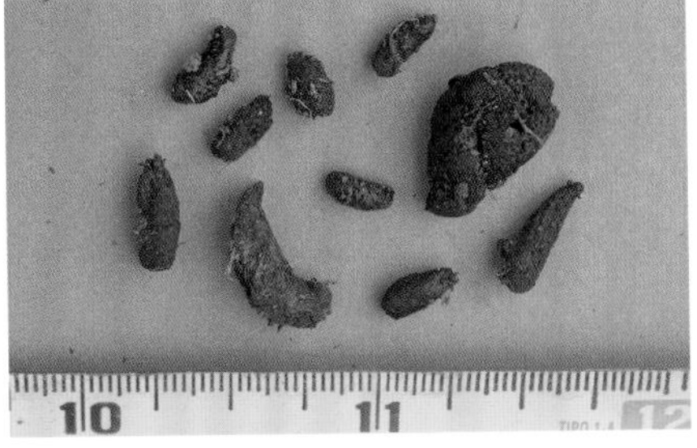

Water vole scats. North Cascades, Washington.

Breeding begins in late spring with the snowpack's retreat and continues through the summer. One to two litters with 5–6 young per litter. Young born early in season may reproduce later in the summer. Animals rarely live through a second winter. Adult males often bear wounds of altercations with other males. **DIET** Herbivorous: grasses, forbs, willow buds, bulbs, and some seeds. **HABITAT & DISTRIBUTION** Along mountain streams and ponds and up to the alpine zone in the Cascades, Selkirks, Rockies, and Blue Mountains. Populations are discontinuous and patchy, as large swaths of forest and other unsuitable habitat often separate usable habitat. **SIMILAR SPECIES** Muskrat, meadow voles.

TRACKS

Museum specimens I have examined resemble other similarly prepared voles but are larger (similar in size to a large chipmunk or a rat). Unlike those of other semiaquatic rodents, water vole toes are neither webbed nor fringed with stiff hairs. Likely prefer to walk and trot, as do other voles.

SIGNS

SCAT 1⁄8–3⁄16 in. (0.3–0.5 cm) D × 3⁄16–1⁄2 in. (0.5–1.2 cm) L. Similar to, but larger than, scats of other voles. Found in and around burrows and feeding areas. **RUNWAYS** Similar to meadow voles but larger (2–2¾ in., or 5–7 cm, wide). Trails often cross or radiate from waterways that are incorporated into travel. Creates subnivean tunnels similar to meadow voles but larger. **FEEDING SIGNS** Clipped vegetation commonly found at feeding sites, along trails, and at water's edge. Diameter of clipped vegetation is larger than that of smaller vole species. Vegetation up to 3⁄8 in. (1 cm) in diameter is cut into lengths up to 4 in. (10.2 cm) long. **BANK BURROWS** Constructed in the banks of streams they inhabit, similar to muskrats. Entrances located around waterline. In winter they create subnivean grass nests similar to those of other voles, but proportionally larger.

RED TREE VOLE

Arborimus longicaudus

This arboreal vole's life revolves around the green needles of the Douglas-fir, their primary (and often sole) food. Nests are usually constructed in Douglas-firs, 15–150 ft. (4.5–45.7 m) off the ground. Nests are found most abundantly in large trees in old forests, associating these voles with late successional (old-growth) forests. In its range, the red tree vole is a principle prey species for the endangered spotted owl (*Strix occidentalis*). It is believed to obtain water by licking dew from the needles and branches of its home tree.

SIZE Length: 15.8–20.6 cm. Weight: 25–47 g Females are larger than males. **FIELD MARKS** Small, reddish-brown back and furred tail. **REPRODUCTION** May breed throughout the year, producing multiple litters of 2–3 young. **DIET** Green needles of Douglas-fir and other conifer needles in small quanti-

Red tree vole. Photo by Chris Maser.

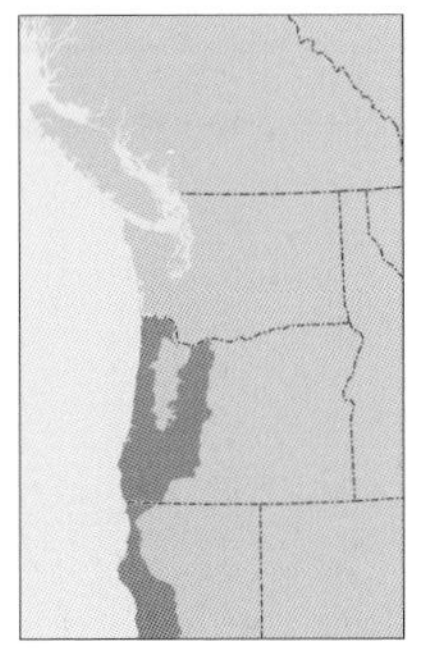

Red tree vole range

ties. **HABITAT & DISTRIBUTION** Only vole to live exclusively in trees. Douglas-fir forests west of the Cascade Crest and in the Coast Range of Oregon. Most abundant in late successional forests but can also be found in younger forests. Sonoma tree vole (*Arborimus pomo*)—Similar habitats along the northern California coast. White-footed vole (*A. albipes*)—Coast Range in Oregon. Little is known about this seemingly rare species, but it is associated with hardwood forests, believed to live in ground burrows, and forages in the trees. **CONSERVATION** Red tree

From left: small debarked Douglas-fir twigs, resin ducts, a single scat, and green Douglas-fir needles. Klamath Mountains, Oregon.

Clipped Douglas-fir branch tips can be found around trees with nests and incorporated into nest structures. Klamath Mountains, Oregon.

A large red tree vole nest wraps around a Douglas-fir at a whorl of branches. Klamath Mountains, Oregon.

voles are a species of interest because of their close association with old-growth forests and because they are an important prey species of the endangered spotted owl.

SIGNS

SCAT Similar in shape to meadow vole scats but smaller and found abundantly in nests. Scats are an important component of the nest structure, helping to secure it to the tree. Scats are not present in the central sleeping chamber. **FEEDING SIGNS** Unique to red tree voles, discards the central, fibrous resin ducts of Douglas-fir needles, which can be found at the bases of trees where voles have been feeding. Fresh ducts are off-white, browning and dulling with age. Small clipped twigs, up to 3⁄16 in. (0.5 cm), are also found around the base of the tree, cut to access their needles before they are discarded. Small twigs are occasionally debarked, as by other voles. **NESTS** Nests low in trees are the most conspicuous signs, but many are constructed higher up and are not visible from the ground. Constructed close to the trunk, nests often sit on a whorl of branches used to secure the structure; they range from football-sized to several feet in diameter. They are constructed from small clipped branches, needle resin ducts, and compacted scat, and often start as simple platforms for feeding that are converted to a covered nest. Nests are occupied by a single individual or a mother with young. In larger trees (believed to be preferred), where the nest may be out of sight high up in the tree, look for resin ducts and clipped branches on the ground.

RED-BACKED VOLES

Southern red-backed vole, *Clethrionomys gapperi*
Western red-backed vole, *C. californicus*

Red-backed voles are exclusively forest dwellers, and volelike signs you find in the interior of a mature forest in the Northwest are likely from one of these two species. Active year-round and both day and night, they are good climbers but are usually encountered around the bases of fallen, rotting logs in mature forests. They do not usually construct the elaborate runs through grasses as meadow voles do, preferring to travel along the overhanging edges of logs. They also use the burrows, tunnels, and trails of other species such as moles and other voles.

SIZE Length: 11.6–17.2 cm. Weight: 6–42 g. **FIELD MARKS** Similar to meadow voles, except for a reddish wash along the back. **REPRODUCTION** Promiscuous. At lower elevations west of the Cascades, may breed throughout the year. At higher elevations, breeds from spring to fall. Up to four litters per year with 4–6 young. Young born in early litters may breed later the same year. Most live less than a year. **DIET** Mainly fruit-

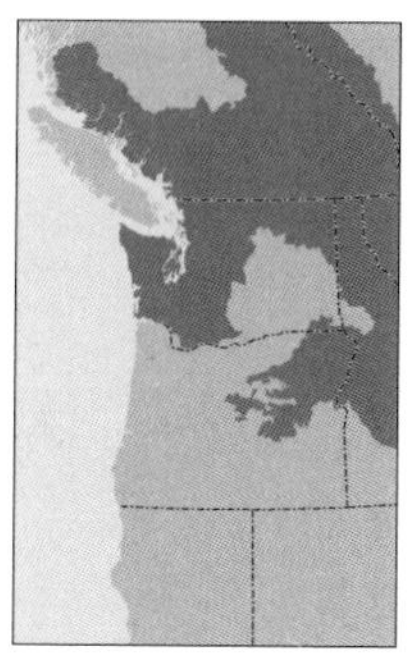

Southern red-backed vole range

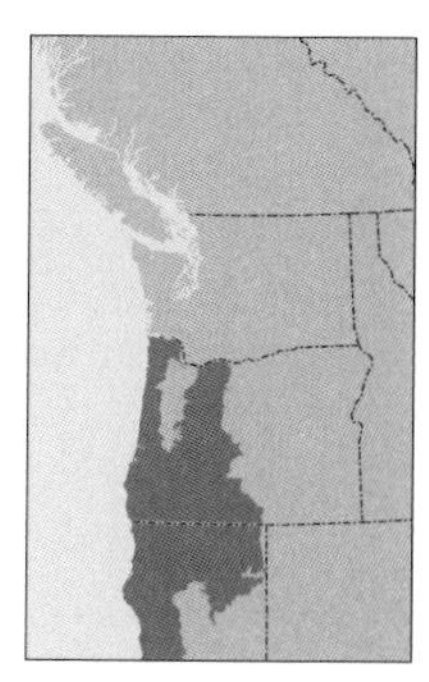

Western red-backed vole range

ing parts of subterranean fungi, or truffles. Also grasses and forbs, conifer seeds, mosses, ferns, insects, and lichens (important in diets at higher elevations). **HABITAT & DISTRIBUTION** Conifer forests; strongly associated with moist sites, fallen logs, and mature stands of trees that provide cover for movement and the fungi they eat. These species do not overlap in range anywhere in our region and are found from sea level to treeline in appropriate habitat. **SIMILAR SPECIES** Other voles, deer mice. **CONSERVATION** Highest abundance of red-backed voles are found in old-growth forests. As with flying squirrels, foraging patterns on truffles are believed to spread spores of these fungi, which increase the forests' productivity. Conversely, these voles are rare in areas of recent clearcuts, because the truffles that are dependent on the trees' root systems die shortly after forests are logged.

TRACKS

Tracks are similar to those of meadow voles.

SIGNS

SCAT Similar to that of other voles. **NESTS** Globular nests made of leaves, mosses, and grasses, 3–4 in. (7.5–10.0 cm) in diameter. Found in natural cavities, under logs, at the base of dense ferns, and occasionally above the ground in a tree. Often surrounded by a series of radiating trails or tunnels. **RUNWAYS** Look for well-worn paths along the overhanging edges of fallen logs (also used by other species such as moles, shrews, and deer mice), along with feeding signs and scats. **FEEDING SIGNS** Most subterranean fungi that are consumed are inconspicuous. Clips shrubs into small sections and debarks, as with meadow voles. Distinguishable only by habitat, if at all.

WESTERN HEATHER VOLE

Phenacomys intermedius

The western heather vole inhabits the subalpine and alpine zones of all of our high mountains. It is active year-round, spending much of its time under deep snow. In winter, several individuals may share a nest.

SIZE Length: 12.2–15.5 cm. Weight: 30–40 g. **FIELD MARKS** Very small; appears similar to meadow voles such as the montane vole, with which it shares parts of its habitat. **REPRODUCTION** Breeding starts in May and continues into the fall. Despite the short summer season, they often produce two litters during the warm weather months. Gestation takes about 3 weeks. Litter size averages 3–4. **DIET** Herbivorous. Bark and berries of huckleberries, bearberry (*Arctostaphylos uva-ursi*), mountain heathers (*Cassiope mertensiana* and *Phyllodoce empetriformis*), and other high-elevation shrubs as well as forbs. May cache food. **HABITAT & DISTRIBUTION** Subalpine and alpine meadows and brushy forest/meadow edges, heather fields, and open high-elevation forests with a thick huckleberry understory. I have frequently found sign in the slight overhangs created from hiking and horse trails cut into high mountain slopes, a location easily observed by backcountry travelers. **SIMILAR SPECIES** Other voles, deer mice.

TRACKS

Tracks are similar to those of meadow voles.

SIGNS

SCAT 1/32–1/16 in. (0.1–0.2 cm) D × 3/16–1/4 in. (0.5–0.7 cm) L. Similar to scats of other voles. Often found in latrines, especially prominent in early summer when sub-

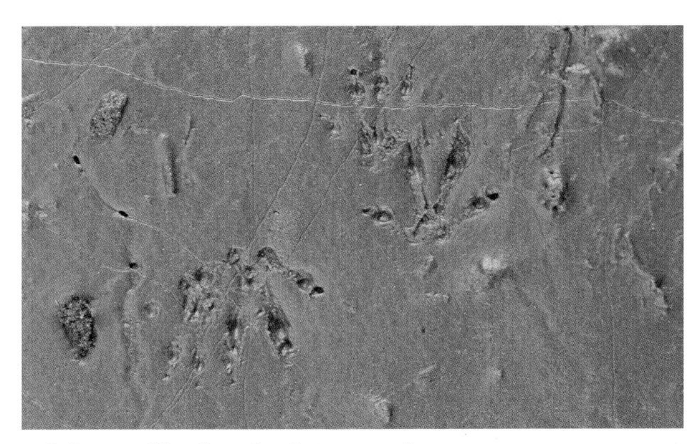

Left front and hind tracks (facing up) of a western heather vole. Tracks in the opposite direction are right front and hind. North Cascades, Washington.

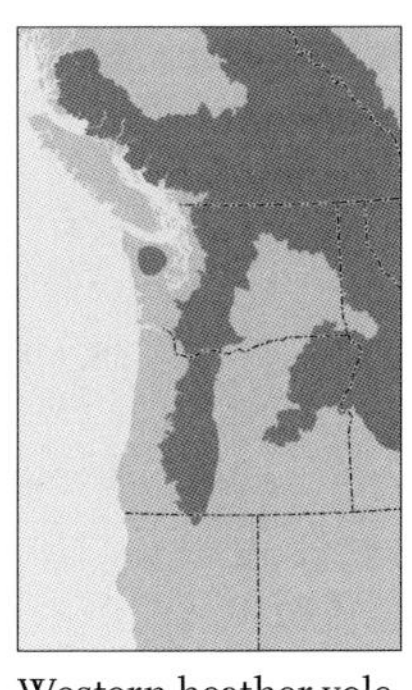

Western heather vole range

These heather vole scats were found along the base of a trail cut in an alpine meadow. North Cascades, Washington.

Heather vole nest and large quantities of scat exposed shortly after snowmelt. North Cascades, Washington.

nivean trails are first exposed. **BURROWS & NESTS** Burrow entrances often found around bases of boulders in meadows, sometimes with radiating trails. Entrance diameter about 2 in. (5.1 cm). Does not construct runways under grass as do meadow voles, but may use them when they are available. Winter nests are subnivean grass spheres with a diameter of about 4¼ in. (11 cm). **FEEDING SIGNS** Cambium feeding on blueberry and huckleberry shrubs and mountain ash (*Sorbus* spp.) occurs in the subnivean zone. Similar to other voles, it clips thin branches into short segments and methodically debarks them; larger branches are debarked in place.

SAGEBRUSH VOLE

Lemmiscus curtatus

This small vole is found in more arid landscapes than other voles. It is apparently semicolonial, with burrows often found in clusters. Its dependence and association with its namesake, sagebrush, is now thought to be less important than was originally believed.

Sagebrush vole burrow entrance at the base of a sagebrush shrub. Columbia Plateau, Oregon.

SIZE Length: 10.3–14.2 cm. Weight: 17–38 g. **FIELD MARKS** Small, light ash-gray in color. **REPRODUCTION** May breed any time of year. Litter of 4–6 young with up to three litters per year. Young reach sexual maturity after 2 months. **DIET** Herbivorous: mainly grasses and forbs. Sagebrush may be most important food in some locations during parts of the year. **HABITAT & DISTRIBUTION** Arid landscapes east of the Cascades. Associated with low-growing, open sagebrush, rabbitbrush (*Chrysothamnus* spp.), and dense, dry grasses. Not usually found in wet meadows as is more common with meadow voles, though they do share some habitats with montane voles (*Microtus montanus*)

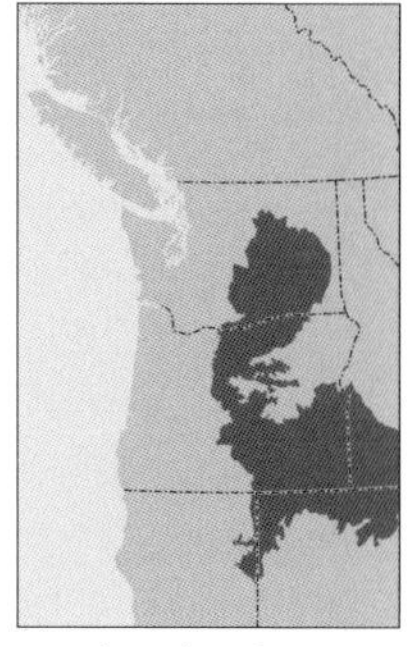

Sagebrush vole range

Tiny burrow entrance of a sagebrush vole. Columbia Plateau, Oregon.

and long-tailed voles (*M. longicaudus*). Uses rock walls, old boards, and even domestic cattle feces for cover. In eastern Washington, sagebrush voles are usually found above elevations of 1000 ft. (300 m). **SIMILAR SPECIES** Other voles, deer mice, harvest mouse, Great Basin pocket mouse.

TRACKS

I have not found their tracks in the field but would assume that they are similar to those of other voles.

SIGNS

SCAT Small but similar to other vole scats. **BURROWS & RUNWAYS** Entrances are small (1 1⁄16 in., or 2.7 cm), found both in the open and under cover of shrubs. Look for clustered burrows with vague trails emanating from entrances and weaving beneath shrubs, but no clearly covered runways as in meadow voles. Nests are subterranean and often include shredded sagebrush bark.

NORTHERN BOG LEMMING

Synaptomys borealis

This boreal species reaches the southern end of its range along the Washington–British Columbia border. Little is known about this species in our region, but it is active year-round.

SIZE Length: 11–14 cm. Weight: 27–35 g. **FIELD MARKS** Short tail. **REPRODUCTION** Farther north in their range, lemmings go through dramatic swings in population density. Whether this occurs in our region is not known. Females can produce multiple litters annually, with 2–8 young per litter. **DIET** Sedges and grasses. **HABITAT & DISTRIBUTION** Mid- to high-elevation bogs and wet meadows, with characteristics that resemble the more expansive habitat of this species much farther north in the continent. **SIMILAR SPECIES** Meadow voles.

TRACKS

Tracks are similar to those of meadow voles. In Washington, where bog lemmings overlap with the larger *Microtus pennsylvanicus*, I have been unable to distinguish their tracks definitively from one another.

SIGNS

Lemmings produce scats, feeding signs, nests, burrows, and runways that are similar to those of meadow voles, with whom they are often found.

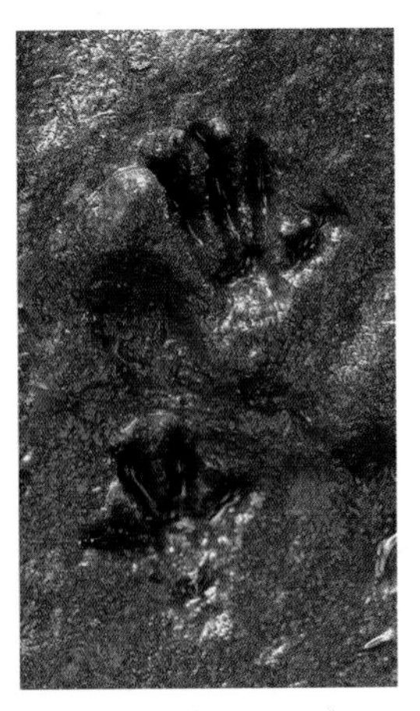

Right front (lower) and hind track of either a northern bog lemming or *Microtus pennsylvanicus*. Both species have been reported at the location where these tracks where discovered. Okanagan, Washington.

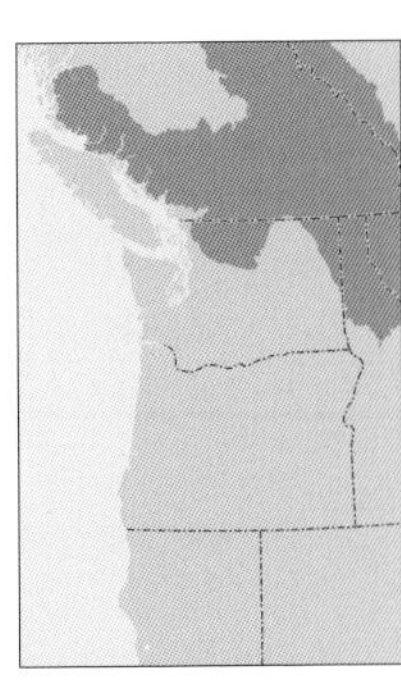

Northern bog lemming range

DESERT-ADAPTED MICE AND RATS

Family Heteromyidae

Heteromyidae is one of the most fascinating families of rodents in our region and includes kangaroo rats, kangaroo mice, and pocket mice. These animals have developed a wide range of amazing adaptations for life in dry conditions. They never need to drink because they get all the water they need by metabolizing their food. They have specially adapted kidneys that produce concentrated urine. They sleep underground, often sealing their burrows to prevent moisture loss. Like pocket gophers (with whom they are closely related), these animals have fur-lined external cheek pouches that can be accessed while their mouths are closed. This helps reduce moisture loss while they collect food. All are nocturnal and have excellent jumping abilities.

Tracks. Foot structure is atypical. Hind feet are enlarged, much larger than fronts, and an excellent clue for distinguishing members of this group from mice and rats (except jumping mice). Claws are long and toes are thin. On the hind foot, toe 1 is greatly reduced or absent.

Track patterns. Kangaroo rats use a unique bipedal hop, with front feet rarely registering during locomotion. Pocket mice bound.

Burrows. All members of this group excavate burrows, often in sandy soils. Size, shape, and location vary among species.

Dust baths. Shallow depressions in the sand are used for dusting for hygiene and scent-marking.

GREAT BASIN POCKET MOUSE

Perognathus parvus

The Great Basin pocket mouse is the most common small mammal in most of our desert landscapes. Their signs are abundant if not conspicuous in areas they occupy. They are nocturnal, reported to be less active on strongly moonlit nights, and inactive in

Great Basin pocket mouse. Photo by B. Moose Peterson/WRP.

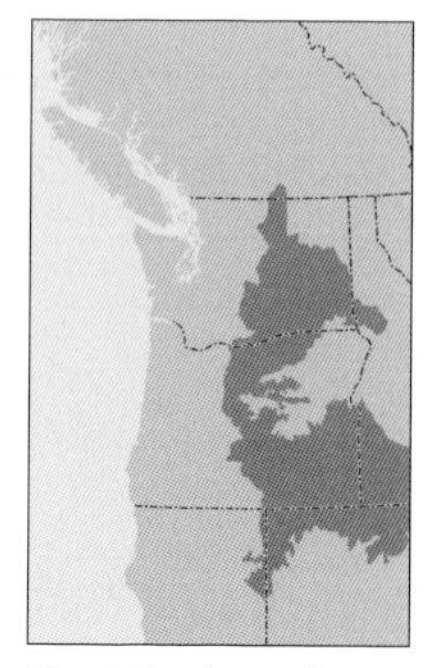

Great Basin pocket mouse range

winter (length of torpor varies from year to year). They excavate burrows in loose, sandy soils, often at the base of a shrub. Two similar species inhabit a limited area in southeastern Oregon—the little pocket mouse, *Perognathus longimembris*, and the dark kangaroo mouse, *Microdipodops megacephalus*. Both are smaller than the Great Basin pocket mouse, whose aggressive behavior may prevent these species from occupying areas it inhabits.

SIZE Length: 16–19 cm. Weight: 21–29 g. **FIELD MARKS** Long, bicolored tail with a slight tuft at the tip. An olive-colored strip may appear between the brown back and light belly. **REPRODUCTION** Breeds from April to August. In years with a relative abundance of precipitation, they can produce 1–3 litters of 2–8 young. During significant drought, reproduction is diminished with many females producing no young. **DIET** Seeds, grasses, forbs, and occasionally insects, especially in the spring before seed crops are available. **HABITAT & DISTRIBUTION** Arid environments east of the Cascades. Associated with sandy soils and often sagebrush and other arid shrub communities. Occasionally found in and around talus and other rocky areas. **SIMILAR SPECIES** Deer mice, harvest mouse.

TRACKS

Tracks are abundant in the sandy soils of their preferred habitat in arid regions, but clear tracks in loose, dry sand are rare (page 59). **FRONT** 1/4–3/8 in. (0.6–1.0 cm) L × 3/16–1/4 in. (0.4–0.7 cm) W. Significantly smaller than hinds. Overall shape is round. **HIND** 5/16–1/2 in. (0.8–1.3 cm) L × 3/16–5/16 in. (0.5–0.8 cm) W. Hinds show significant aberrance from the typical rodent structure. Longer than wide. Toe 1 is reduced and lower on the foot than in other mice. Toes 2–4 are long and thin. Claws often register prominently in tracks. **TRACK PATTERNS** Distinctive bounding pattern. Trail width is narrower and group length is more elongated than in deer mice. Front tracks reg-

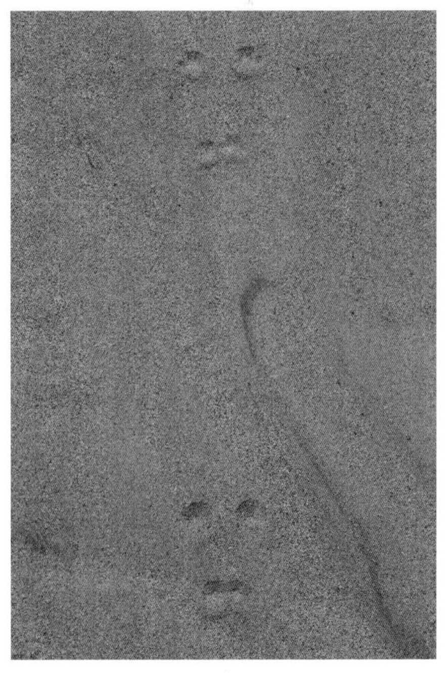

Typical narrow track pattern in sand, with front tracks nearly overlapping. Columbia Plateau, Washington.

Great Basin pocket mouse bound

ister either overlapping each other or nearly so, whereas the front tracks of deer mice are slightly or greatly separated. The significant difference in size between fronts and hinds is also apparent in their track pattern, another good diagnostic feature. **BOUND** Stride: 3 3⁄8–7 13⁄16 in. (8.5–19.9 cm). Trail width: 13⁄16–1 1⁄8 in. (2.0–2.9 cm). Group length: 1 1⁄4–1 7⁄8 in. (3.2–4.8 cm).

A dust bath area used by Great Basin pocket mice reveals numerous marks from their long tails. Columbia Plateau, Washington.

Burrow with throw mound of a Great Basin pocket mouse. Columbia Plateau, Washington.

SIGNS

DIGS Along with storing grains in burrows (larder hoarding), pocket mice also scatter hoard, burying small quantities of seeds in numerous locations. Look for small digs scattered around areas with other pocket mouse signs. These digs may indicate the retrieval of caches or a foraging behavior for insects. **DUST BATHS** Like other members of Heteromyidae, pocket mice engage in extensive dust bathing. This likely serves three purposes: reducing populations of parasites such as fleas, removing naturally produced oils from hair that otherwise becomes matted, and facilitating scent-marking for intraspecies communication. Baths are shallow, sandy depressions, often with distinct trails radiating from them. Recently used baths often contain tracks and tail marks. I have found Ord's kangaroo rat tracks in the same bath areas with Great Basin pocket mouse tracks. **BURROWS** 1 in. (2.5 cm) diameter. Found under shrubs or in the open. May be plugged during the day, and small throw mounds may be apparent. Areas with hard soils are not usually inhabited by this species. In rocky areas, they live in natural cavities created within rock piles and talus.

KANGAROO RATS

Ord's kangaroo rat, *Dipodomys ordii*
Chisel-toothed kangaroo rat, *D. microps*

The Ord's kangaroo rat is the most widely distributed kangaroo rat in our region. It is nocturnal and active year-round. Males and females demonstrate strong territorial behaviors and are solitary. The chisel-toothed kangaroo rat occupies arid desert basins in southeastern Oregon. It is similar in size to the Ord's kangaroo rat but has distinct burrowing and feeding behaviors. It uses its chisel-shaped teeth to strip the outer layers of shadscale (*Atriplex* spp.) leaves to access their moisture-rich centers. Kangaroo rats can significantly alter their surroundings by increasing the amount of open ground and loose soil.

SIZE Length: 21–36 cm. Weight: 40–70g. **FIELD MARKS** Large hind feet and smaller front feet and legs. Tail is long with a bushy tip. The chisel-toothed kangaroo rat has prominent white spots around the eyes, and bicolored ears. Color varies in Ord's kangaroo rats. **REPRODUCTION** The Ord's kangaroo rat produces 1–2 litters annually with 1–6 young per litter. Breeding occurs in the late winter to early spring and sometimes in the fall. The chisel-tooth kangaroo rat produces only one litter (1–8 young) in the spring. Emergence of young for the first time has been correlated with the peak of moisture content in shadscale, their primary food source. Except for breeding, both species are solitary. **DIET** Herbivorous. Mostly grass and shrub seeds. Green vegetation and insects are rarely consumed. Shadscale leaves are a mainstay of the chisel-toothed kangaroo rat. It collects them by climbing into the shrub's branches, a behavior that distinguishes it from Ord's. When shadscale leaves are not available, the chisel-toothed kangaroo rat consumes seeds. **HABITAT & DISTRIBUTION** Ord's kangaroo rat—shrub-steppe, juniper woodlands, arid grasslands, and desert basins with sandy soils. Range may be expanding in Washington. Chisel-toothed kangaroo rat—Range in our region limited to desert basins in southeastern Oregon. California kangaroo rat (*Dipodomys californicus*)—Occupies a small area close to the Oregon–California border in the slopes and foothills of the Klamath

Ord's kangaroo rat. Photo by Alan D. St. John.

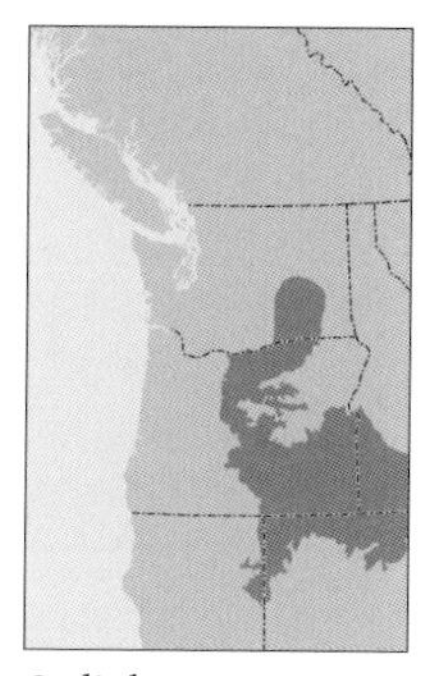

Ord's kangaroo rat range

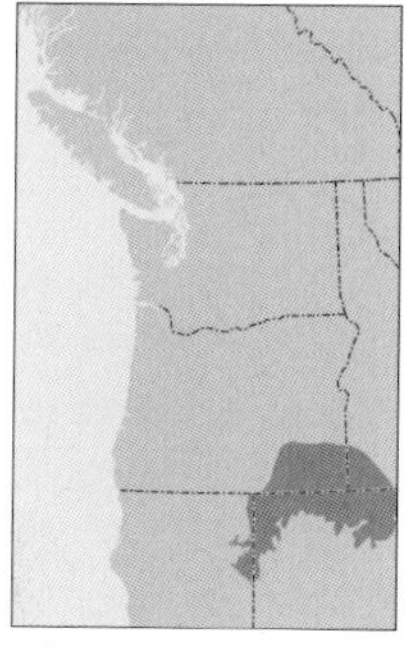

Chisel-toothed kangaroo rat range

Mountains and East Cascades ecoregions and does not overlap with areas inhabited by the other two species. **SIMILAR SPECIES** Great Basin pocket mouse.

TRACKS

Unique tracks and track patterns are apparent in their preferred sandy-soiled habitat. Measurements are for Ord's kangaroo rat only (page 60). **FRONT** Oblong in shape. Much smaller than hind feet. Rarely register, as the animal usually does not place them on the ground when traveling. **HIND** 5⁄8–1 7⁄16 in. (1.6–3.7 cm) L × 7⁄16–7⁄8 in. (1.1–

Ord's kangaroo rat burrow under a tumbleweed. Northern Great Basin, Oregon.

A large kangaroo rat dust bath in sandy soil. Northern Great Basin, Oregon.

Hind tracks of an Ord's kangaroo rat with tail drag between them. Columbia Plateau, Oregon.

At left, two trails of hopping Ord's kangaroo rats (moving toward the viewer); the larger tracks at right are from a Nuttall's cottontail bounding away from the viewer. Columbia Plateau, Washington.

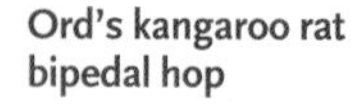

Ord's kangaroo rat bipedal hop

2.2 cm) W. When the entire heel registers (common), the track's overall shape is similar to that of an ice cream cone. Five toes on both species: toe 1 is greatly reduced and rarely registers distinctly. Toes 2–5 are long and thin with large claws that usually register distinctly. Toes 2–4 are parallel and relatively even in size, while toe 5 attaches to the palm lower on the outside of the foot. The foot is furred and the palm is indistinct. **TRACK PATTERNS** Bipedal hop is distinctive (also used by the smaller dark kangaroo mouse). Tail drag may appear, especially when the animal is moving slowly. Hind feet land side by side. **BIPEDAL HOP** Stride: 6⅛–14⅜ in. (15.5–36.5 cm). Trail width: 1⅛–1⅞ in. (2.9–4.8 cm).

SIGNS

SCAT Small and pointy. Rarely encountered. **BURROWS** Both species live in burrows with entrances that are usually plugged during the day to conserve moisture (often below the surface and not apparent from the outside). Ord's kangaroo rat—Creates burrows in the open, under shrubs, and at the base of fallen juniper logs. Some mounding of soil under shrubs, likely refuse from excavation. Entrance diameters average 2⅜ in. (6 cm). With use, entrances grow in size and especially width. Well-defined trails often radiate from burrows. Chisel-toothed kangaroo rat—Produces more massive mounds of soil at the base of shrubs, often shadscale, with multiple entrances. Mounds average 12 in. (30 cm) tall and 4 ft. (1.2 m) across. Some have been recorded at up to 13 ft. (4 m) across! Burrow entrances average 2⅜–3⅛ in. (6–8 cm) in diameter. Each mound is reported to be occupied by only one kangaroo rat. **DUST BATHS** Like pocket mice, kangaroo rats create shallow depressions in sandy soil for dust bathing, which serve both hygienic and communication functions. Along with tracks, look for marks from their long tail within the dust bath. **FEEDING SIGNS** Similar to pocket mice, Ord's kangaroo rats scatter hoard and larder hoard seeds. I have found cut grass leaves and stems in uncovered shallow digs. In southeast Oregon, shadscale shrubs close to mounds and burrows show abundant chisel-toothed feeding signs.

Chisel-toothed kangaroo rat mound at the base of a desert shrub. Northern Great Basin, Oregon.

Two digs and a small buried food cache were created by an Ord's kangaroo rat. Tail marks appear on the throw mound at left. Columbia Plateau, Washington.

ODDBALL RODENTS
Porcupine, Mountain Beaver, and Pocket Gophers

Families Erethizontidae, Aplodontiidae, and Geomyidae

Each of these three families have distinct body shapes, behavior patterns, and tracks and signs. Porcupines are covered with quills for defense. They are slow moving, and inhabit a variety of habitats throughout our region. Pocket gophers are fossorial creatures that occasionally travel above ground. Throw mounds of soil from underground excavations are common pocket gopher signs. The mountain beaver is a distinctive Northwest species that is semifossorial and lives in wet forests in the Cascades and west to the coast. Burrows and feeding signs are commonly encountered.

Tracks. Vary widely among the three groups, all with significant variations from the typical rodent structure.

Track patterns. All primarily use a walking gait, with pocket gophers and mountain beavers speeding to a trot.

PORCUPINE
Erethizon dorsatum

Except for their bellies and faces, porcupines are covered with quills, specialized hollow hairs that serve as an important defense mechanism for this slow-moving creature. Quills dislodge easily, becoming embedded in an enemy. Porcupines have poor vision but good senses of smell and hearing. They often climb trees for refuge or to feed. Most activity is nocturnal or crepuscular, though they occasionally emerge during the day. Porcupines are active year-round, less so during the coldest parts of winter. They are mainly solitary but several may share a winter den.

SIZE Length: 60–130 cm. Weight: 3.5–18.0 kg. Males are larger. **FIELD MARKS** Body quills are distinctive. Face is dark. **REPRODUCTION** Courtship and premating behavior is complex and involves vocalizations and occasional fighting between males. Breeding occurs September to November. One precocial young is born between April and June; they can climb trees soon after birth and start eating plant matter within a week. **DIET** Tree cambium and evergreen needles in winter; plant leaves, roots, and seeds in summer. In arid forests, pine is a favored food. May gnaw on human items, such as wooden handles, shoes, car wiring, and plywood in search of salts, which they crave. **HABITAT & DISTRIBUTION** Found throughout our region. In arid landscapes they are associated with riparian vegetation, and in forests they are most abundant in younger conifer stands. **SIMILAR SPECIES** Scat and feeding signs could be confused with those of beavers, muskrats, and mountain beavers.

TRACKS

Front and hind feet have large, relatively flat pads with a pebbly appearance that is apparent in good substrate. The palm and heel are fused into a single pad. A slightly raised mound often appears in the center of the track. Toes often do not register in tracks but the long, curved claws register far ahead of the palm pad in an arc. Length measurements do not include toes or claws (page 65). **FRONT** 1 5/16–2 in. (3.3–5.1 cm) L × 1 1/8–1 5/8 in. (2.8–4.1 cm) W. Four toes. Claws are longer than in the hinds but the foot is smaller. The heel is narrower than

Porcupine

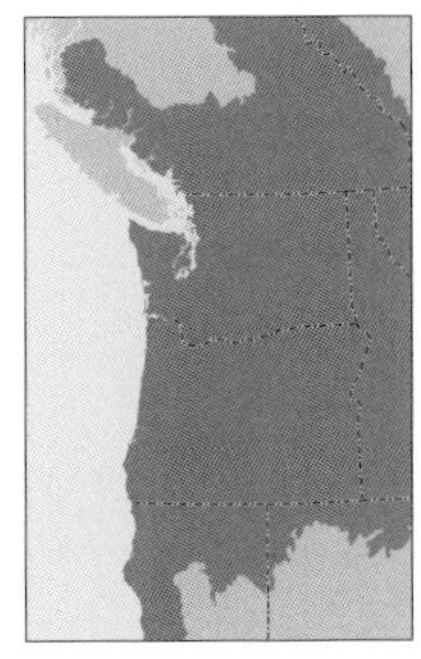
Porcupine range

the palm. Toe and claw length average 1 7/16 in. (3.6 cm). **HIND** 1 7/16–2 3/4 in. (3.7–7.0 cm) L × 1 1/16–2 1/16 (2.7–5.2 cm) W. Five toes, all clawed. Hind heel is broader than front heel. Toe and claw length average 1 3/16 in. (3 cm). **TRACK PATTERNS** Walks, leaving a pigeon-toed direct register or an overstep pattern when moving quickly, though porcupines are rarely in a hurry. Quill drag marks are often seen with tracks. **WALK** Stride: 8 7/8–19 1/2 in. (22.5–49.5 cm). Trail width: 3 9/16–6 11/16 in. (9–17 cm).

This left hind track shows the pebbly appearance typical of porcupine tracks. Columbia Plateau, Washington.

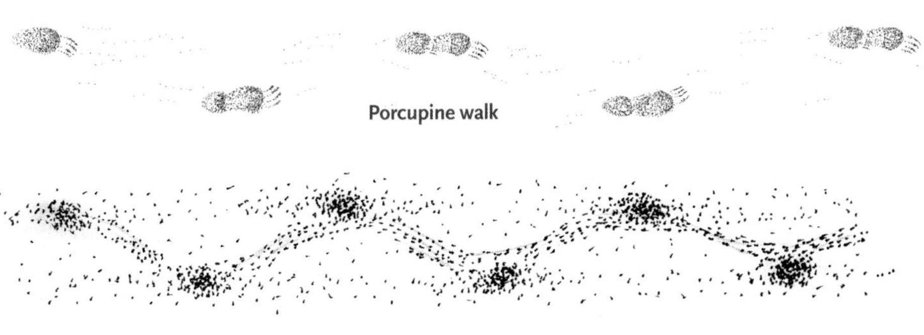
Porcupine walk

Porcupine walk in deep snow

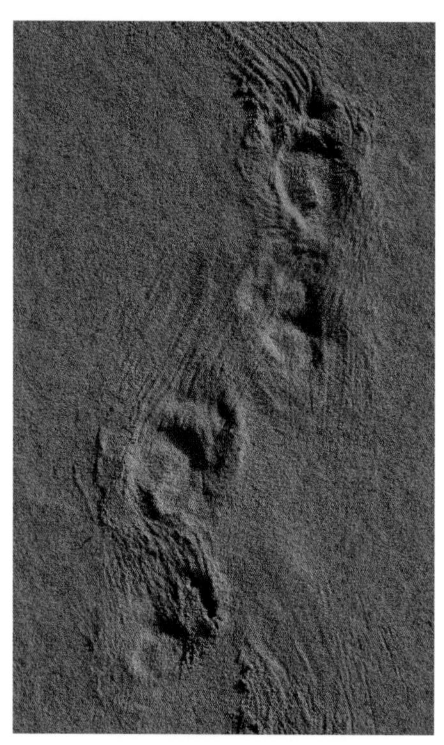

Overstep walk pattern showing quill drag marks. Northwest Coast, Oregon.

Typical porcupine scats. Canadian Rockies, Montana.

Claw marks of a porcupine on a quaking aspen tree (*Populus tremuloides*) are even smaller than those of a bear cub. Columbia Plateau, Oregon.

A natural cavity at the base of a basalt cliff is used as porcupine refuge, with scats and quills at the entrance. Columbia Plateau, Washington.

Porcupine cambium feeding signs on a small shore pine (*Pinus contorta* var. *contorta*). Northwest Coast, Oregon.

SIGNS

SCAT ¼–½ in. (0.6–1.3 cm) D × 9⁄16–7⁄8 in. (1.5–2.3 cm) L. Oblong pellets, deposited individually or loosely attached in a string of several. **FEEDING SIGNS** Cambium feeding signs found at ground level or high up in a tree, either on the trunk or along branches. Incisor marks are substantially larger than those of squirrels and smaller rodents but smaller than those of beavers. Scats often found around the base of the tree in which it has been feeding. **DENS** Uses rock crevices when available. Preferred dens can be used for generations and may be overflowing with years of scats and shed quills, along with an overpowering odor. Where rock outcroppings are unavailable, porcupines use abandoned beaver lodges and hollows under fallen trees for dens.

MOUNTAIN BEAVER

Aplodontia rufa

Mountain beavers, nearly endemic to our region, are burrowing animals, but they are also capable tree climbers and are active year-round from low elevations to near treeline. Their primitive kidneys require a high level of water intake, which restricts their range to areas close to a water source. Many other species use their extensive burrow and trail systems. Mountain beavers are solitary, defending their burrows and nests from other mountain beavers. Tracks are occasionally found on muddy roads and trails and along the edges of wetlands. Burrows and feeding signs are more common to find than tracks.

SIZE Length: 23.8–47.0 cm. Weight: 0.81–1.33 kg. **FIELD MARKS** Medium sized, brown pelage, small eyes, and long whiskers. **REPRODUCTION** Breeding occurs January to June, usually in March. After about a month, 2–3 altricial young are born and are weaned at about 7 weeks. **DIET** Herbivorous. Sword fern (*Polystichum munitum*), bracken fern (*Pteridium aquilinum*) (toxic to many other species), red alder, vine maple, salal (*Gaultheria shallon*), conifer bark, grasses, and forbs. **HABITAT & DISTRIBUTION** Forest environments from sea level to treeline. Selects well-drained slopes near water for burrow construction, around which all activity is based. Does well in second-growth forests and is considered a pest on tree plantations because it can be destructive to saplings. Absent from lower elevations east of the Cascade Crest. **SIMILAR SPECIES** Ground squirrels, hoary marmot, pika.

Mountain beaver. Photo by Tom and Pat Lesson.

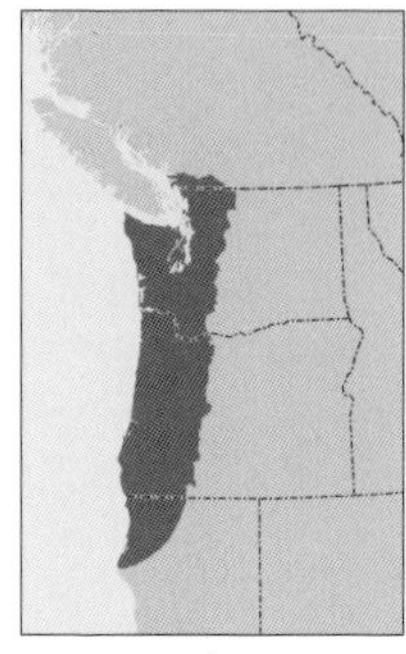

Mountain beaver range

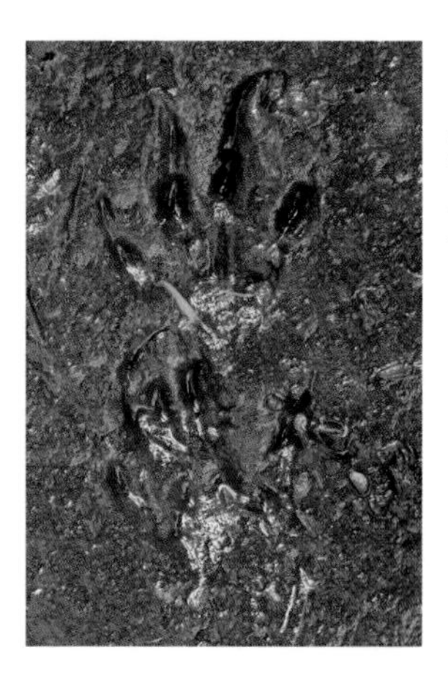

Left front (top) and hind of a mountain beaver. West Cascades, Washington.

Right hind on top of right front. West Cascades, Washington.

TRACKS

Minor variation from typical rodent structure. Fingerlike toes connect to the palm pad in tracks. Front and hinds are similar in size (page 61). **FRONT** 13⁄16–1 1⁄8 in. (2.0–2.8 cm) L × 7⁄8–1 1⁄8 in. (2.3–3.0 cm) W. Four toes register with prominent long claws. Toes 2–4 curve inward. Proximal pads often do not register. **HIND** 11⁄16–15⁄16 in. (1.8–2.4 cm) L × 13⁄16–1 3⁄16 in. (2–3 cm). Handlike in appearance. Claws often show but are smaller than on front foot. Heel rarely registers. **TRACK PATTERNS** Commonly walks (direct or indirect register). Trots when moving quickly. **WALK** Stride: 7 5⁄16–11 7⁄16 in. (18.5–29.0 cm). Trail width: 2 5⁄8–3 3⁄4 in. (6.7–9.5 cm).

SIGNS

SCAT Elongated pellets, deposited in latrines inside burrows and rarely found. May surface in throw mounds around burrows after excavation. Caprophagic. **BURROWS** At lower elevations, look for burrows in and around old

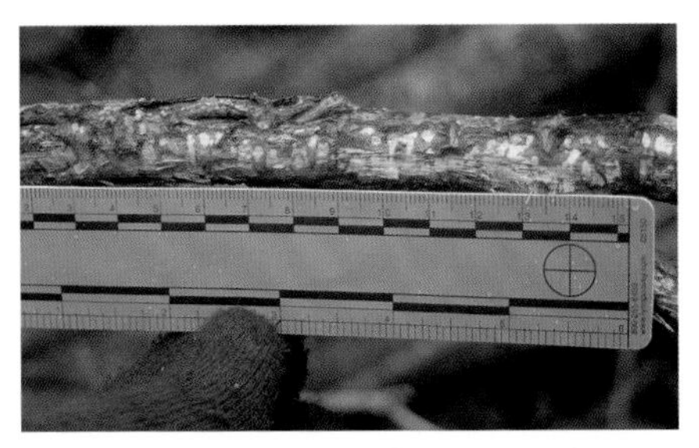

Cambium feeding signs of a mountain beaver on a slide alder. North Cascades, Washington.

Mountain beaver burrow and throw mound, with clipped sword fern fronds and a Douglas-fir branch collected by the animal. West Cascades, Washington.

Vine maple showing stunted growth from mountain beaver feeding activity. West Cascades, Washington.

Mountain beaver walk

slash piles and at the bases of rotting stumps and logs. At higher elevations after snowmelt, look for signs of winter activity—well-worn trails under the snow. Look for burrows at higher elevations near large boulders and on steep slopes in thickets of slide alder (*Alnus crispa*). Burrow entrance diameters average 5¼ in. (13.3 cm), though I have found them as large as 8 in. (20.3 cm). Some entrances are accompanied by a throw mound of loose soil. **FEEDING SIGNS** Climbs shrubs and saplings to a height at which the stem diameter is up to ⅝ in. (1.6 cm) and cuts it off, leaving a distinctive shape as the shrub grows around the injury. Bark may be removed directly from larger branches and may look similar to cambium feeding by voles or porcupines, distinguished in part by the size of marks left by the incisors (of medium size). Mountain beavers often collect piles of clipped vegetation (with distinctive 45-degree angle cuts) at burrow entrances.

POCKET GOPHERS

Northern pocket gopher, *Thomomys talpoides*
Western pocket gopher, *T. mazama*

Pocket gophers are fossorial creatures that create extensive tunnel systems and numerous highly visible throw mounds above their subterranean homes. Their excavation and soil aeration activities likely provide an important ecological function in areas where they reside. The northern pocket gopher is solitary and defends its tunnel system from others. During the breeding season, a greater degree in overlapping use of burrow systems is apparent. These species move aboveground at night, and their remains are regularly found in owl pellets. You may catch a glimpse of the animal as it briefly pokes its head out of a tunnel entrance to collect food or expel dirt. They transport food in their external, fur-lined cheek pouches.

Northern pocket gopher

SIZE Length: 16.5–26.0 cm. Weight: 60–160 g. Males are larger. **FIELD MARKS** Oblong shape, short limbs, short tail. Fur color varies widely. Small ears with black patch behind them. Western pocket gopher ears are pointy with a larger black patch than on the northern pocket gopher, whose ears are rounded. **REPRODUCTION** Breeding starts in March, but pregnant females have been captured as late as August. Gestation is 18 days, with 4–7 young per litter. May produce more than one litter per year. **DIET** Herbivore: grasses and forbs, leaves, stems, and roots. They often pull the plant down into their burrow from below. **HABITAT & DISTRIBUTION** Northern pocket gopher—Cascades and east in Washington. In Oregon, its range begins east of the Cascades. Meadows, alpine meadows, agricultural fields, open forests, and arid grasslands. Prefers soft, deep soils but also uses harder clay soils and gravelly areas. Townsend's pocket gopher (*Thomomys townsendii*)—In a few desert basins in southeast Oregon and adjacent states, in areas with deep soils created by ancient lake beds. Western pocket gopher—the Cascade Crest and west in Oregon. In Washington, it is found only in the Olympics and in a few limited locations associated with oak woodlands and now endangered prairie habitat in the Puget lowlands. Range overlaps with the Botta's pocket gopher (*T. bottae*) in the southwest corner of our region. The much larger Camas pocket gopher (*T. bulbivorus*) is

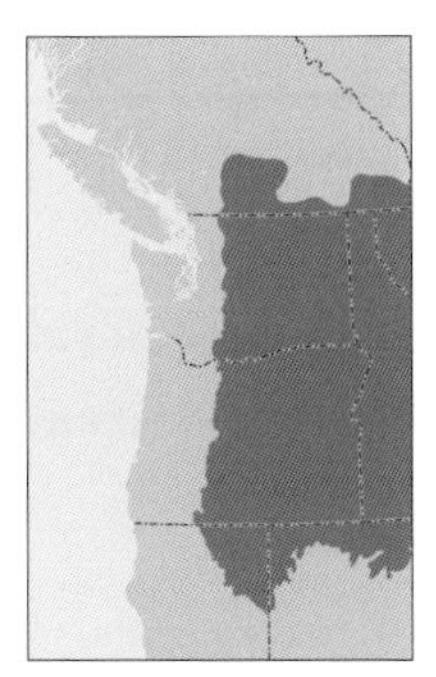
Northern pocket gopher range

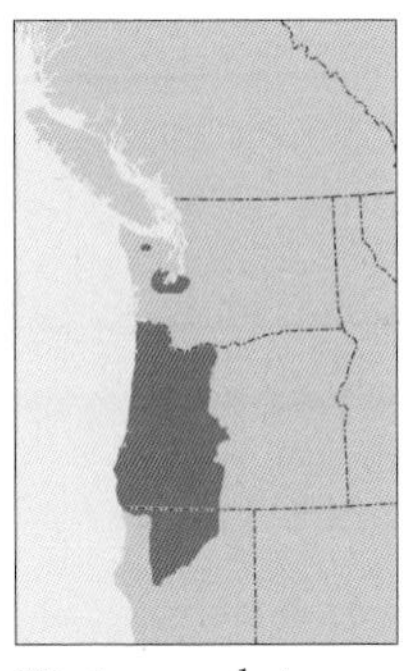
Western pocket gopher range

Tracks of a northern pocket gopher; track at lowest right is a hind, and the next track up and left is a front. Canadian Rockies, British Columbia.

Northern pocket gopher trot

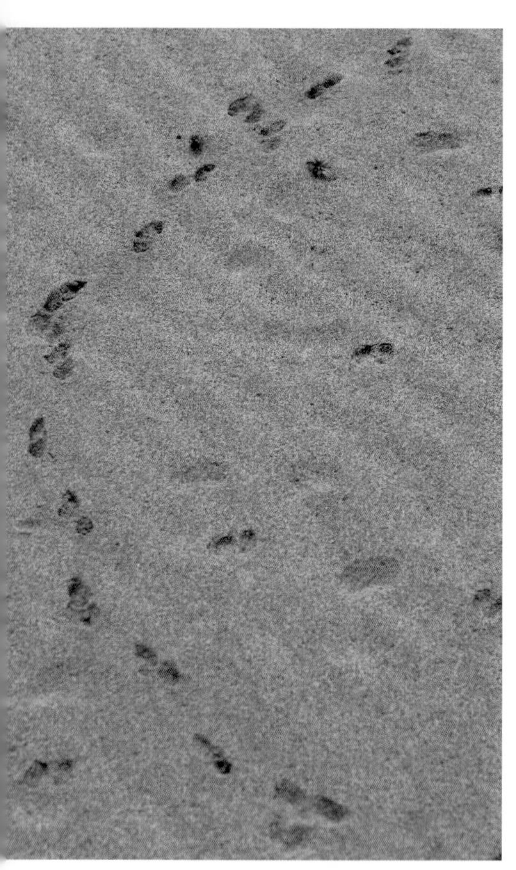

Typical overstep trotting pattern of a northern pocket gopher arcing through the frame and bisected by the bipedal hopping trail of an Ord's kangaroo rat. Columbia Plateau, Washington.

A northern pocket gopher mound in fine soil shows the plugged burrow entrance hole. Columbia Plateau, Washington.

I startled a northern pocket gopher that was in the middle of expelling earth at this mound, causing it to retreat without plugging its hole. Columbia Plateau, Washington.

endemic to Oregon's Willamette Valley. **SIMILAR SPECIES** Moles. **CONSERVATION ISSUES** Isolated populations of western pocket gopher in Washington are protected.

TRACKS

Atypical foot structure, as feet are highly adapted to its fossorial lifestyle. Front and hind tracks are similar in size, with five clawed toes that are thin and without bulbous tips (page 59). **FRONT** 3⁄8–5⁄8 in. (1.0–1.6 cm) L × 3⁄8–7⁄16 in. (0.9–1.1 cm) W. Overall shape is wedgelike, widest at the toes. Toe 1 is smaller and lower in the track than toe 5. Toes 2–4 are farther forward in the track, with toe 3 the farthest forward. All claws are long, but those on the center three toes are longest. Heel has two distinct pads. **HIND** 3⁄8–9⁄16 in. (0.9–1.4 cm) L × 5⁄16–9⁄16 (0.8–1.4 cm) W. Prominent hind claws are about half the length of those on fronts. Toes radiate more evenly than on fronts. Palm is an indistinct continuation of the toes. Narrow heel has no distinct pads and rarely registers. **WALK & TROT** Stride: 2 5⁄8–5 11⁄16 in. (6.6–14.5 cm). Trail width: 1 1⁄8–1 7⁄8 in. (2.8–4.7 cm).

SIGNS

MOUNDS Produces conspicuous, fan-shaped mounds of soil that are often confused with mole mounds. The gopher burrows to the surface and pushes soil out of the tunnel system, radiating away from the hole. It then plugs the entrance, which is to one side of the mound (or under one edge). Plug may be apparent or difficult to find. You can often push open the plug and see that the hole angles out of the ground rather than emerging straight up (a feature more characteristic of moles). East of the Cascades, gopher mounds are more common to find than those made by moles. **ESKERS** Width: 2 1⁄8–3 9⁄16 in. (5.4–9.0 cm). Pocket gophers are active throughout the winter. In areas with consistent snowpack, their tunnels are alternately subnivean and subterranean. As they dig forward in these conditions, they use new material to backfill the tunnel behind them. When soil is deposited in snow tunnels, serpentine eskers are produced on the surface and are apparent when the snow melts. **PLUGS** Pocket gophers create holes to the surface to access food aboveground. They usually plug these entrances, which average 1 7⁄8 in. (4.7 cm) in diameter.

Northern pocket gopher eskers radiating from a central point. Middle Rockies, Idaho.

CARNIVORES Order Carnivora

The order Carnivora comprises five families in our region. All members of this order consume at least some animal tissue, although many are omnivores rather than strictly carnivores. Each family is covered separately.

FELINES

Family Felidae

Charismatic carnivores with global distribution, felines have excellent vision and are primarily visual hunters that stalk or ambush prey. Sharp, retractable claws are used for climbing trees (primarily for seeking refuge) and killing prey. All three native species are obligate carnivores and will not consume fruit or other vegetation, as will many canines. Felines are sexually dimorphic, and mature males are larger than females.

All felines in our region are solitary and avoid direct interactions with others of the same species, except for juveniles traveling with their mothers and mating pairs traveling together for brief periods. Males occupy home ranges that overlap with multiple females with whom they mate. Timing of breeding and birth of young varies among species. Felines scent-mark extensively, using urine, scat, and secretions from specialized scent glands to communicate their presence and breeding condition.

Tracks and Signs

Feline tracks can be confused with those of canines. Their sharp, retractable claws rarely register in tracks. If claws do appear, they often register less deeply than other parts of the track. The palm pad makes up 50% or more of the track's overall volume, which is significantly larger than in canines. The overall shape of the palm pad is trapezoidal, and it is trilobed on the posterior edge and bilobed on the anterior edge. The negative space is smaller than in canines and forms a downward-facing C-shape. The toes are laid out asymmetrically with toe 3 most forward. Canada lynx differ from this general description because of foot adaptations that accommodate cold and snowy environments.

Front. Overall round shape. Four toes reliably register. Toe 1 is greatly reduced and almost never registers. A single proximal pad registers only when the animal is running or moving in deep substrates. The palm pad often cants outward.

Hind. Overall shape is oval to round. Four toes register and are asymmetrically oriented, although less so than on the front foot. The palm pad is relatively smaller than on the front but similar in shape.

Track patterns. Most often walk, either in a direct register or an overstep pattern. They also trot, bound, and gallop.

Scent-marking. Both adult males and females engage in scent-marking behaviors. Resident males appear to leave marking signs, such as the scrapes described for bobcats and mountain lions, more frequently than other sex or age classes.

Scat. Scats are tubular in shape and clearly segmented, often with blunt ends. The contents are often densely packed. Scats may be buried or associated with scent-marking behavior. The scents of wild feline scat and urine are similar to those of a domestic cat.

Feeding signs. Prey may be moved for caching and is often meticulously covered. Hair may be sheered off the hide of large prey

DETERMINING SEX USING TRACKS AND SIGNS

As with tracks of other sexually dimorphic species in which males are larger than females, tracks of mature male felines are generally larger. Tracks at the upper end of the size ranges provided for bobcats and mountain lions are likely males. Global studies on several large cat species have attempted to determine whether tracks can be used to determine the sex of felines (Liebenberg 1990b; Sager and Singh 1991; Stander and Ghau 1997).

Mark Elbroch studies felines in North America and has developed general criteria for evaluating tracks of mountain lions and bobcats. Adult male mountain lion and bobcat tracks often show several morphological differences from those of adult females. Specifically, a male's front tracks are relatively wider than long, its toes are thicker and rounder on both feet, and the palm pads are relatively larger, less angular, and broader on the leading edge, especially on the hind feet, in comparison to tracks of an adult female.

Elbroch points out that these criteria are accurate, with some exceptions. Many tracks show several but not all of these characteristics, and the tracker must weigh various observations of the track in developing a hypothesis about the maker's sex. It is unclear whether Canada lynx, which are less sexually dimorphic and have a foot structure substantially different from other native felines, share these distinctions. Furthermore, tracks in deep substrates, such as loose snow, where the animal's toes have splayed widely, are impossible to analyze effectively.

These track tracings of two known-sex mountain lions collected by Casey McFarland and Mark Elbroch illustrate characteristic differences between male and female tracks. Tracks were collected from track plates: the animal stepped on a sooted sheet of aluminum and then stepped on contact paper that recorded the track.

(deer and other ungulates), especially in winter when pelage is thickest. Small tufts of hair, clearly sheered at their base, may be discovered near the carcass. This is a sign distinctive of felines (Morse 2004).

BOBCAT

Lynx rufus

Bobcats are adaptable predators found in most habitats in our region, and tracks and signs of this exciting feline are relatively common. They can be active at any time of day, alternating periods of 4–8 hours of activity with rest periods of about 8 hours.

SIZE Length: 47.5–125.0 cm. Weight: 4.1–18.3 kg. Males are larger. **FIELD MARKS** Mottled coat and short tail, with black on top of the tip and white on lower parts. Small tufts on ear tips. **REPRODUCTION** Females breed from January to August. Development takes about 2 months, and most litters are born in May. Average litter size is 2–3 kittens. Females raise young without assistance from males. Young disperse as early as their first fall and typically no later than spring, before their mother's next litter is born. **DIET** Obligate carnivores. In most of the bobcat's range, rabbits and hares are their main prey, along with other small mammals and birds. In western Washington, mountain beavers are a primary food source. **HABITAT & DISTRIBUTION** Widely distributed and adaptable from the coast to eastern arid lands. Missing from the highest elevations and some eastside areas heavily impacted by agriculture. Often uses rock outcrops for refuge and bedding locations. Hunts along forest roads, wetlands, and thickets. Male home ranges of 3–42 square miles (7–108 km^2) expand during the breeding season and overlap with those of multiple females, whose ranges are 3–17 square miles (8–45 km^2). **SIMILAR SPECIES** Mountain lion, coyote, Canada lynx.

TRACKS

Typical feline shape. Foot pads are hairless, leaving clear tracks in good substrate. Claws are retractable and rarely show, but

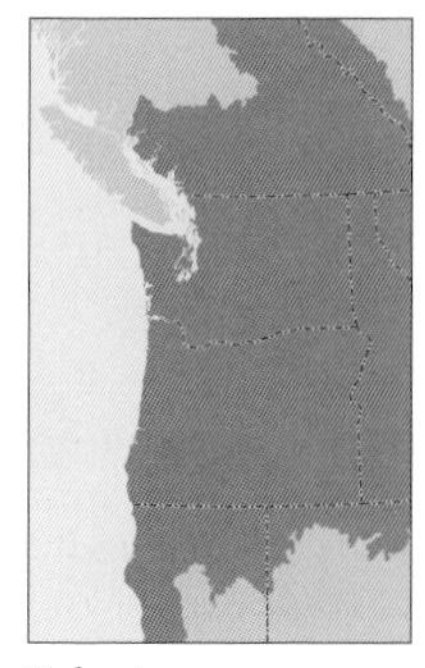

Bobcat range

when they do, they are often more shallow than foot pads. Toes can splay widely in deep snow, leaving a larger track (page 64). **FRONT** 1 9/16–2¼ in. (4.0–5.7 cm) L × 1⅝–2⅜ in. (4.1–6.0 cm) W. Round in shape. Broader than hind. **HIND** 1⅝–2⅜ in. (4.1–6.0 cm) L × 1⅝–2¼ in. (4.2–5.7 cm) W. More symmetrical than front, often longer than wide. The palm pad is smaller than on the front foot. **DIRECT REGISTER WALK** Stride: 19–31½ in. (48.3–80.0 cm). Trail width: 2¾–8 in. (7.0–20.3 cm). **OVERSTEP WALK** Stride: 22⅝–33½ in. (57.5–85.1 cm). Trail width: 4¼–5½ in. (10.8–14.0 cm).

Direct register walking track pattern in snow with characteristic drag marks leading into and out of tracks. In deep snow, bobcat trails may resemble coyote trails. East Cascades, Washington.

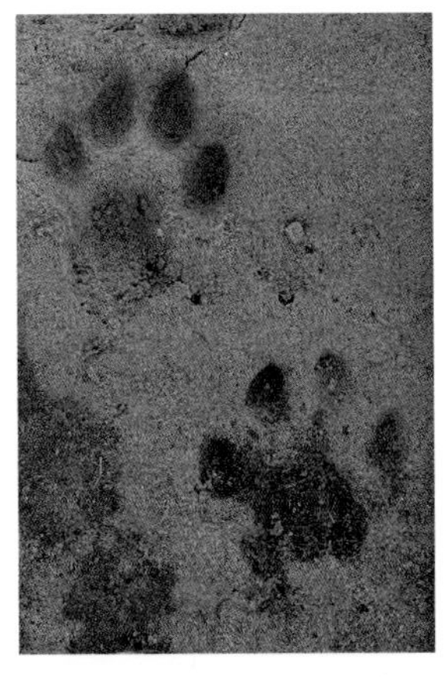

Left front (lower) and hind tracks of a large (probably male) bobcat. Columbia Plateau, Washington.

Right front (lower) and hind tracks of a smaller bobcat. Puget Trough, Washington.

Bobcat overstep walk in snow

Typical bobcat scat next to a small scrape. Photo by Jason Knight.

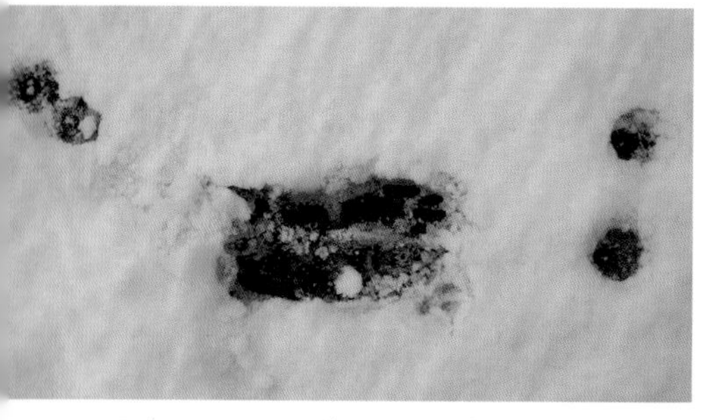

Bobcat scrape without scat. The cat was facing right and scraped backward with its two hind feet. West Cascades, Washington.

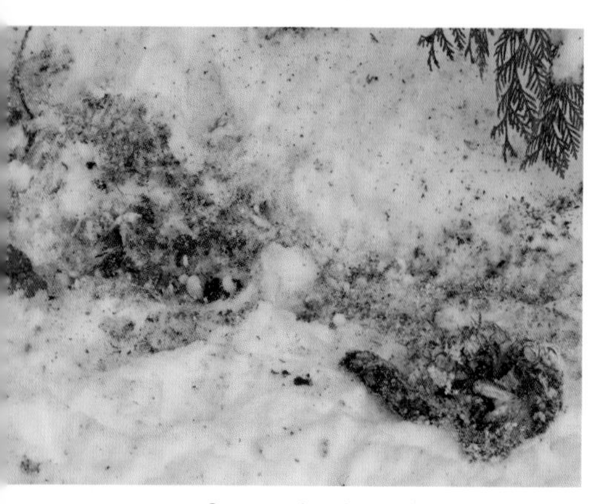

Remains of a snowshoe hare consumed by a bobcat—hair, stomach contents, and one hind leg. East Cascades, Washington.

SIGNS

SCAT ½–1 in. (1.3–2.5 cm) D × 2¼–9⅝ in. (5.7–24.5 cm) W. Usually segmented with blunt ends. Compact, with less hair and smaller bone fragments than similar-sized coyote or mountain lion scat. Scats may be deposited in or by a scrape on game trails, or at the edges of a prominent trail junction. Like domestic cats, bobcats sometimes bury scat in mounds. **SCRAPES** Bobcats make conspicuous scrapes for intraspecies communication at prominent or important features within their home range. Most scrapes I have encountered appeared to have been made by the animal's two hind feet (which is about the width of the scrape). The animal pushes dirt or forest debris into a pile, leaving a noticeable rectangular trough with a mound at the end. The bobcat may urinate or defecate on the mound and/or trough. Its urine has a distinctive scent, similar to that of a domestic cat. **KILL SITES** Prey may be consumed where it is killed or carried to a protected location such as the base of a tree with low-hanging branches. May remove feathers of large birds, with some quills sheared cleanly and others plucked (quills intact). Larger prey may be partially covered.

CANADA LYNX

Lynx canadensis

Lynx are exceptionally well adapted to harsh winters; their large, furred feet allow them to travel easily through deep, loose snow. In the boreal forests farther north on the continent, the lynx population fluctuates radically and is linked to populations of snowshoe hare, their primary prey. These fluctuations apparently do not occur in our region. Both hares and lynx maintain pop-

Canada lynx. Photo by Richard Day/Daybreak Imagery.

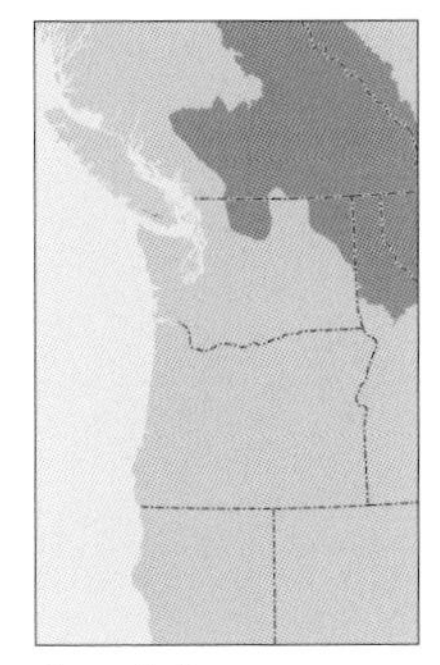

Canada lynx range

ulation densities similar to population lows in more northern regions, likely because of marginal habitat conditions on the southern end of their range and increased competition (for both hares and lynx) from species that do not exist in more northern latitudes. Trailing a lynx through fresh snow as it weaves through dense forest from one area of snowshoe hare activity to another is truly a window into the elemental world of predator and prey.

SIZE Length: 67–107 cm. Weight: 4.5–17.3 kg. Males are slightly larger. **FIELD MARKS** Short tail completely tipped with black, grayish fur without distinct mottling; distinctive long tuft of hair at the tip of each ear; and a "beard." Long hind legs make the back slope forward toward the head. Large front and hind feet. **REPRODUCTION** Breeding occurs in March or April, with parturition in late May or June. Natal dens are under fallen trees or similar landscape features. Litter sizes of 1–6 can vary depending on prey density, increasing in times of abundance. Kitten survival also varies considerably, depending on prey availability. **DIET** Mainly snowshoe hares, especially in winter (up to 99% of the total biomass consumed in one study). Also consumes red squirrels, voles, and birds. **HABITAT & DISTRIBUTION** Limited distribution in our region; restricted to areas with cold winters and deep snow. Strongly associated with high-elevation mature subalpine fir (*Abies lasiocarpa*) and Engelmann spruce (*Picea engelmannii*) forests, which are used for denning, and younger lodgepole pine forests, which are used for hunting because of high densities of snowshoe hare. The historic range of this species in areas south of northern Washington is unclear. Juvenile Canada lynx can disperse great distances (310 miles, 500 km, or more), so sightings at great distances from breeding populations occasionally occur. Home range boundaries often correspond to terrain features such as ridgelines and major streams. In north-central Washington, female home ranges average 15 square miles (39 km^2), and males' ranges average 26.5 square miles (69 km^2). **SIMILAR SPECIES** Mountain lion, bobcat,

Typical walking trail of a Canada lynx in snow. Notice the foot's large size in comparison to the stride. Okanagan, Washington.

Right front (lower) and hind tracks of a Canada lynx on firm snow. Okanagan, Washington.

Left hind track in mud. Photo by Brian McConnell.

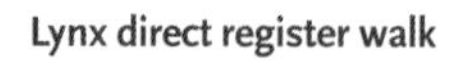

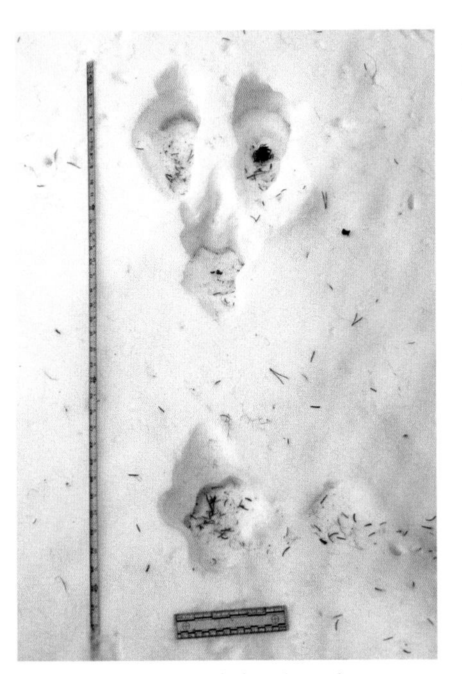

Typical track pattern of a lynx bounding in snow. This lynx was chasing a snowshoe hare. North Cascades, Washington.

wolverine. **CONSERVATION ISSUES** Federally endangered in our region. Fires have heavily impacted large portions of their prime range regionally.

TRACKS

Tracks are usually observed in the snow. Feet are heavily furred, with fur often completely covering the small toe and palm pads, making detail in tracks indistinct. Tracks are exceptionally large relative to the animal's size, and because of this, tracks are often shallow, even in loose snow. Unlike tracks of other felines, the anterior edge of the overall track shape has a distinct "stair-step" appearance made by the relationship between the middle and outer toes. (Bobcats and mountain lions have a more even arc to the placement of their toes.) Toe and palm pad may appear as small depressions within the overall track shape. Palm pad is smaller than in other felines and has a distinctive shape in tracks. Claws rarely show (page 71). **FRONT** 2 11⁄16–4½ in. (6.8–11.4 cm) L × 2¾–4½ in. (7.0–11.4 cm) W. **HIND** 2 3⁄8–4¾ in. (6.0–12.1 cm) L × 2½–5 in. (6.4–12.7 cm) W. Hind tracks are often longer than front tracks. **TRACK PATTERNS** In snow, almost always uses a direct register walk. On firm substrates, oversteps, as in other felines. Look for a relatively small stride compared to the foot size. When running (such as in pursuit of prey), leaves a distinctive bounding pattern. **WALK** Stride: 18½–37 in. (47–94 cm). Trail width: 3½–11 in. (8.9–27.9 cm). **BOUND** Stride: 39–112¼ in. (99–285 cm). Trail width: 6 3⁄8–12 in. (16.2–30.0 cm). Group length: 15–55 1⁄8 in. (38.1–140.0 cm).

SIGNS

SCAT Usually found and definitively identified from trailing an animal in snow. Otherwise impossible to distinguish from bobcat scats where their ranges overlap. **SCENT-MARKING** Lynx urinate or rub their cheeks on logs and stumps to mark their territory. When trailing lynx in the snow, you may see them approach objects and mark them. Examine the object at about knee height to detect hairs that may have been rubbed off. Urine has a characteristic catlike odor.

MOUNTAIN LION

Puma concolor

The largest felines in our region specialize in hunting hoofed mammals. Rarely seen by people, these big cats are shy and usually flee after being spotted by a human, though

Mountain lion. Photo by Mark Elbroch.

they can live and hunt close to human populations. Although rare, attacks on humans have occurred. Mountain lions can be active at any time of the day, but they are usually most active at dawn and dusk. In areas with lots of humans, nocturnal activity increases.

SIZE Male length: 1.0–1.5 m. Weight 36–120 kg. Female length: 0.86–1.3 m. Weight: 29–64 kg. **FIELD MARKS** Large in size and tawny in color. Tail is as long as the rest of the body. **REPRODUCTION** Females commonly breed every other year, at any time of the year. Gestation is 90 days, with litters of 1–5 kittens. Juveniles stay with their mother for about a year before being abandoned and dispersing to establish their own home ranges. **DIET** Mostly deer and elk. West of the Cascades, beavers are also important prey. They opportunistically feed on rabbits, hares, and other small mammals. **HABITAT & DISTRIBUTION** Adaptable to diverse habitats, including dense forests, mountainous terrain, and arid and brushy landscapes. Found in areas that support deer or elk. Some animals shift territories seasonally, following prey to higher elevations in the summer months. Populations include resident and transient animals. Residents are mature animals that maintain a defined territory. Males occupy larger areas than females, overlapping with several females with which they breed. Residents often share the use of many parts of their range but avoid direct contact with other individuals through scent-marking activities. Transients are subadult animals that have not yet established their own territories. **SIMILAR SPECIES** Bobcat, lynx, domestic dogs, black bear.

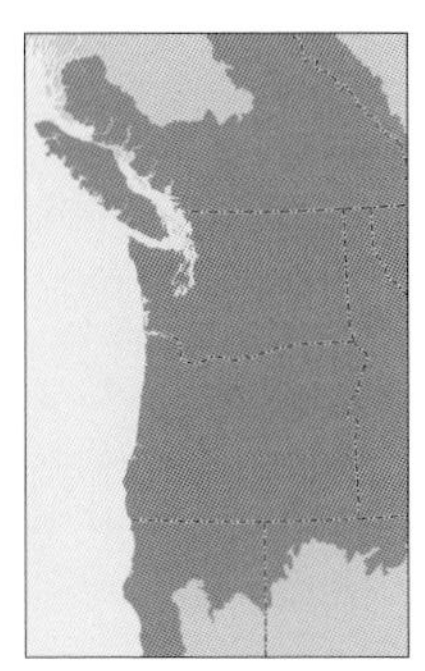

Mountain lion range

TRACKS

Typical feline structure. Tracks are larger than those of bobcats. Toe and palm pads are not densely furred and register clearly (page 70). **FRONT** 3 1/16–4 in. (7.8–10.1 cm) L × 2¾–4 5/16 in. (7–11 cm) W. Round. Larger than hind. **HIND** 2 15/16–4 1/8 in. (7.5–10.5 cm) L × 2 9/16–4¾ in. (6.5–12.1 cm) W. Slightly smaller than front and less asymmetrical—often longer than wide. **TRACK PATTERNS** Walks in an overstep or direct register pattern depending on speed of travel and substrate. Runs in a gallop. **DIRECT REGISTER WALK** Stride: 27 9/16–42¼ in. (70.0–107.3 cm). Trail width: 4¾–11 in. (12.1–27.9 cm). **OVERSTEP WALK** Stride: 40 5/8–46 7/16 in. (103.2–118.0 cm). Trail width: 6¾–12 5/8 in. (17.1–32.0 cm).

SIGNS

SCAT ½–1 7/16 in. (1.3–3.6 cm) D × 7 1/16–21½ in. (18.0–54.6 cm) L. Tubular and segmented with at least one blunt end. Usually in large quantity, attesting to the animal's size. Obligate carnivore, so vegetation in scat is unlikely. Scats often left on travel routes in conjunction with other scent-marking signs, or they may be buried. **SCRAPES** Used for marking territory and communicating with other mountain lions in the area. Uses its hind feet to create a rectangular depression (about the width of its two hind feet), ending in a pile of debris—similar to bobcats but proportionally larger. May urinate on or deposit a scat in or by the scrape. Often found atop needles under prominent large conifers and along edges of wetlands and ridgelines. **SCRATCH MARKS ON TREES**

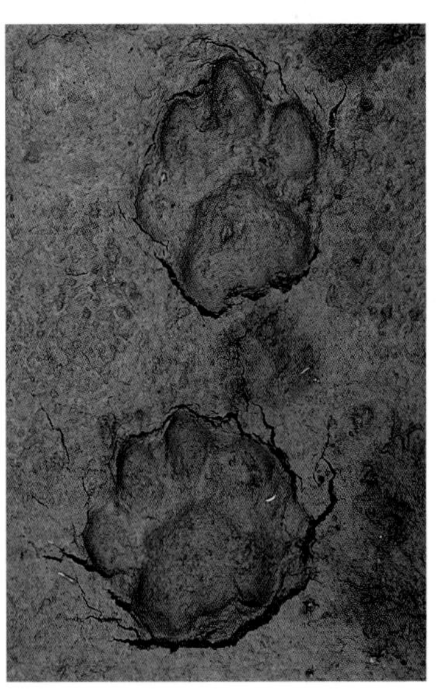

Left front (lower) and hind tracks of a large (likely male) mountain lion. Puget Trough, Washington.

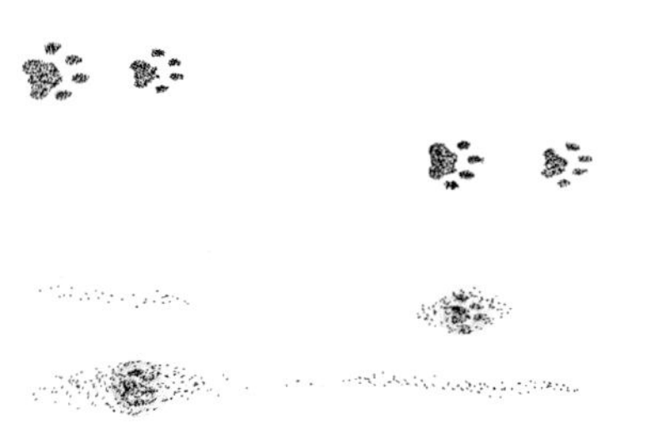

Mountain lion overstep walk

Mountain lion direct register walk in snow

Typical scrape of a mountain lion. The knife's tip points to a scat deposited by the mound of debris. This scrape, along with several others, were discovered under an overhanging rock outcrop. Columbia Plateau, Oregon.

Three or four different scats were deposited in the same location under a prominent tree along a ridgeline. Leaving multiple scats in one location is common for this species. Northern Great Basin, Oregon.

A mountain lion killed and partially covered this black-tailed deer. The white patch on the hind quarter shows where the lion sheared the deer's hair. West Cascades, Washington.

These claw marks on a bigleaf maple (*Acer macrophyllum*) were found close to the carcass of a recently killed deer. West Cascades, Washington.

Mountain lions use their sharp claws on tree trunks, possibly as a scent-marking behavior. **KILL SITES** Often drags large prey to protected location, such as a brushy area or bottom of a ravine, and uses forepaws to drag branches and debris over the carcass after initial feeding. A lion may return to a carcass to feed for a week or more, and often beds nearby. Carcass may be moved between feedings. Deer and elk carcasses are often entered at the base of the rib cage. Hair may be meticulously sheered off the prey, leaving distinctive tufts of hair around the feeding site.

HOUSE CAT

Felis catus

Tracks are commonly found in rural and urban locations, less often in deeper wildlands. Our region has both truly feral cats and free-roaming domestic cats that return home, both of which compete with native predators and significantly impact songbird populations. The domestic cat was bred from the African and European wildcat (*Felis silvestris*).

SIZE Length: 73–79 cm. Weight: 3.3–4.5 kg. **REPRODUCTION**: Females can have two litters per year with an average of four kittens in each. **DIET** Wild foods include insects, rodents, amphibians, reptiles, and ground-nesting or foraging birds. **SIMILAR SPECIES** Bobcat, gray fox, ringtail. **CONSERVATION ISSUES** Feral cats have been linked to the decline of endangered bird species in North America and globally. In North America, feral and free-roaming cats likely kill more than 1 billion small mammals and several million birds each year.

TRACKS

Typical feline track shape, with more variation in track shapes than in our region's other felines, perhaps as a result of domestication (page 62). **FRONT** 1⅛–1⁷⁄₁₆ in. (2.9–3.7 cm) L × 1⅛–1¾ in. (2.9–4.5 cm) W. Round in shape. Four toes show regularly. Toe 1 rarely shows, but may be split with two claws on some individuals. **HIND** 1¼–1⅝ in. (3.1–4.2 cm) L × 1¹⁄₁₆–1⅝ in. (2.7–4.2 cm) W. Oval to round. Often longer than front foot. Toes are longer and thinner than those of front. Palm pad smaller than on front and in some tracks may not appear bilobed on the leading edge. **TRACK PATTERNS** Direct register walk is the most common gait. **DIRECT REGISTER WALK** Stride: 14³⁄₁₆–24 in. (36–61 cm). Trail width: 2¹⁄₁₆–4¹⁵⁄₁₆ in. (5.2–12.5 cm).

SIGNS

SCAT Tubular and segmented with at least one blunt end. Sometimes creates a latrine with multiple scats that are often buried. **SCRAPES** Feral cats create scrapes that sometimes include a scat. Many scrapes appear to be made with the front feet, with the cat moving around a central location and piling up soil and debris into one mound.

Left front (top) and hind track of a feral cat. Klamath Mountains, Oregon.

CANINES

Family Canidae

Canines display a variety of social organizations including strongly hierarchical pack structures. Males are larger then females in all species. Hostility between coyotes and red foxes and between wolves and coyotes is well documented, with the larger species actively killing the smaller, possibly because they are considered competition. Wolves have been reported to tolerate red foxes, but not always. Distinguishing domestic dog tracks from native wildlife is covered in the respective section.

Tracks and Signs

Canine tracks are far more symmetrical than those of cats. The palm pad is triangular shaped and relatively small compared to feline tracks. The palm pad's anterior edge is rounded, and the posterior edge often has two distinct lobes, one on either side of the pad's central subpad (most apparent on hind feet). Except for the gray fox, canine claws are not retractable and often, but not always, register in tracks. More space appears between the palm pad and the oval toes than in felines. Red fox tracks show several notable variances from this general description.

Front. Four toes register reliably. Toe 1 is greatly reduced, vestigial, and rarely registers in tracks. Front tracks are larger than hinds. Palm pad is proportionally larger in the front track than in the hind. A single proximal pad may show in deep substrate or when the animal is running.

Hind. Four toes register (there is no toe 1). Palm pad is small and can resemble the combined shape of the middle two toes in loose substrates.

Track patterns. Canines are excellent distance runners. Trotting is a preferred method of travel. All species use a gait called a side trot, which is used regularly only by canines. The animal is turned slightly obliquely to its direction of travel, with its head to one side and its hind end to the other. This gait is apparently used to avoid interference between the front and hind legs while running.

Scent-marking. All canines use scat and urine for scent-marking. Urine has a distinctive scent for each species, which some people can differentiate. As is common with domestic dogs, wild canines scrape their front and/or hind feet after urinating or depositing a scat, a sign commonly found for coyotes and wolves. Estrus occurs in the late winter and is preceded by blood in the female's urine. Scats are often left in prominent locations.

GRAY WOLF

Canis lupus

The largest canines in the world are social animals that form highly structured packs. Wolves are a keystone species—that is, many other species benefit from their presence, including smaller carnivores that scavenge their kills. In some instances, even plant communities benefit from reduced browsing pressure by ungulates that change feeding patterns in the presence of wolves.

SIZE Length: 0.9–1.3 m. Weight: 23–80 kg. Males are larger. **FIELD MARKS** Most wolves in our region are gray, but several different color phases include black or white. Larger than a coyote, with a shorter and blunter snout and ears. **REPRODUCTION** In a stable population only the dominant female in a pack will breed, but in expanding or

Gray wolf

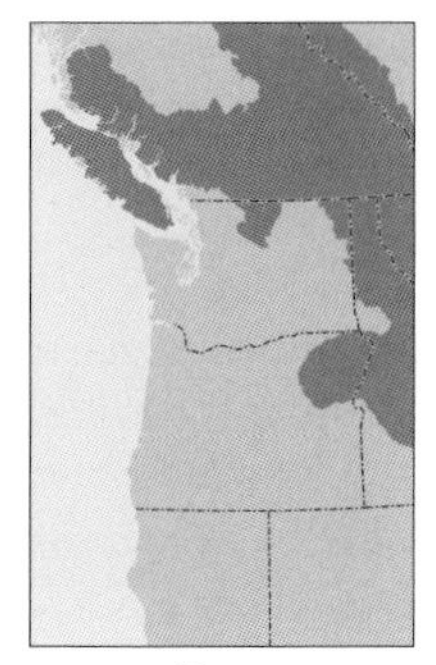

Gray wolf range

exploited populations multiple females may breed and rear young. Breeding occurs February to April, and 3–11 pups are born in late April to July. Dens, often used repeatedly, are dug into hillsides close to water, rarely in a natural cavity. By late summer, pups travel with adults and their tracks are adult-sized. **DIET** Obligate carnivores. Mostly ungulates (deer and elk), but also beavers and smaller mammals. Some packs kill livestock. **HABITAT & DISTRIBUTION** Formerly ranged throughout our region. Well established in central Idaho and northwestern Montana, their range is now expanding in northern Washington and northeastern Oregon. **SIMILAR SPECIES** Domestic dogs, coyote. **CONSERVATION ISSUES** Wolves were systematically extirpated from the region in the first half of the 20th century and then reintroduced into Idaho and Montana in the 1990s. Wolves naturally reestablished themselves in parts of northwestern Montana prior to this. They have expanded into Washington and Oregon from Idaho and British Columbia. The return of wolves to the contiguous United States has been a highly political issue, and the future of wolves in the Northwest continues to be a topic of debate.

TRACKS

Typical canine shape. Larger than coyote tracks, and could be confused only with tracks of the largest breeds of dogs. Tracks often cant forward with toes being deeper than the palm pad. Claws usually register. Adult male tracks are distinctly larger than all other age and sex classes (page 70). **FRONT** 3½–5 1⁄16 in. (8.9–12.8 cm) L × 2 15⁄16–4 7⁄8 (7.5–12.4 cm) W. Significantly larger than hinds. Oval to round. Outer toes may spread in deep substrate or in larger individuals. Negative space either X-shaped or H-shaped depending on toe splay. **HIND** 3 3⁄8–4 9⁄16 in. (8.6–11.6 cm) L × 2 11⁄16–3 7⁄8 in. (6.8–9.8 cm) W. Smaller and narrower than front. Palm pad is smaller than front with the central portion often registering deeper than the outer “wings.” **TRACK PATTERNS** Most commonly trots (direct register or side trot). Walks in a direct register or overstep pattern. **DIRECT REGISTER TROT** Stride: 51 3⁄16–61 in. (130–155 cm). Trail

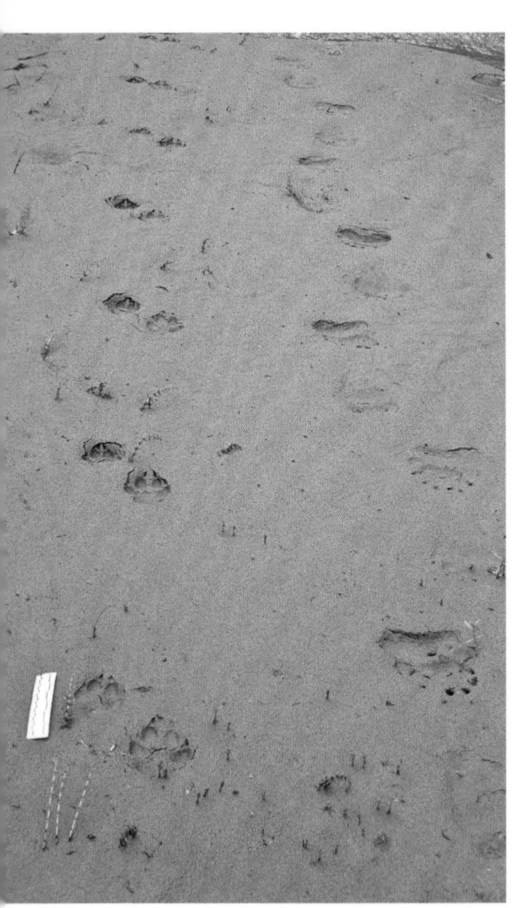

At left, a wolf's side trot heads away from the viewer. At right, an overstep walk of a grizzly bear heads toward the viewer.

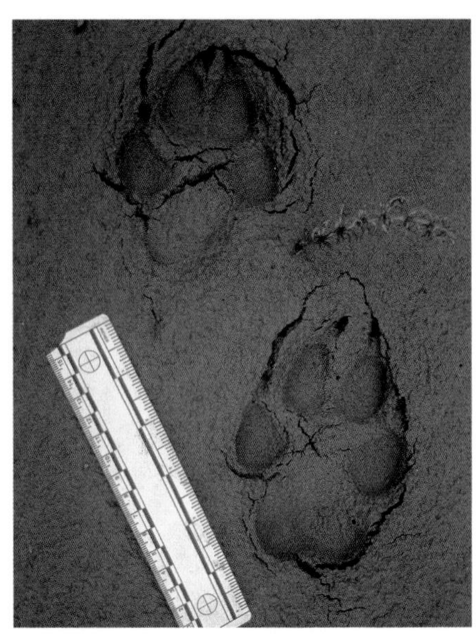

Right front (lower) and hind tracks of a gray wolf

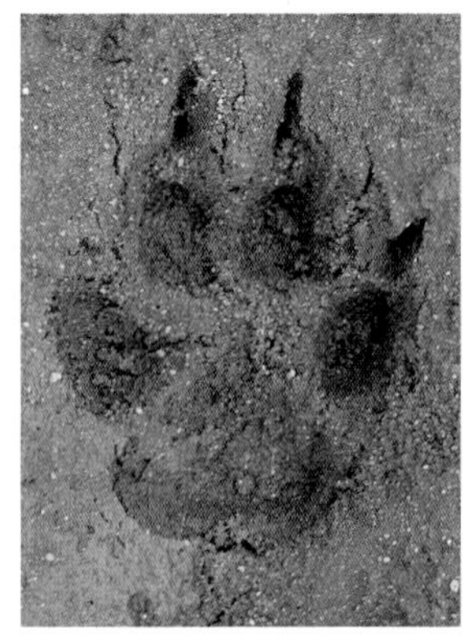

Hind track with outer toes splayed in mud

Wolf direct register trot

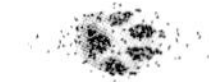

Wolf direct register walk

width: 3 15⁄16–8½ in. (10.0–21.5 cm). **SIDE TROT** Stride: 46 1⁄16–84¼ in. (117–214 cm). Trail width: 6½–10 5⁄8 in. (16.5–27.0 cm). **WALK (DIRECT REGISTER AND OVERSTEP)** Stride: 43 7⁄8–48 5⁄8 in. (111.5–123.5 cm). Trail width: 4 15⁄16–9 1⁄16 in. (12.5–23.0 cm).

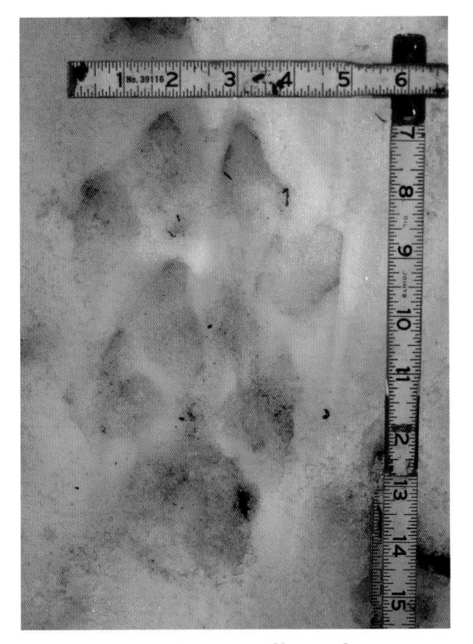

Smaller hind track on top of larger front track. North Cascades, Washington.
Photo by Brandon Sheely.

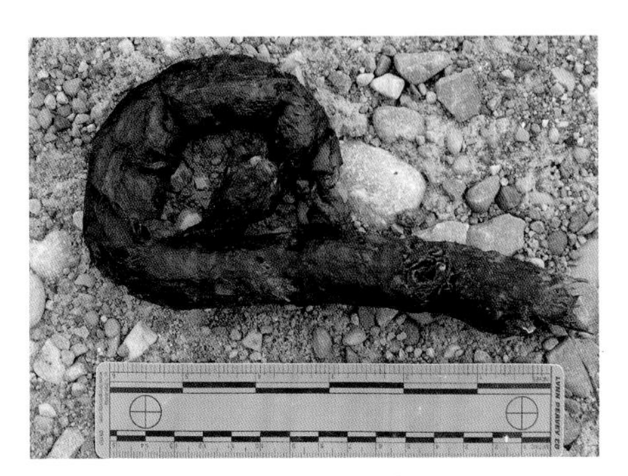

Average-sized wolf scat. Canadian Rockies, Montana.

SIGNS

SCAT 1 1⁄16–1½ in. (1.8–3.8 cm) D × 2¾–16 15⁄16 in. (7–43 cm). Large and tubular, often with at least one tapered end. Not twisted as in coyote or mustelid scats. Often dark and strong smelling. Occasionally may be made up entirely of undigested grass. Scats vary in size depending on recent diet and amount consumed. **FEEDING SIGNS** A pack of wolves kill a hoofed mammal about once a week. Prey are usually brought down by the coordinated effort of multiple animals, and feeding usually begins around the animal's belly area. Much of the prey is consumed at the kill site. Eventually pieces

Runny wolf scat, typical when the animal has been feeding on a fresh kill. Middle Rockies, Idaho.

These bone fragments are from a mule deer that was killed and consumed by wolves. Middle Rockies, Idaho.

The far edge of this large meadow is used as a wolf rendezvous location. A vague wolf trail cuts through the grass, heading to the line of green trees in the distance. Middle Rockies, Idaho.

are dragged off and cached. Bones are consumed, and bone fragments are common signs at wolf-consumed carcasses. Adults regurgitate consumed meat for young at the den or take parts of the animal to the den.

DENS Wolves usually excavate their own dens but may renovate another animal's burrow or expand a natural cavity such as under the root ball of a fallen tree. Dens are usually well concealed and close to water (within ¼ mile, or 0.5 km). The immediate vicinity is often littered with bones and scats. Den sites usually have sandy or loose soil, facilitating excavation. Multiple holes are often dug in the immediate vicinity. Den entrances are 14–25 in. (35–64 cm) in diameter. Tunnels are usually 6–14 ft. (1.8–4.3 m) long and end with an enlarged chamber where pups are born. Dens are barely large

Typical wolf den entrance dug into a hillside. Canadian Rockies, Alberta.

Rocky Mountain elk carcass shortly after being killed by wolves. Feeding began at the belly below the rib cage. This carcass was completely dismantled and dispersed by wolves and other carnivores within 48 hours, at which point only the elk's stomach contents and a few tufts of hair remained. Canadian Rockies, Montana.

enough to accommodate a curious human. **RENDEZVOUS SITES** When pups are young, a pack of wolves meets at a particular location to socialize upon returning from hunting or before heading out. Midsummer, when pups are more mobile, the rendezvous locations may shift away from the den site. Such locations may be littered with tracks and scats. Locations vary but include floodplains and the edges of large meadows.

COYOTE

Canis latrans

The most abundant wild canine in our region, coyotes have expanded their range in the wake of extirpation of other carnivores in large parts of North America and have adapted well to anthropogenic changes to the environment. The animals' social structure varies depending on terrain and food sources, ranging from fairly large pack structures to small family groups. They are wide-ranging and can cover many miles in a single day of hunting and foraging.

SIZE Length: 0.75–1.00 m. Weight: 7–20 kg. Males are larger. **FIELD MARKS** Pointy ears and nose. Tail often black-tipped. **REPRODUCTION** Pairs are monogamous and often strongly bonded. Mating occurs from January to March, with an average of six pups born between March and May. Den is dug by the female into a stream bank or hillside. Brush piles or hollow logs are also used. Den may be reused in subsequent years. Depending on social organization of particular group and habitat conditions, young may disperse or cohabitate with parents. **DIET** Highly omnivorous and adaptable: berries, mushrooms, insects, salmon, small mammals, and fawns. Occasionally kill adult deer, but also scavenge road-killed animals and other carcasses. **HABITAT & DISTRIBUTION** Occupies all habitats in our region, including urban and suburban settings. **SIMILAR SPECIES** Red fox, gray wolf, domestic dogs, bobcat. **CONSERVATION ISSUES** Because of breeding behaviors and removal of other predators, decades of attempts to extirpate coyotes have failed, leaving a larger population than prior to

Coyote

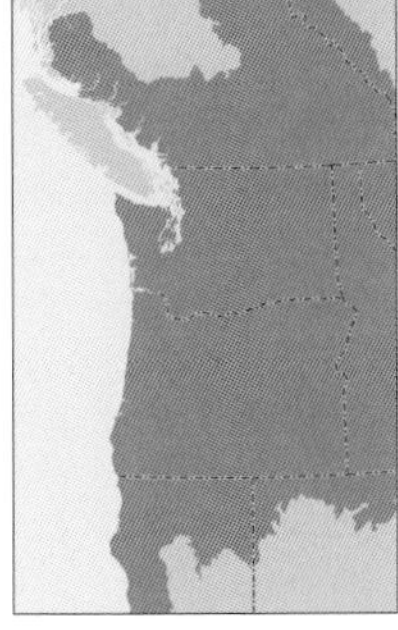

Coyote range

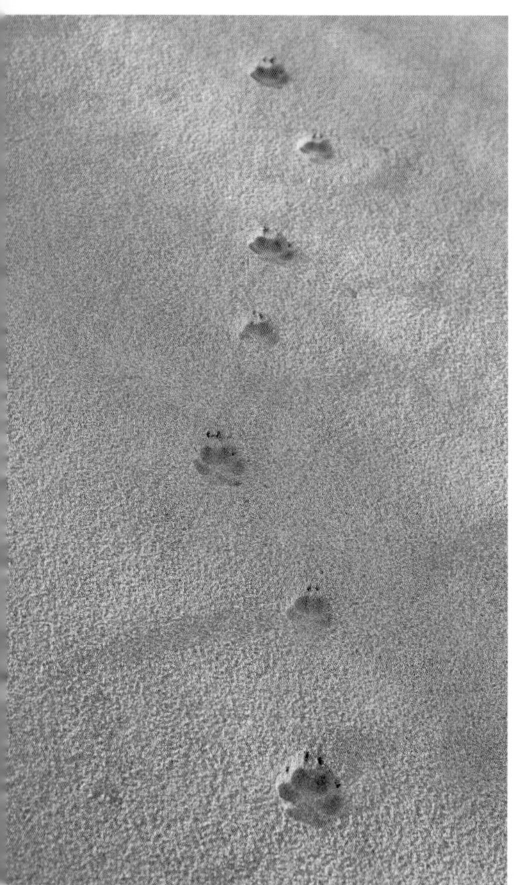

Overstep pattern common for a walking coyote. Columbia Plateau, Washington.

Trail of a coyote in a side trot. From bottom to top: left front, left hind, right front, right hind, left front, left hind. Columbia Plateau, Oregon.

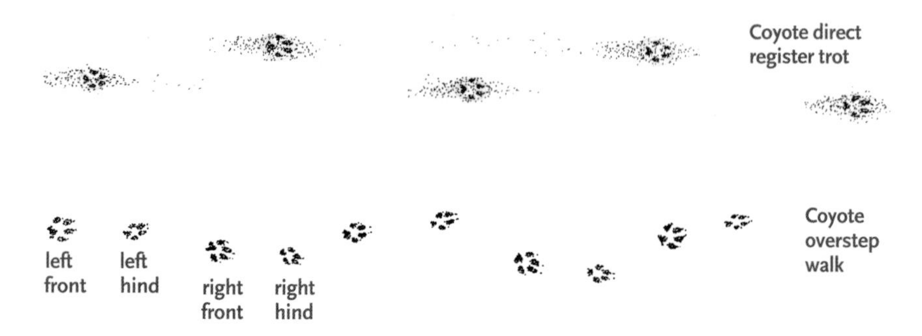

European settlement of the Northwest. In some areas, coyotes compete with rare carnivores such as Canada lynx.

TRACKS

Typical canine structure. Strongly digitigrade, often showing a slant forward with the toes deeper than the palm pad. Space between toes and palm pad forms an X-shape with a raised mound in the center (page 67). **FRONT** 2 1⁄8–3 1⁄16 in. (5.4–7.8 cm) L × 1 5⁄8–2½ in. (4.2–6.3 cm) W. Oval shaped. Wider and more robust than hind foot. Palm pad is triangle shaped, rounded on the leading edge. Two distinct lobes on the outer edge of the pad's posterior may register less deeply than the pad's central portion. Toes splay in deep substrate; otherwise oriented forward. Claws on outer toes register close to the central toes. **HIND** 1¾–2 15⁄16 in. (4.4–7.5 cm) L × 1 3⁄8–2 1⁄8 in. (3.5–5.4 cm) W. Smaller than front. Oval, with outer two toes tucked tightly behind center toes. Palm pad is small with two lightly registering "wings." Central portion of the palm pad is often the only part that registers, making an impression similar in size to a toe. Sharp claws often show. Outer claws sit close to inner toes, while inner claws are oriented in line with track or curve toward the track midline. **TRACK PATTERNS** Usually trots (direct register or side trot), but also walks (overstep or direct register). Flees or pursues prey in an extended gallop. Bounds in deep snow. **OVERSTEP WALK** Stride: 24–38 9⁄16 in. (61–98 cm). Trail width: 2¼–7 11⁄16 in. (5.7–19.6 cm). **DIRECT REGISTER TROT** Stride: 32 7⁄8–46 7⁄16 in. (83.5–118.0 cm). Trail width: 2 5⁄16–4½ in. (5.8–11.5 cm). **SIDE TROT** Stride: 38 3⁄8–53 9⁄16 in. (97.5–136.0 cm). Trail width: 4 1⁄8–6½ in. (10.5–16.5 cm).

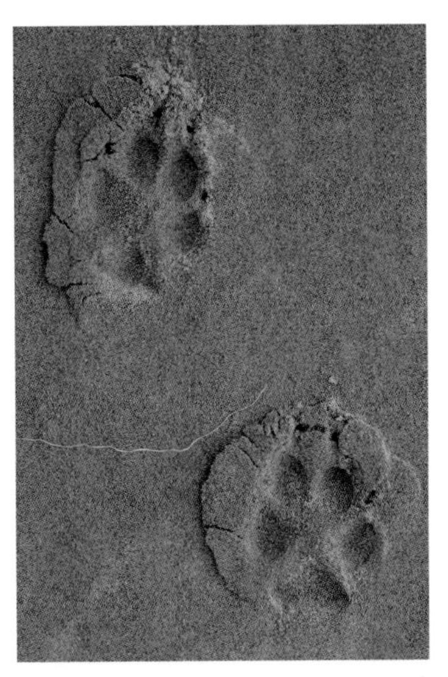

Right front (lower) and hind tracks of a large coyote. Northwest Coast, Oregon.

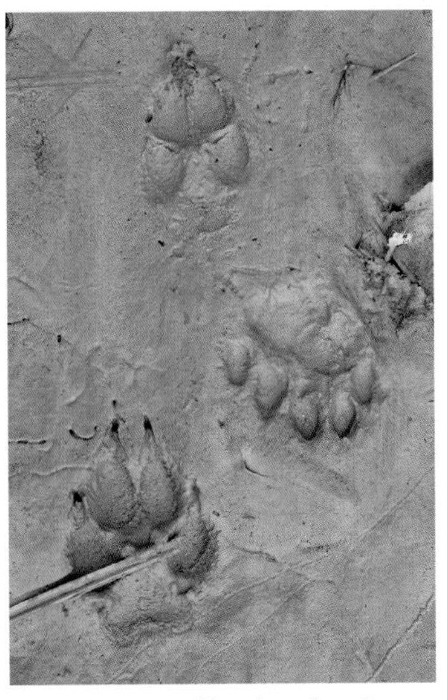

Front (lower) and hind tracks of a coyote. The tracks of a bobcat (hind over front) face the opposite direction. Puget Trough, Washington.

A coyote scat containing small mammal remains. West Cascades, Washington.

A coyote scat containing huckleberries. West Cascades, Washington.

A coyote ate only the nutritious brain from this chum salmon that died and was stranded on a riverbank after spawning. Wolves and bears leave similar feeding signs. Puget Trough, Washington.

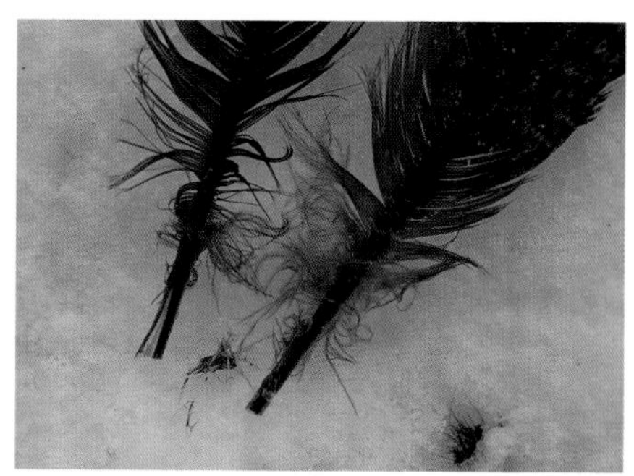

Coyotes will roughly sheer many of the quills of turkeys (shown here), grouse, ducks, and other large birds they consume. Okanagan, Washington.

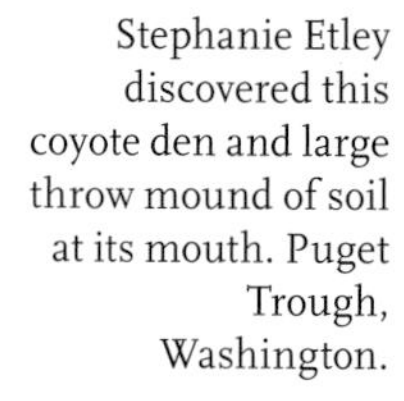

Stephanie Etley discovered this coyote den and large throw mound of soil at its mouth. Puget Trough, Washington.

SIGNS

SCAT ½–1 3⁄16 in. (1.3–3.0 cm) D × 3¾–13 in. (9.5–33.0 cm) L. Tubular, often twisted with a pointy end. Occasionally segmented. Large variety of contents from omnivorous diet. Often deposited on well-used game trails or dirt roads and at other prominent locations on the landscape. **SCENT-MARKING** Scats and urine are often left along regularly traveled roads and trails, and are used to mark territory and communicate with other coyotes. Urine has a distinctive musky odor. **FEEDING SIGNS** Coyotes hunt, but also scavenge carcasses killed by other animals or from other causes. Large carcasses, such as deer, are eaten beginning at the rear. Over time, parts of the carcass are dragged off and cached. Look for an increasing abundance of scats on trails near the carcass. **DIGS** Digs in snow, sand, and forest debris to look for insects or fungi or to bury or retrieve a cache. Triangle shaped, similar to those of gray foxes but proportionally larger. **DENS** Often excavated in hillsides with loose, sandy soil, but may also renovate existing burrows or use natural cavities such as the space under a fallen log. Excavated dens average about 12 in. (30 cm) in diameter and are 5–25 ft. (1.5–7.5 m) long.

This mule deer was killed and partially consumed by coyotes. The area was surrounded by coyote trails. East Cascades, Oregon.

This example of scent-marking by a mated pair of coyotes is called "double marking." The higher urine mark is from the male, and the lower one is from the female. The small amount of blood in her urine indicates the onset of estrus. The alpha male and female in a wolf pack also double mark. East Cascades, Oregon.

RED FOX

Vulpes vulpes

Three subspecies of this adaptable omnivore are found in our region—two natives and at least one introduced species. The Cascade red fox (*Vulpes vulpes cascadensis*) and the Rocky Mountain red fox (*V. vulpes macroura*) occupy high elevations in the Cascades and Blue Mountains/Rockies, respectively. At lower elevations both east and west of the Cascades, non-native foxes can be found, escapees from fur farms or intentionally introduced as a game species. It is uncertain whether introduced populations overlap with native foxes or if they have interbred anywhere in our region.

SIZE Length: 0.83–1.10 m. Weight: 3–7 kg. Males are larger. **FIELD MARKS** Red to dark gray (several forms); large, bushy tail with white tip; dark feet and legs. **REPRODUCTION** Breeding usually occurs in January or February, and a litter of 2–10 pups is born between mid-March and mid-May, peaking in April. Dens are dug into hillsides in dense timber or with a southern exposure. May dig its own den or use an existing burrow system. Pairs are usually monogamous, and the male assists with young. Young travel with parents extensively starting in midsummer. **DIET** Omnivorous; varies with location and season. Small mammals, fruits and berries, carrion, and insects. **HABITAT & DISTRIBUTION** Native subspecies occupies high-elevation forests, subalpine meadows, and edge habitat in the Cascades and Rocky Mountains. West of the Cascades, introduced red foxes are found in farmlands and suburban habitats, less so in eastern Washington. They are absent from desert areas except when water is present. A related species, the kit fox (*Vulpes macrotis*), is well adapted to arid environments and has been documented in the far southeastern corner of Oregon. It is unclear if a resident population inhabits the area. **SIMILAR SPECIES** Gray fox, coyote, bobcat. **CONSERVATION ISSUES** Little research has been done on our native red fox, so its status remains a mystery. Elsewhere, introduced foxes have been linked with declines in songbirds and ground-nesting birds.

Red fox range

TRACKS

Some variance from typical canine shape. Large feet relative to body size. Tracks slightly smaller than those of coyotes. Feet are heavily furred (becoming thickest in winter), which can blur details within the track. X-shaped negative space is prominent (page 64). **FRONT** 1⅞–2½ in. (4.7–6.4 cm) L × 1 7/16–2 5/16 in (3.7–5.9 cm) W. Oval to round, often as wide as long (a good distinction from coyotes). Toe pads may be obscured by fur. Sharp claws often register. Palm pad is furred except for a small chevron, which registers alone or deeper than the rest of the pad. **HIND** 1¾–2 5/16 in. (4.4–5.9 cm) L × 1⅜–1⅞ in. (3.5–4.7 cm) W. Oval. Smaller than front, especially in width. Outer toes tucked in tightly behind front toes. Palm pad is small and often appears as a single, small, oval depression. Fur may obscure details of all features. **TRACK PATTERNS** Commonly trots (direct register or side trot) or walks (direct register or overstep). **DIRECT REGISTER WALK** Stride: 22–26½ in. (55.8–67.3 cm). Trail width:

Right front track of a red fox shows the deeper chevron-shaped impression in the palm pad and the wavy impression from hair on the heavily furred foot. Puget Trough, Washington.

Front (at left) and hind tracks of a red fox.
Photo by Jonah Evans.

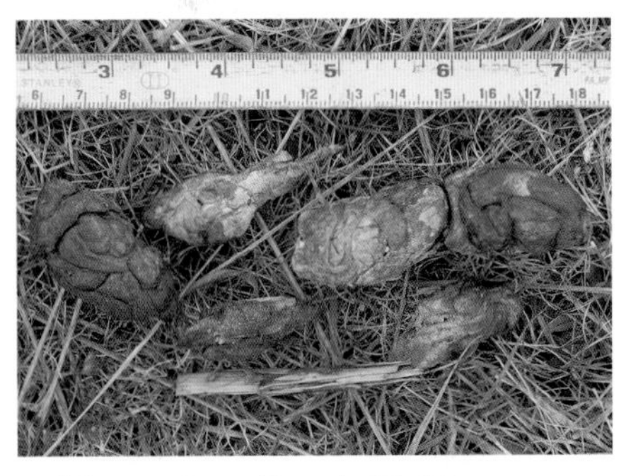

Red fox scat. Puget Trough, Washington.

Red fox side trot

Red fox walk

2 7⁄16 – 6 in. (6.2–15.2 cm). **TROT (DIRECT REGISTER AND SIDE)** Stride: 31–40 in. (78.7–101.6 cm). Trail width: 2¼–6 5⁄16 in. (5.7–16.0 cm).

SIGNS

SCAT 7⁄16 – 7⁄8 in. (1.1–2.2 cm) D × 1–7 7⁄8 in. (2.6–20.0 cm) L. Tubular, often pointed at one end. Contents varied. Often left at scent posts along well-traveled trails and roads. Urine scent-marking occurs at territorial and social scent posts and in conjunction with foraging activities. **DENS** Dug into hillsides and banks with loose soil, in rock outcroppings, or in refurbished burrows excavated by other species. Entrances average 8 in. (20 cm) wide and 12 in. (30 cm) high. Tunnels can be up to 40 ft. (12.2 m) deep but are not usually more than 15–20 ft. (4.6–6.1 m).

GRAY FOX

Urocyon cinereoargenteus

The gray fox is the only native canine that climbs trees. It is mainly nocturnal but is seen occasionally by day. Its legs are shorter relative to its body size than all of the other canines in our region.

SIZE Length: 0.8–1.1 m. Weight 3–7 kg. **FIELD MARKS** Mainly gray, with black-tipped hair along back. Sides are reddish-brown. Throat and underside are white. **REPRODUCTION** Male and female pair travel together throughout the year, breeding in mid-February to March, with 2–7 pups born in April or May. Natal dens are usually located in thick brush. Young disperse in late summer or fall. **DIET** Omnivorous; varies depending on location and season. Berries, insects, and small mammals. **HABITAT & DISTRIBUTION** Woodlands and brushy areas in western Oregon and south. **SIMILAR SPECIES** Red fox, house cat, ringtail.

TRACKS

Typical canine structure, with the exception of semi-retractable claws that rarely register in tracks except when the animal is running or climbing steep slopes. Tracks are smaller than those of a red fox, more similar in size

Gray fox. Photo by Kristi Dranginis Photography.

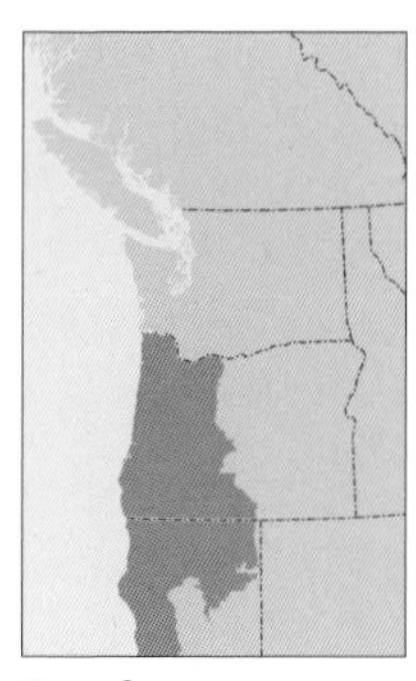

Gray fox range

Front (lower) and hind tracks of a gray fox. Northwest Coast, Oregon.

When a gray fox walks, its track pattern shows a shorter stride and wider trail width than in a direct register trot. Northwest Coast, Oregon.

Gray fox scat containing huckleberries (*Vaccinium* spp.). Northwest Coast, Oregon.

A gray fox's direct register trot across a coastal sand dune. Northwest Coast, Oregon.

A gray fox created this dig, probably to find insect larvae burrowing under the sand. Northwest Coast, Oregon.

to a house cat (page 63). **FRONT** 1 7/16–2 in. (3.6–5.1 cm) L × 1 3/16–1 3/4 in. (3.0–4.5 cm) W. Oval to round. When they register, claws appear sharp. Negative space makes a well-defined X. Toe 1 and proximal carpal pad may show in deep substrate or when the animal is running. Palm pad is triangle shaped and a bar may register along the track's posterior edge, which may be confused with the arched chevron that is central to the palm pad of a red fox. **HIND** 1 1/4–1 3/4 in. (3.2–4.5 cm) L × 1–1 7/16 in (2.6–3.6 cm) W. Oval to round. Palm pad often registers lightly, sometimes no larger than toes. Distinctly smaller than front foot. **DIRECT REGISTER WALK** Stride: 14 3/8–17 11/16 in. (36.5–45.0 cm). Trail width: 2 15/16–4 13/16 in. (7.5–12.2 cm). **DIRECT REGISTER TROT** Stride: 21 5/8–30 3/8 in. (55.0–77.2 cm). Trail width: 1 11/16–3 3/4 in. (4.2–9.5 cm). **STRADDLE TROT** Stride: 30 1/2–38 3/8 in. (77.5–97.5 cm). Trail width: 3 1/8–4 15/16 in. (8.0–12.5 cm).

SIGNS

SCAT 7/16–5/8 in. (1.1–1.6 cm) D × 3 1/8–8 1/16 in. (8.0–20.5 cm) L. Left on prominent objects along travel routes and at trail junctions. During late summer, contents are almost exclusively fruit in many areas. **DIGS** Creates small, triangle-shaped digs, often looking for insects. Appear similar to those made by coyotes but are proportionally narrower.

DOMESTIC DOG

Canis domesticus

Domestic dogs were bred from wolves, and come in many shapes and sizes. Their tracks can be confused with other animals, from gray foxes, to wolves or mountain lions, depending on the breed.

SIZE Varies depending on breed. **HABITAT & DISTRIBUTION** Signs are ubiquitous throughout the region. Free-roaming dogs are encountered in rural and semiwild locations throughout the Northwest. **SIMILAR SPECIES** Other canines or felines, depending on size.

TRACKS

Domestic dogs come in all shapes and sizes, and tracks can be confused with those of any of our native canines. The broad variation in their tracks is an excellent clue to differentiating them from wild canine tracks. Get familiar with the tracks and signs of wild canines, and then look for features that vary from those to identify dogs.

Several features are often useful in separating dog tracks from those of wild canines. Dog tracks often appear flatter than native canine tracks, which tend to be considerably deeper in the toes than the palm pad. Larger dog breeds' large and blunt claws almost always show and splay outward. **FRONT** Oval to round, often squatter in appearance than those of wild canines. Often as wide as long. Palm pad triangular and often flat across the posterior edge and in the track floor. Toes splay, and negative space is often more compressed and less X-shaped than in wild canines. **HIND** Smaller and more oval than the front. Negative space often has more of an X-shaped appearance than the front foot. Palm pad often registers deeper and larger than in a similar sized wild canine, whose pad is often not much larger than a toe. The middle two claws often appear blunt and are oriented straight ahead (rather than sharp and slightly convergent, as in wild canines). Outer claws register more frequently than in wild canines. The outer two toes of wild canines are often tucked tightly into the track, a trait not as consistently found in domestic dogs. In similar-sized animals, the hind often resembles the front foot of a coyote. **TRACK PATTERNS** Erratic and energetic track patterns are excellent clues in

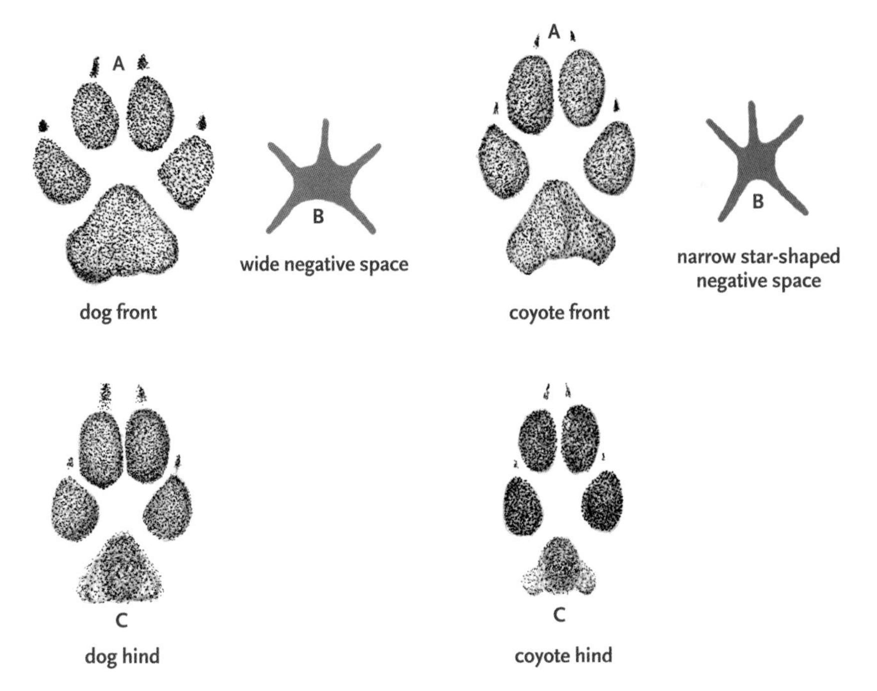

Distinguishing coyote from similar-sized dog tracks: **A** Blunt claws and splayed toes in dog; in coyote thinner, sharper claws, toes and claws generally not splayed. **B** In dog, wider negative space between toes and palm. **C** In coyote, palm pad is relatively smaller and shallower than in dog; in coyote hind track, outer subpads are shallower than central pad.

helping you identify tracks of domestic dogs. Where wild canine tracks and trails are directed toward accomplishing survival tasks, domestic dogs run back and forth aimlessly. Common gaits include walks, trots, lopes, and gallops. In walks, look for a sloppy pattern, with hind feet placed less precisely on fronts. Another clue may be the size of the feet compared to the stride. Many dogs have shorter legs relative to their foot size when compared with similar-sized wild canines. Such a dog's stride is correspondingly shorter as well. Dogs that spend large amounts of time in the wild can demonstrate trail characteristics more typical of wild canines.

BEARS

Family Ursidae

Bears have excellent senses of smell and hearing but relatively poor eyesight. Intelligence, adaptability, and strength allow these omnivores to access a wide variety of food sources. Because of their large size, feeding habits, and marking behaviors, bears leave fascinating and prolific signs on the landscape. Sexually dimorphic, adult males are larger than females.

In winter, bears may be dormant for up to 4 months. Bears are not true hibernators, however, as their core temperatures decrease only slightly and their heart rates are only about half of normal. They are easily aroused during warm periods and may become active briefly throughout the winter. Adult females hibernate for up to 6 weeks longer than males.

In both Northwest bear species, males and females are promiscuous, and a litter may include cubs sired by different males. Breeding usually occurs in early to midsummer. Once impregnated, females experience delayed implantation, and fetus development stops until late fall. Bears are solitary except for females with cubs. At productive feeding locations, bears may tolerate the presence of other bears.

Tracks and Signs

Bears are plantigrade, leaving large, flat tracks (not canted forward toward the toes as with many digitigrade species). Five toes on the front and hind feet, with toe 1 being the smallest (the opposite of a human foot, with which bear tracks can be confused) and not always clearly registering in tracks. Claws usually register on front and hind tracks.

Front. Single proximal pad occasionally registers.

Hind. Significantly longer than front track when the full heel registers.

Track patterns. For both bear species, most common gait is a lumbering overstep walk. Tracks often appear pigeon-toed, canting inward toward the trail's midline. When moving quickly, bears lope or gallop.

Signs. Diverse and often conspicuous. Both species feed on forbs, grasses and sedges, and insects but also consume meat when they can. They occasionally cache large carcasses of animals they have killed or, more commonly, appropriated from another carnivore. Unlike most other carnivores, bears sometimes consume the stomach contents of ungulate carcasses, possibly to acquire bacteria that aids in the digestion of the plant matter they consume. Cached carcasses may be covered with local materials.

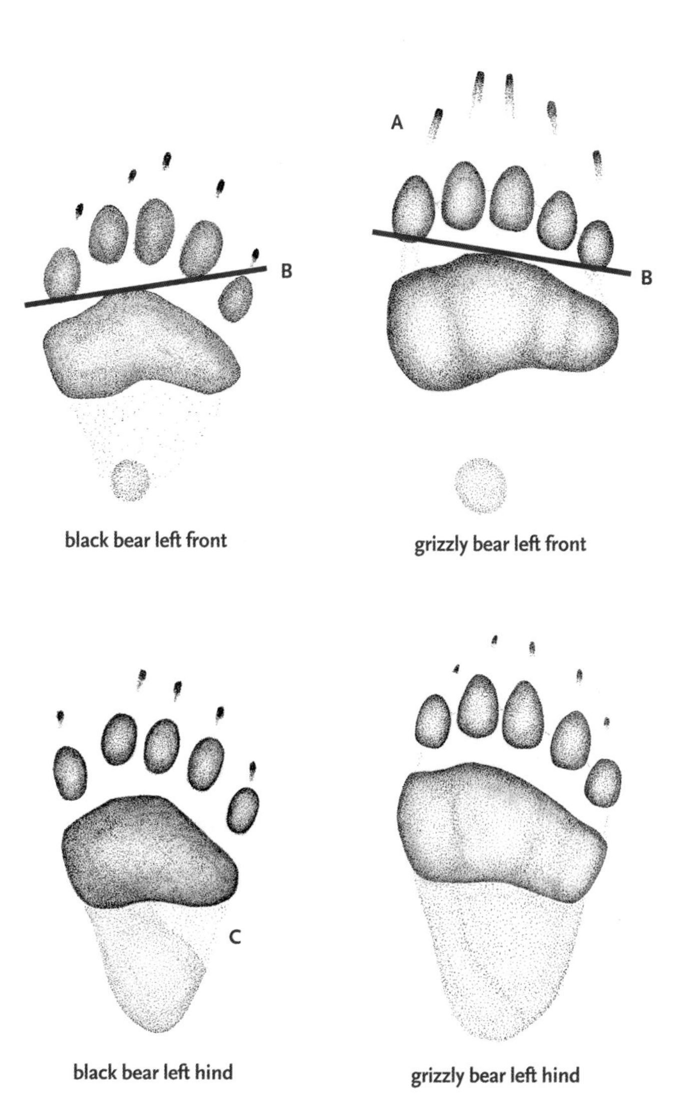

Key features distinguish black and grizzly bear tracks: **A** Claws on front tracks of grizzly bears are usually longer than the length of associated toe and significantly longer than hind claws. **B** In black bear tracks, inner toe falls mainly below a line drawn from bottom of outermost toe and across the top of palm; inner toe falls above line in grizzly bear tracks ("Palmisciano test" most reliable on front feet). **C** Black bears have a wedge of hair on the inside of their hind foot that can leave a distinctly raised area in this portion of the track.

BLACK BEAR

Ursus americanus

Black bears are the most common bears in our region. Well-adapted for life in a forested environment, both cubs and adults can climb trees for safety and to forage. They are active irregularly through the winter, more so west of the Cascades. They are diurnal or crepuscular but can also be active at night, especially when attempting to avoid interactions with people. Extensive overlap exists in the home ranges of both males and females. Females with cubs are the least tolerant of other bears in their vicinity.

SIZE Male length: 1.44–2.0 m. Weight 70–280 kg. Female length: 1.2–1.6 m. Weight: 45–120 kg. **FIELD MARKS** Convex profile to top of skull, with a flat back. Despite their name, black bears can also be gray or cinnamon in color. Color is not a reliable criteria for distinguishing this species from grizzly bears. **REPRODUCTION** Breeds in June or July. Young are born in late January or early February, while mother is still semidormant in the den. Young may den with mother their first winter, dispersing at around 17 months. Females breed every other or every third year. **DIET** Omnivorous; varied depending on specific location and season. Berries, fruit, sedges, grasses, and insects are all eaten abundantly. Depredates on deer fawns in the spring. Commandeers deer and elk carcasses killed by mountain lions. **HABITAT & DISTRIBUTION** Common in all forested and mountain environments in our region. Use of the landscape changes with the season as food sources change. Look for signs around wetlands and early green-up areas in the spring. Areas with large quantities of berries or tree mast crops attract bears in the late summer and fall. Often use human trails, old logging roads, and ridgeline game trails for travel. **SIMILAR SPECIES** Grizzly bear.

TRACKS

Tracks are large. Toes are significantly arched on both front and hind feet. Palm pad is large and flat, often registering more clearly than any other part of the track.

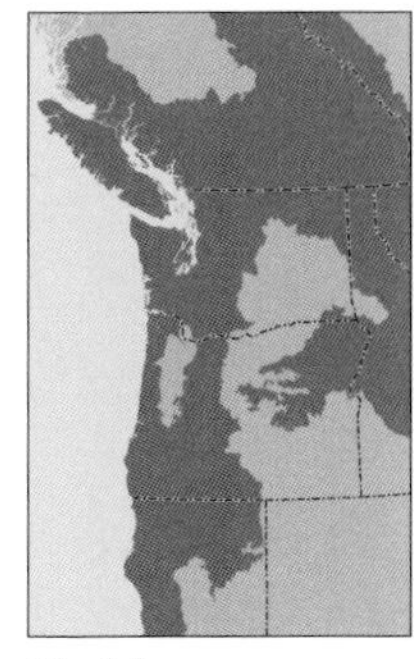

Black bear range

Claws are similar in length on the front and hind feet. Unlike grizzly bears, toes are unwebbed for the entire length of the toe pad. This allows them to separate slightly in deep substrate, giving tracks a less crowded appearance than grizzly tracks in similar conditions (page 71). **FRONT** 3 11⁄16–5 1⁄16 in. (9.4–12.8 cm) L × 3 7⁄8–5 3⁄16 in. (9.8–13.1 cm) W. Claws are sharp and 1.25 in. (3 cm) or less in length, slightly longer than claws of hind foot. A line drawn from the base of toe 5 across the leading edge of the palm pad bisects the middle of toe 1. **HIND** 3 3⁄4–4 3⁄8 in. (9.5–11.1 cm) L without heel or 5 11⁄16–7 1⁄2 in. (14.5–19.1 cm) L with heel × 3 5⁄8–4 15⁄16 in. (9.2–12.5 cm) W. Palm and heel pads partly separated with a wedge of hair on medial portion of the foot that may vaguely register in track (not present in grizzly bear hind feet). **TRACK PATTERNS** Overstep and direct register walk. Tracks are often slightly pigeon-toed. Runs in a lope or gallop. **OVERSTEP WALK** Stride: 34 7⁄8–51 1⁄8

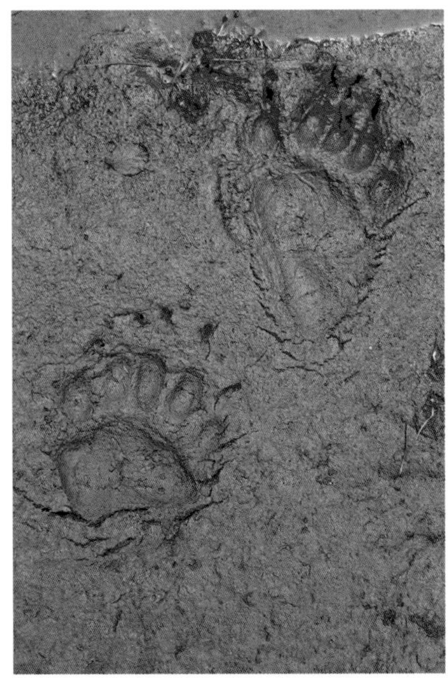
Left front (lower) and hind tracks of a black bear. North Cascades, Washington.

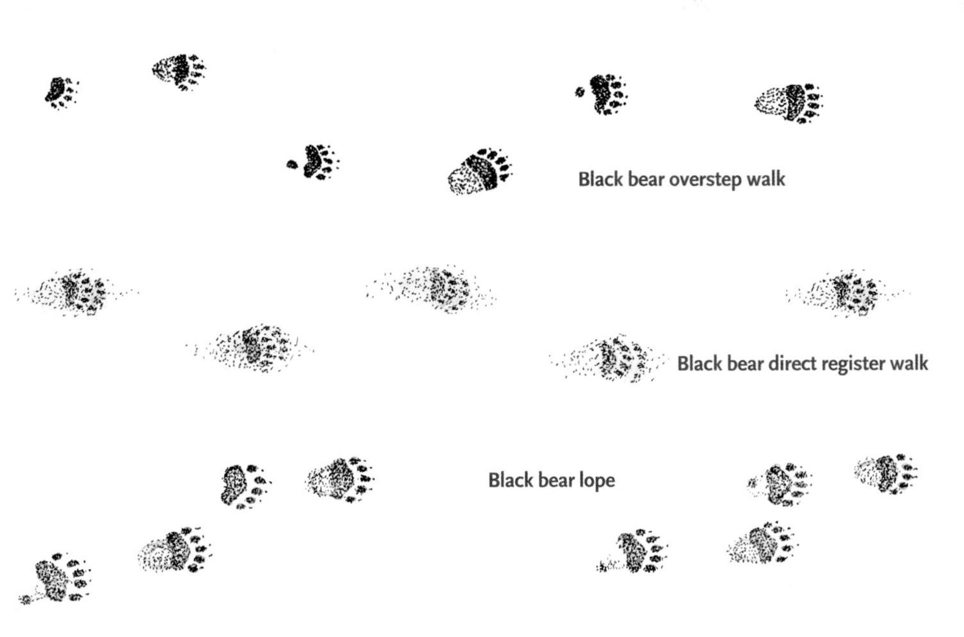

Black bear scat containing grass or sedges. East Cascades, Washington.

Black bear scat containing salmonberries (*Rubus spectabilis*). Puget Trough, Washington.

These rocks were overturned by a black bear in search of insects. North Cascades, Washington.

A black bear tore this log apart searching for insect larvae. Foraged logs often show claw or tooth marks. Middle Rockies, Idaho.

A subalpine fir is scarred from a bear's cambium feeding that occurred many years earlier. North Cascades, Washington.

in. (88.5–129.9 cm). Trail width: 6⅞–12⅜ in. (17.5–31.5 cm). **DIRECT REGISTER WALK** Stride: 30–43¹¹⁄₁₆ in. (76.2–111.2 cm). Trail width: 5½–12⅝ in. (14–32 cm).

SIGNS

SCAT 1–2⁹⁄₁₆ in. (2.5–6.5 cm) D. Highly varied in size and shape depending on the animal's size and diet; tubular to amorphous. It is often easy to deduce what a black bear was eating from its scat. **FEEDING SIGNS** Many foraging activities leave substantial marks on the landscape. **CAMBIUM FEEDING** In many parts of our region, bears feed on the cambium of trees, including Douglas-fir, western hemlock (*Tsuga heterophylla*), western red cedar (*Thuja plicata*), true firs (*Abies* spp.), and black cottonwood. They peel down the outer bark with their claws and scrape off the cambium with their lower incisors. In some forests, these signs are prolific yet often overlooked. Though this type of sign is visible year-round, this behavior is most common in the spring. **SCENT-MARKING & TREE RUBS** Claw, bite, and rub marks can be found on prominent trees along bear travel routes and feeding areas. They often mark conspicuous smaller trees or large trees along ridgelines or edges of wetlands. They also break off the leaders of saplings. Bite marks often leave a distinctive rectangular scar. Hairs in rub marks are wavy and may sun bleach blonde with age. **TREE CLIMBING** Black bears climb trees for refuge or to access food such as alder or aspen catkins or mast in fruit and nut trees, and they leave distinctive claw marks on the bark. Occasionally, a black bear will perch in a fruit tree and break off multiple branches, pulling them back toward its perch to access the fruit. This behavior creates a large clump of branches in the tree, sometimes called a "bear nest." **BEDS & DENS** Bears often bed

The bark from this western red cedar was stripped and fed on by a black bear. West Cascades, Washington.

A black bear chewed off parts of a trail sign in Olympic National Park. Bears often mark human objects, especially those that are conspicuous, such as fence posts, buildings, and power poles. Northwest Coast, Washington.

under large conifer trees with low-hanging branches on the edge of a feeding area such as a wetland or on a mountainside with lots of berries. Bears favor certain trees and habitually bed there. Look for an oval depression worn into the earth, marking where it curls up. Scats are often found around the site. Winter dens are located in more sheltered locations, such as in excavated spaces under fallen logs.

A black bear climbed this aspen tree many years ago, leaving claw marks all the way up the trunk. The scarred bark on the lower portion of the tree is probably from cambium feeding and/or scent-marking activities of elk. East Cascades, Washington.

This subalpine fir was the only tree in the middle of a large meadow and slide alder patch, making it an attractive target for marking attention from bears. North Cascades, Washington.

The shallow depression in front of this large ponderosa pine is a black bear bed that has been used repeatedly. Notice the two scat piles. Canadian Rockies, Montana.

Jeff Rosier discovered this black bear den in a large log that had been hollowed out by a fire many years earlier. Klamath Mountains, Oregon.

GRIZZLY BEAR

Ursus arctos

Grizzly bears are the largest carnivores in our region. Unlike black bears, grizzlies evolved to make their living away from the forest in open terrain. They are excellent diggers and rarely climb trees. They are known for their aggressive behavior, which may be a defense mechanism that evolved from life in open terrain where trees were not readily available for refuge.

SIZE Length: 1.0–2.8 m. Weight: 80–500 kg. (avg. 200 kg males, 135 kg females). **FIELD MARKS** Top of head has a concave profile. Large shoulder hump. Silver-tipped hair. **REPRODUCTION** Both sexes are promiscuous. Young are born while mother is still in the den. Juveniles stay with mother for up to 3 years and are actively protected. Low reproduction rates; females breed every 2–3 years and often don't start breeding until they reach 5 years of age. **DIET** Omnivorous; varies depending on location and season. Sedges, spring forbs, grass, berries, small mammals, roots and bulbs, insects, fawns, elk calves, mature ungulate carcasses killed by wolves or mountain lions, whitebark pine (*Pinus albicaulis*) seeds. **HABITAT & DISTRIBUTION** Large chunk of northern Washington and the Idaho panhandle are federally designated grizzly bear recovery zones. Their current distribution in the North Cascades is unknown. Small population along Canadian border with northeast Washington and northern Idaho. Well established in northwest Montana. Need large, contiguous chunks of wildlands because of their large spatial requirements and to help insulate them from negative interactions with humans, their largest source of mortality. **SIMILAR SPECIES** Black bear. **CONSERVATION ISSUES** Range extensively reduced from historic presence because of systematic hunting and poisoning campaigns. Human interactions are problematic because of the bears' aggressive behaviors and occasional livestock depredation. Poaching is also a problem, and some bears are killed by vehicles while crossing roadways. The reestablishment of a stable population in the North Cascades and Okanagan ecoregions will require habitat conservation, protection, and

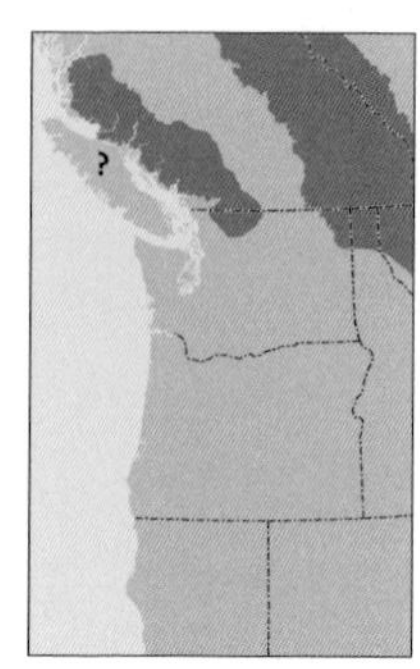

Grizzly bear range

enhancement of wildlife corridors between core habitat areas, plus education of the human population to help reduce negative interactions with bears.

Left front track of a grizzly bear. The proximal pad has lightly registered. Canadian Rockies, British Columbia.

TRACKS

Tracks average larger than those of black bears. Size alone is unreliable for distinguishing these two species. The grizzly's toe arrangement and the palm pad's anterior edge are less arched than those of a black bear. Digits have mesial webbing that connects about a third to two-thirds of the toes pads, often making the toes appear crowded in tracks (page 72). **FRONT** 4 1/8–5 5/16 in. (10.5–13.5 cm) L × 4 1/8–5 11/16 in. (10.5–14.5 cm) W. Long, usually blunt claws used for digging, extending up to 2 1/2 in. (6.4 cm) beyond end of the toes (can be worn

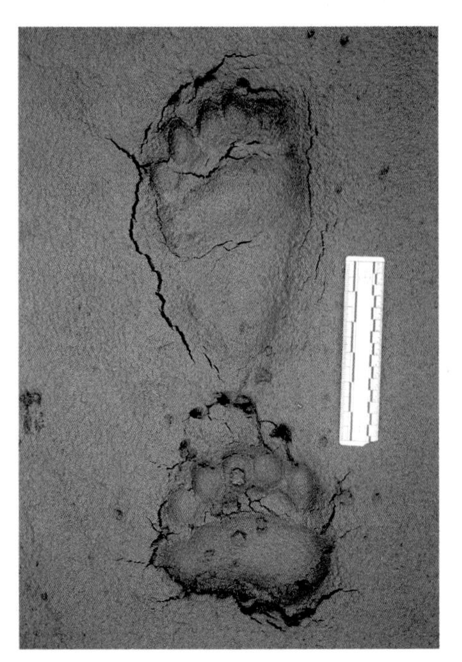

Right front (lower) and hind tracks of a grizzly bear

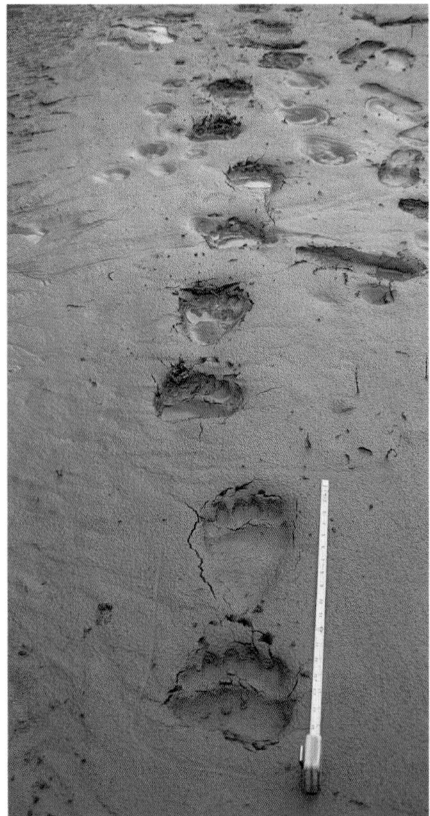

Overstep walking pattern of a grizzly bear (similar to black bear pattern)

down from extensive digging). A line drawn from the base of toe 5 and across the top of the palm pad barely intersects toe 1, with most of the toe above the line. **HIND**: 4 5⁄16–5 1⁄8 in. (11–13 cm) L without heel or 6 5⁄16–8 1⁄16 in. (16.0–20.5 cm) L with heel × 4 1⁄8–5 11⁄16 in. (10.5–14.5 cm) W. Hind claws are shorter than those on the front. Palm and heel pads are fused, with no hair separating them. **DIRECT REGISTER WALK** Stride: 37 3⁄8–42½ in. (95–108 cm). Trail width: 10–12 3⁄16 in. (25.5–31.0 cm). **OVERSTEP WALK** Stride: 40 3⁄16–62 3⁄16 in. (102–158 cm). Trail width: 9 7⁄16–16 15⁄16 in. (24–43 cm).

SIGNS

SCAT Up to 2¾ in. (7.0 cm) D. Varies depending on diet and animal's size. Tubular or amorphous in shape. Usually impossible to differentiate from black bear scat in the field without accessory clues. **RUB POSTS** Rubs on trees, fence posts, telephone polls, and buildings to mark territory. Look for scratch marks from claws; paired bite marks, with the top and bottom canine teeth leaving opposing marks; and wavy, silver-tipped hairs attached to the rub post. **CARCASS FEEDING** May chase wolves and other predators away from kills. Often eat noses off the skulls of ungulate carcasses and break open large bones. Carcass may be covered with material from the immediate vicinity. **DIGS** Use their long claws for digging up roots, bulbs, small rodents, and their food caches—a distinctive sign of a grizzly bear,

Wavy hairs of a grizzly bear are stuck to a tree where bears commonly rub. Canadian Rockies, Montana.

Large grizzly bear scat

Grizzly bite marks on a log cabin. The top row is from the animal's upper canines, the bottom row is from the lower. Similar signs are also found on trees, as with black bears.

as black bears do not typically forage in this manner. Digs often found in grasslands and wet meadows, with sod removed in clumps of about the width of the bear's front foot. Digs are occasionally large and deep. A grizzly bear may return to a favored feeding area year after year to dig for bulbs, leaving the ground lumpy and uneven.

A foraging grizzly bear turned over these large clumps of soil in a wet subalpine meadow. Canadian Rockies, British Columbia.

RACCOONS AND RINGTAILS

Family Procyonidae

Both members of this family in our region are omnivorous, nocturnal, and adept climbers.

Tracks and Signs

Raccoons and ringtails both have five toes on their front and hind feet.

RACCOON

Procyon lotor

Raccoons are both cleaver and dexterous. Agile climbers, they spend the day resting in the safety of trees, foraging on the ground and along bodies of water at night. Raccoon sightings are common in most rural, suburban, and urban areas in our region.

SIZE Length: 60–95 cm. Weight: 4.0–15.8 kg. Males are larger. **FIELD MARKS** Small, bearlike appearance, with dark eye mask, pointy nose, ringed tail (page 51), rump carried higher than the head. **REPRODUCTION** Mating occurs in the spring; 1–7 young born between late May and July. Females that are not impregnated in the spring may come into estrus again in the fall. Young are born in a tree hollow, other natural cavity, or human structure. By midfall, young are semi-independent but may overwinter with mother. **DIET** Omnivorous: fruit, berries, and crayfish are common along with foods pilfered from human sources. **HABITAT & DISTRIBUTION** Strongly associated with both trees and water. Well adapted to urban, suburban, and rural environments. In our region, they are missing only from the highest mountain environments and from arid locations with no water nearby. **SIMILAR SPECIES** Fisher, river otter, bobcat, opossum.

TRACKS

One of the most common medium-sized mammal tracks found in our region (page 66). **FRONT** $1\frac{9}{16}$–$3\frac{1}{8}$ in. (4–8 cm) L × $1\frac{9}{16}$–$2\frac{7}{8}$ in. (3.9–7.3 cm) W. Fingerlike toes (resembling a small human hand) radiate from and connect to the C-shaped palm pad in tracks. Sharp claws often register. **HIND** $1\frac{15}{16}$–4 in. (5.0–10.2 cm) L × $1\frac{7}{16}$–$2\frac{15}{16}$ in. (3.7–7.5 cm) W. Fingerlike toes usually splay less than in front foot. Toe 1 starts noticeably lower than the other four toes. Toes connect to the palm in tracks.

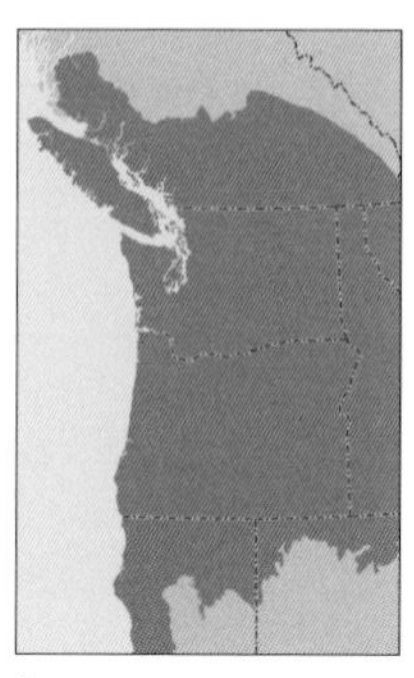

Raccoon range

Palm pad is C-shaped but longer than on front foot. When the heel shows, the hind foot is substantially longer than the front. Claws often show. **TRACK PATTERNS** In one of the most distinctive track patterns for North American mammals, raccoons use an extreme overstep walk, with the front foot from one side of the body falling next to the hind foot from the other side. Runs in a bound or gallop. **WALK** Stride: 18 1⁄8–34 5⁄8 in. (46–88 cm). Trail width: 3¼–7¼ in. (8.3–18.4 cm).

SIGNS

SCAT 7⁄16–1 1⁄8 in. (1.1–2.8 cm) D × 2¾–10¼ (7–26 cm) L. Often an even, blunt-ended tube, but can be loose and amorphous depending on diet. Communal latrines

Right front (at right) and left hind tracks of a raccoon. Puget Trough, Washington.

Typical track pattern of a walking raccoon. The front from one side of the body is paired with the hind from the opposite side. Puget Trough, Washington.

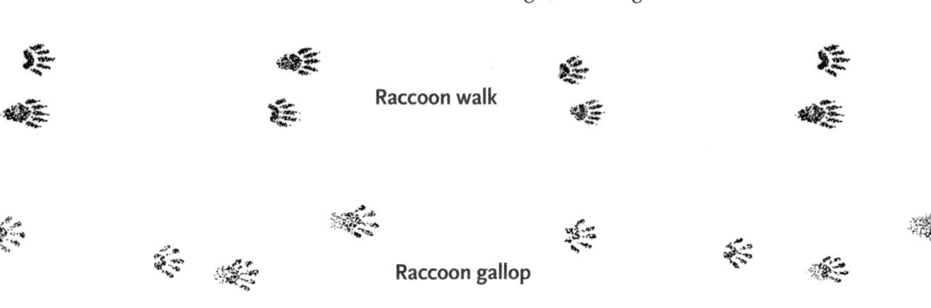

with numerous scats created at base of or in crotch of prominent trees, or occasionally under a bridge. Be careful around raccoon feces, as it may contain parasitic roundworms that can infect humans if inhaled or ingested.

Typical raccoon scat. Puget Trough, Washington. Photo by Tobais Maloy.

Large raccoon latrine under a tree. Puget Trough, Washington.

RINGTAIL

Bassariscus astutus

Ringtails are primarily nocturnal but are occasionally seen in the day during the spring breeding season. They are excellent climbers of trees, rocks, and human structures and can rotate their feet 180 degrees to aid in descending. They have an affinity for crevices and tight spaces. When they are frightened they may secrete a fluid from their anal gland with a sweet, musky scent.

SIZE Length: 61.6–81.1 cm. Weight: 0.75–1.30 kg. **FIELD MARKS** Similar in size to a small house cat; slender, with large ears and a bushy, ringed tail. **REPRODUCTION** Breeding occurs in March or April. Males and females may travel together at this time. Produces a single litter each year in May or June with 1–5 young. Males may assist with rearing young. Young disperse as early as their first fall. Dens are in cliff crevices or in hollow snags, stumps, or brush piles. **DIET** Omnivorous; forages and hunts both arboreally and terrestrially. Diet is seasonal and includes berries, red tree voles, maple seeds, conifer seeds, and insects. **HABITAT & DISTRIBUTION** Broken, rocky, and brushy terrain. Cliffy terrain along rivers is heavily used in southwestern Oregon, which is the northern extent of its range. **SIMILAR SPECIES** House cats, spotted skunk, gray fox.

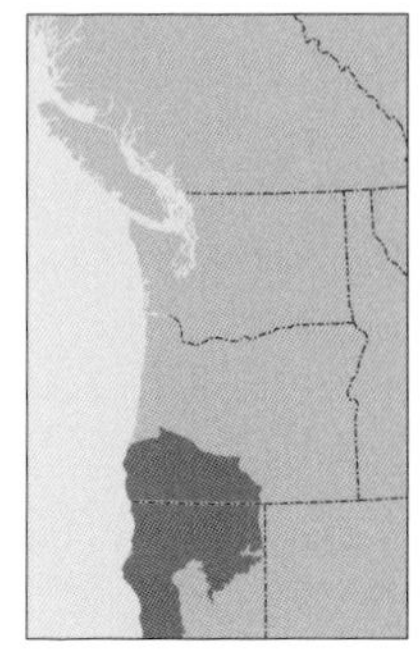

Ringtail range

Ringtail. Photo by George Andrejko, Arizona Department of Fish and Game.

TRACKS

Semiretractable claws rarely register in tracks. Oval; longer than wide. Similar in size to a house cat. Palm pads are not divided as in spotted skunks (page 61). **FRONT** 15⁄16–1 3⁄8 in. (2.4–3.5 cm) L × 3⁄4–1 1⁄8 in. (1.9–2.8 cm) W. Oval toes do not connect to palm pad. Toe 1 registers more often than on hind foot. Palm pad is longer than wide and makes up a larger amount of the overall track volume than in a feline. Leading edge is rounded and bottom has two lobes. A single proximal pad usually registers in tracks. **HIND** 1–1 3⁄4 in. (2.5–4.4 cm) L × 11⁄16–1 5⁄16 in. (1.8–3.4 cm) W. Similar in size to front foot with a longer palm pad. Toe 1 rarely shows in tracks when walking. **TRACK PATTERNS** Walk, bounds, lopes, and gallops. Direct register walking stride is shorter than for house cats and greater than in spotted skunks. **DIRECT REGISTER WALK** Stride: 11–15 7⁄8 in. (28.0–40.3 cm). Trail width:

Ringtail walk

Ringtail gallop

2 3/16–3 9/16 in. (5.5–9.1 cm). **LOPE** Stride: 17½–21 in. (45.5–53.5 cm). Trail width: 2 15/16–3 3/8 in. (7.5–8.5 cm).

SIGNS

SCAT Varies widely depending on diet. May be twisted and ropey (as in mustelids) when feeding on small mammals, or may resemble small raccoon scats when diet is more omnivorous.

Right front (above) and hind tracks of a ringtail. The inner toe on the hind foot registers lightly and could be easily overlooked. Photo by Jason Knight.

All four tracks of a ringtail in a track pattern characteristic of a gallop.

WEASELS AND THEIR KIN

Family Mustelidae

Mustelidae includes a highly energetic and inquisitive group of carnivores. Sexual dimorphism is pronounced in this family, with males in some species 60% larger than females. Their family name refers to their strong-smelling anal gland secretions. Their senses of hearing and smell are well developed; their vision is less so.

All members of this family in our region exhibit delayed implantation, some for up to 10 months. They produce one litter annually in the late spring or early summer. With some notable exceptions, they are solitary.

Tracks and Signs

All members of this family in our region have five toes on their front and hind feet, and their claws generally register in tracks. The toes are oriented asymmetrically around a palm that is thin and arced, being lower and thinner on the inside of the track. Palms of both front and hind feet have four distinct subpads. Two species with highly specialized lifestyle adaptations, river otters (swimming) and badgers (digging), show significant differences from the general descriptions presented here.

In addition to otters, which have obvious webbing, several other mustelids have been reported to share the characteristic. My examination of mink feet revealed mesial webbing on front and hind feet. I suspect that proximal or mesial webbing appears between the toes of other species as well.

In species with heavily furred feet whose tracks are often encountered in snow, track length ranges include measurements both with and without the heel registering. (The heel is difficult to separate from the palm in tracks in loose snow.) Sexual dimorphism within mustelids produces large ranges in track measurements. Measurements that fall on the upper end of a species' range are probably from male tracks. These large ranges make it difficult to differentiate several similarly sized species, and this is compounded by the fact that tracks are often discovered in snow, which can easily distort the size of the track-making foot.

Front. A single proximal pad on the front foot of all species shows occasionally. Toes are oriented in an arc, with toe 1 the lowest. Toe 1 does not always register in tracks, though it does so more reliably than the toe on the hind foot in many species.

Hind. Toes are oriented asymmetrically, with toe 1 lower than the other four toes. Toes 3 and 4 are often strongly paired in several species. Toe 1 and the associated palm pad (smallest and lowest on the inside of the track) do not always register clearly. If toe 1 does not register, the toe orientation on the hind track can resemble that of a canine.

Track patterns. Most mustelids use a distinctive form of bounding or loping that resembles a Slinky moving across the landscape. They all also walk, and badgers often trot. With the exception of badgers, all species create a distinctive track pattern in deep snow that appears to be an angled pattern of two, but is actually made by all four feet. Squirrels also leave a similar pattern of two but will not produce the angled pattern of weasels and other mustelids. Knowing the trail width ranges and averages for various species can help you sort out which mustelid left the tracks you find in the field. (See the track measurement graphs in the back of the book.)

Scat. Often twisted or ropey. Form can change with diet, at times appearing tubular or loose and amorphous. A few species

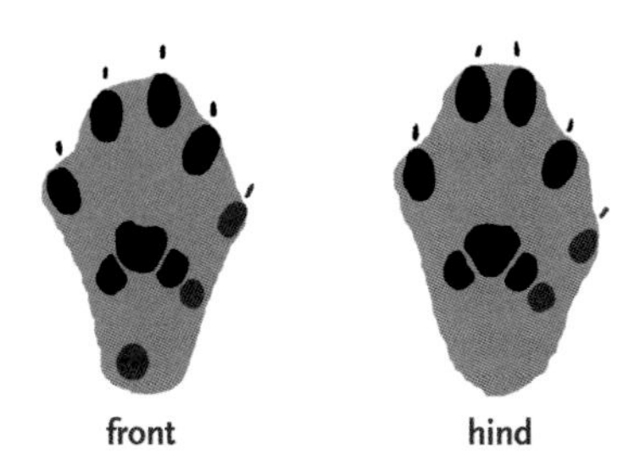

Generic structure of the left front and left hind tracks of a mustelid.

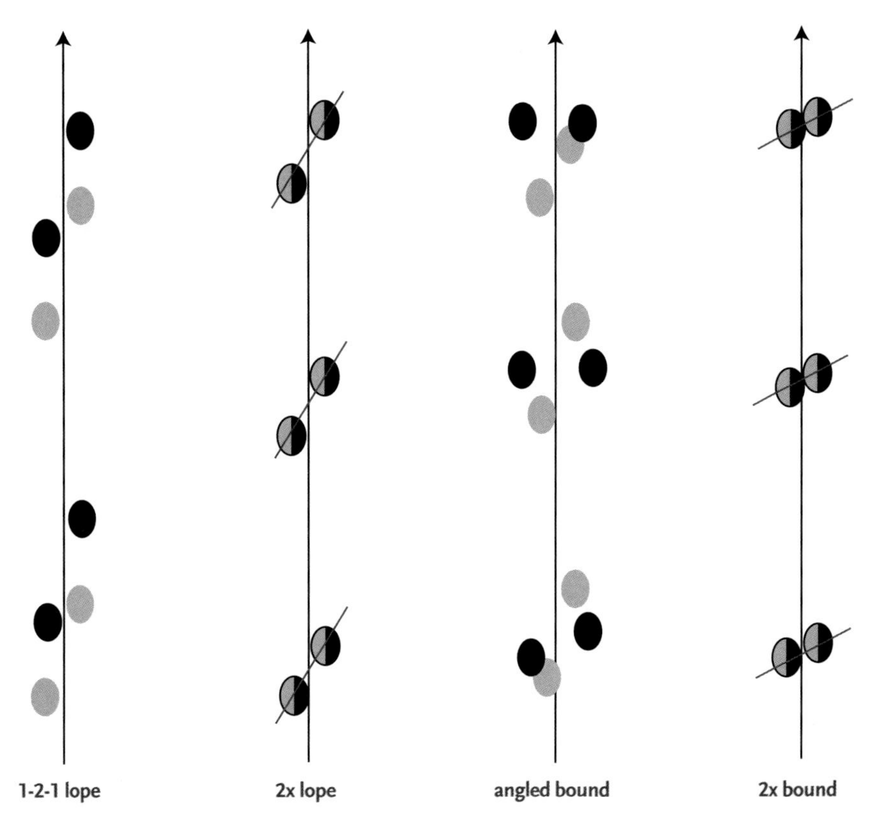

Common mustelid track patterns

consume fruit when in season, which is apparent in scats.

Feeding signs. Several species readily take birds while they are on the ground or at water's edge. Some feather quills are sheered rather than plucked.

WEASELS

Short-tailed weasel, *Mustela erminea*
Long-tailed weasel, *M. frenata*

Weasels are the smallest mammalian predators in our region. Their long, thin body shape allows them to explore burrows, hollow logs, and the subnivean zone during the winter. This body shape also poses a problem: it is inefficient at conserving heat. Weasels have a high surface-area–to–body-mass ratio and cannot role up in a ball like a canine or feline to conserve heat as they sleep. To make up for this, weasels have a high metabolic rate and high caloric needs in the winter, and they need well-insulated bedding locations such as the nest of a rodent upon which they have preyed.

These two species can be difficult, if not impossible, to distinguish in the field,

Short-tailed weasel in winter coat. Photo by Leonard Lee Rue III.

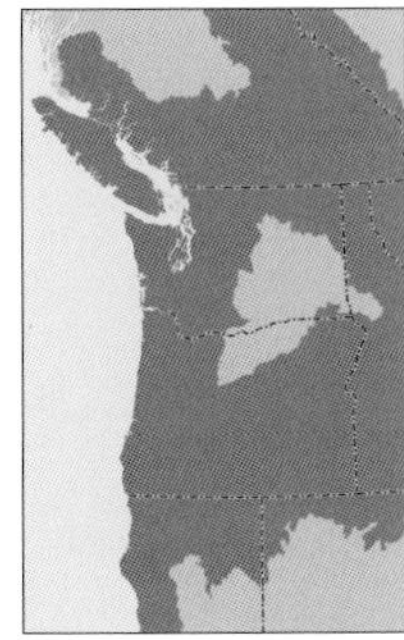

Short-tailed weasel range

Long-tailed weasel in summer coat

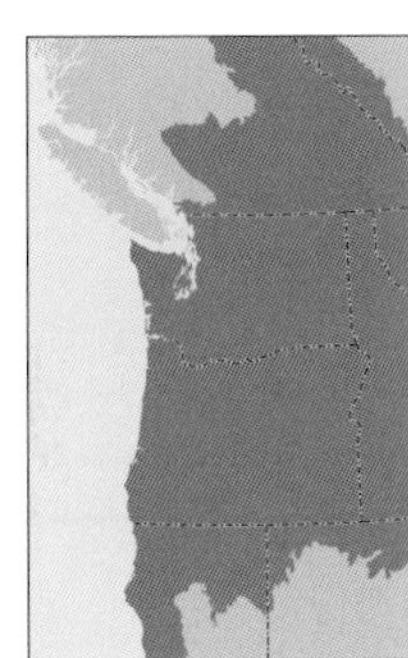

Long-tailed weasel range

either from live sightings or tracks. While long-tailed weasels are usually larger than short-tailed weasels, because they are both sexually dimorphic, a slight size overlap exists between large male short-tailed weasels and small female long-tailed weasels. A few minor habitat use distinctions between the two species are also not absolute.

SIZE Short-tailed weasel—Length: 19–34 cm. Weight: 25–116 g. Long-tailed weasel—Length: 28–42 cm. Weight: 80–450 g. Males are larger in both species. **FIELD MARKS** The long-tailed weasel's tail is about half the length of the rest of it body or longer (44–70%), and a short-tailed weasel's tail is about a third (30–45%) of its body's length. In the Cascades and the mountains to the east, both species can turn white in the winter to blend in with the snow. **REPRODUCTION** Similar behavior in both species. In spring, a male often lingers around the female for several weeks, leading up to and following the birth of young. He may bring food to the den. Breeding occurs after the birth of young. Delayed implantation occurs so that birth does not occur until the following spring. The male can mate with a female and females in her litter. Females not pregnant in the spring come into estrus and breed earlier than pregnant females. **DIET** Carnivorous, with occasional consumption of fruit. Resource partitioning occurs between the two species and between males and females within a species, with the smaller animals in each category focusing on smaller prey. Short-tailed weasels specialize in mouse-sized prey. Both enter and hunt in the prey's burrow. Long-tailed weasels do this with ground squirrels and mountain beavers, two species larger than usual for a short-tailed weasel's prey. The brain case of some prey may be consumed first, with the rest of the carcass abandoned or consumed later. **HABITAT & DISTRIBUTION** Regionally, long-tailed weasels are absent only from the most arid locations and coastal British Columbia. Short-tailed weasels are present in the Cascades and west; east of the Cascades, they are found only in higher mountain environments. Short-tailed weasels might be more common at higher elevations in the region than long-tailed weasels, and some research suggests that they are more effective predators in the subnivean zone than their larger cousins. Both species frequent brushy areas, talus fields, and other areas with abundant small mammal populations. While not aquatic, weasels and their tracks are often seen in riparian habitats. Long-tailed weasels are more likely to travel in the open, through fields and clearcuts. West of the Cascades, long-tailed weasels often use mountain beaver runs and burrows, likely because of high prey densities there. Long-tailed weasel home ranges vary in size seasonally and with prey abundance from as little as 24 acres (0.1 km^2) in summer with abundant prey to nearly 400 acres (1.6 km^2) in winter with scarce prey. Males have larger home ranges than females. **SIMILAR SPECIES** Mink, red and Douglas's squirrels, spotted skunk.

TRACKS

Typical mustelid structure. Feet heavily furred year-round, thickest in winter when fur can completely obscure the toe and metacarpal pads. Tracks often found in snow, where details are sparse because of

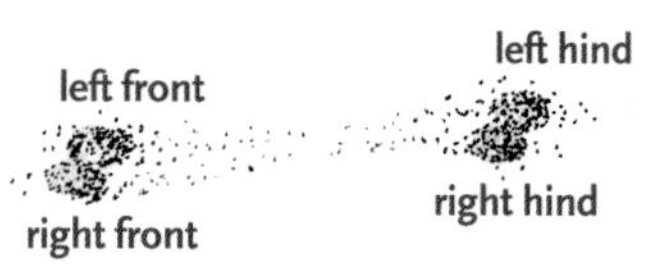

Weasel 4x bound (dumbbell pattern)

Weasel bound variations

the substrate and fur. In mud, claws are often the most prominent feature in tracks. Front and hind feet are round or oblong and similar in size. Smaller tracks are likely from short-tailed weasels, and larger tracks are from long-tailed weasels. Front or hind tracks longer than 13⁄16 in. (2.0 cm) are likely from long-tailed weasels (pages 60 and 61). **FRONT** 3⁄8–1 3⁄8 in. (1.0–3.5 cm) L × 7⁄16–1 1⁄8 in. (1.1–2.9 cm) W. Wider and more symmetrical than hind. Palm pad is thin and drops lowest on the inside of the track, corresponding to toe 1, which registers lightly or not at all. Single proximal pad registers in the clearest tracks. **HIND** ½–1¼ in. (1.2–3.2 cm) L × 7⁄16–1 3⁄16 in. (1.1–3.0 cm) W. Distinctly asymmetrical. If toe 1 doesn't register (common), remaining four toes oriented similar to a canine, though smaller. Meta-

Right front (at left) and hind tracks of a short-tailed weasel. Okanagan, Washington.

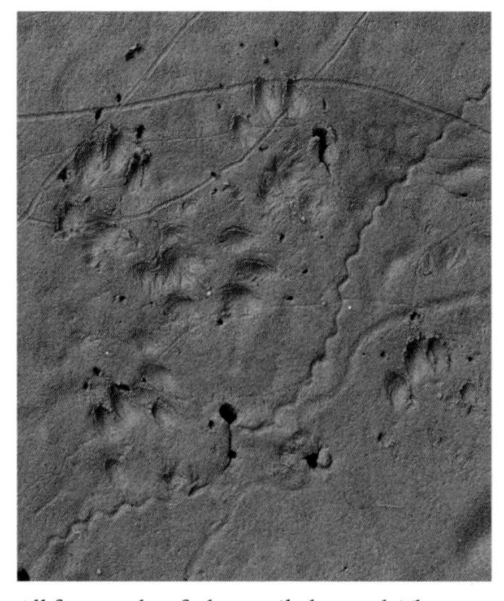

All four tracks of a long-tailed weasel. The two uppermost tracks are fronts and the two lowest are hinds. The track in the center, facing right, is another right front track. The wavy impressions in the palm pad are from fur. Puget Trough, Washington.

At right, a typical oblique pattern created by a short-tailed weasel. Each apparent group of two is actually all four feet. The larger tracks to the left are from a bounding red squirrel.

Dumbbell track pattern of a weasel. The first (lower) depression in each dumbbell is from the front feet, the second depression is from the hinds. Canadian Rockies, Washington.

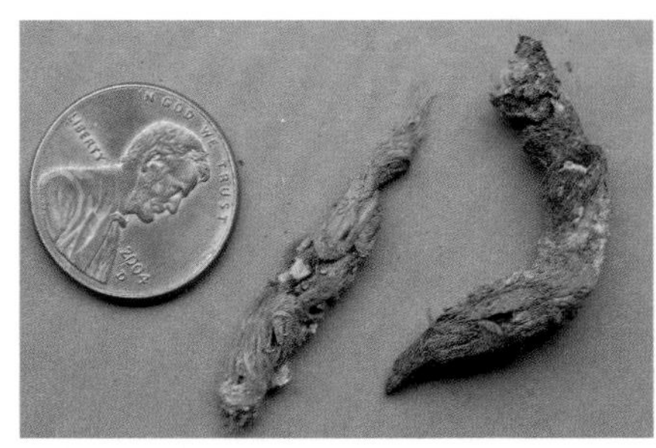

Long-tailed weasel scats. Columbia Plateau, Washington.

Short-tailed weasel scat. Canadian Rockies, British Columbia.

carpal pad is thin and arced as in front foot. **TRACK PATTERNS** In snow the most common pattern is a 2x bound. At faster speeds, hind feet pass the front feet, often leaving drag marks connecting the fronts and hinds, creating a "dumbbell" pattern. In substrates other than snow, long-tailed weasels usually leave a pattern of four that is more boxy in appearance than the irregular pattern left by mink. **2x BOUND** Stride: 7 1⁄16–35 5⁄8 in. (18.0–90.4 cm). Trail width: 1–2 5⁄8 in. (2.5–6.7 cm). Group length: 1¼–3½ in. (3.2–8.9 cm). **4x BOUND (DUMBBELL PATTERN)** Stride: 21 5⁄8–46 in. (55.0–116.8 cm). Trail width: 1 1⁄8–2 5⁄16 in. (2.9–5.9 cm). Group length: 10¼–18 7⁄8 in. (26–48 cm).

SIGNS

SCAT 1⁄8–1⁄4 in. (0.3–0.6 cm) D × 1–3 in. (2.5–7.6 cm) L. Thin, ropey, and often twisted. May be left on logs over travel routes, along trails, or inside or at the entrance to a burrow used as a den or for hunting.

MINK

Neovison vison

While not truly aquatic, mink are strongly associated with water and riparian habitats. They swim readily but can venture far from water in their foraging endeavors. Most activity is nocturnal. The first time I saw

Mink. Photo by Tom and Pat Lesson.

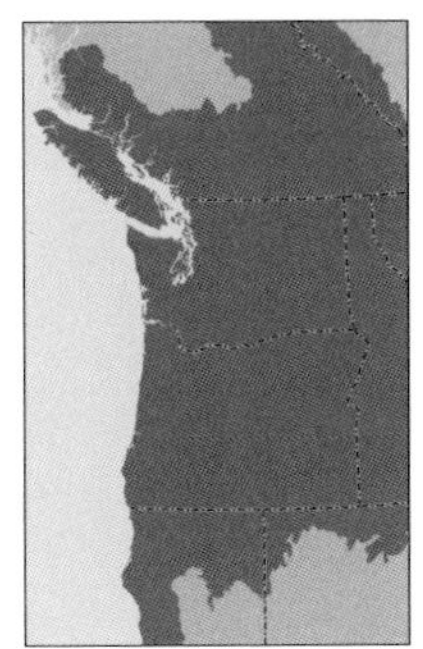

Mink range

a mink, it was running off with a sparrow fledgling in its teeth, being chased by the mother bird. The mother's frantic chirps alerted me to its presence.

SIZE Male length: 55–70 cm. Weight: 0.55–1.25 kg. Female length: 47–60 cm. Weight: 0.55–1.00 kg. **FIELD MARKS** Brown to black fur, often with a white patch on chin or chest. Slender tail. Larger than long-tailed weasels and similar in size to American martens. **REPRODUCTION** Estrus occurs as early as late February and is induced in part by increasing daylight. Polygamous (occasionally promiscuous). A brief period of delayed implantation (greater if the female breeds earlier in the winter) is typical. Young are born in May. Natal den is a muskrat burrow or deep in a log jam. Young reach adult size by November when they disperse. **DIET** Diverse carnivorous diet includes terrestrial and aquatic species. Mouse-sized mammals and muskrats are often important prey, along with small fish, birds, eggs, insects, crayfish, and garter snakes. **HABITAT & DISTRIBUTION** Throughout the region in riparian, wetland, and coastal marine habitats. **SIMILAR SPECIES** Long-tailed weasel, American marten, Norway and roof rats.

TRACKS

Typical mustelid structure. Feet are much less furred than those of weasels and American martens, making clear toe and metacarpal pad registration in tracks more common than in these species. Toes have mesial webbing that occasionally shows in deep tracks where the toes are widely spread. Claws often blend into the toe pad's print. Toe 1 does not always register, especially on hind track. Fronts are slightly larger than hinds, a simple distinction from rats (page 62). **FRONT** 1¹⁄₁₆–1⅞ in. (2.7–4.7 cm) L × 1³⁄₁₆–1¾ in. (3.0–4.4 cm) W. The proximal pad often registers clearly. Toes often splay in a characteristic star shape. **HIND** ⅞–1¾ in. (2.3–4.4 cm) L × ¾–1¾ in. (1.9–4.4 cm) W. Toes may splay as in fronts, but less so. Toe orientation is more asymmetrical than on front foot. **TRACK PATTERNS** Most common gait is a lope. Typical track pattern in substrates other than deep snow is a group of four tracks, often overlapping and usually more elongated than the pattern of long-tailed weasels. **2x BOUND & LOPE** Stride: 1 1/8–30 11/16 in. (28.2–78.0 cm). Trail width: 1 9/16–4 9/16 in. (3.9–11.6 cm). Group length: 3 3/8–6 11/16 in. (8.5–17.0 cm). **ELONGATED LOPE** Stride: to 52½ in. (133.4 cm). Group length: to 11¼ in. (28.5 cm).

SIGNS

SCAT 3/16–9/16 (0.5–1.4 cm) D × 1½–4 in. (3.8–10.2 cm) L. Thin, twisted, and ropey. Left on logs, under overhangs along water's edge, or in latrines outside a den site. **DENS** Various locations are used for resting, and these are often similar to structures used for natal dens. Resting dens are rarely used for more than a few days before the animal moves on, but it may return to the den in the future. Tracks found entering or exiting a muskrat or other burrow may also indicate hunting activity.

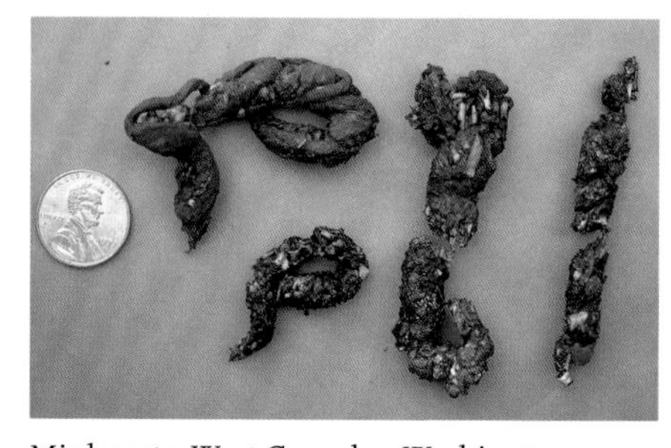

Mink scats. West Cascades, Washington. Photo by Paul Houghtaling.

All four feet show in a common track pattern of a mink. The topmost tracks are the hinds. The mesial webbing of the front feet and right hind foot has registered distinctly. Puget Trough, Washington. Photo by Jason Knight.

Mink tracks, from bottom to top: left front, left hind, and right hind has indirectly registered on the right front. North Cascades, Washington.

Mink lope

AMERICAN MARTEN

Martes americana

Martens are nocturnal forest animals that travel on the ground and in the trees. In winter, they use the subnivean zone extensively. They avoid humans, although individual martens can be incredibly bold after they learn that a human camp may be a source of food. Marten tracks are difficult to find in the summer in their preferred habitat of dense forest litter. With the winter snows, however, their tracks zigzag across the forest, leaving a fascinating story of encounters with other forest creatures, hunting adventures, dens, and food caches.

SIZE Male length: 56–68 cm. Weight: 0.47–1.25 kg. Female length: 50–60 cm. Weight: 0.28–0.85 kg. **DIET** Variety of small animals, including meadow voles, squirrels, snowshoe hares, birds, fish, and insects. Also carrion of larger animals such as deer and elk. Avid berry eaters in the late summer. Can hunt in trees and explore burrows and rock piles, and they are light enough to travel and hunt over deep, loose snow. **REPRODUCTION** Polygamous. Breeding occurs as late as July or August. Delayed implantation occurs and young are not born until March or April. **HABITAT & DISTRIBUTION** Most abundant in undisturbed, high-elevation conifer forests with complex structure of standing, leaning, and fallen trees that creates spaces under the snowpack; the marten depends on these spaces for hunting and thermal regulation when resting. They also use rock outcroppings, are active around tree lines, and are occasionally found at lower elevations in ponderosa pine/Douglas-fir forests on the east side of the Cascades. Male home ranges are almost twice the size of female ranges. Ranges vary widely, from one-third to 6 square miles (0.8–15.7 km^2), and change with prey species density. **SIMILAR SPECIES** Mink, fisher. **CONSERVATION** Clearcuts that fragment forest landscapes not only reduce the amount of usable habitat, but pose obstacles to dispersal of juveniles. Martens have a low reproductive rate for their size. This and home range requirements have slowed population rebound following intensive trapping in parts of our region.

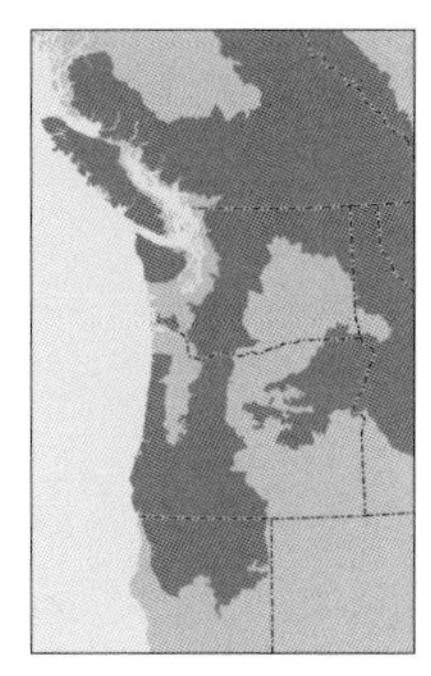

Marten range

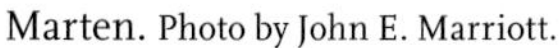

Marten. Photo by John E. Marriott.

TRACKS

Typical mustelid structure. Round to oblong overall. On average, larger than a mink and smaller than a fisher. Claws are semiretractable. Toes tend to splay less than mink toes. Foot size is large compared to body size (an adaptation for snow travel). Feet are furred year-round, becoming thickest in winter. In snow, tracks register less deeply than those of fishers, which are up to four times heavier. Fronts are slightly larger than hinds (page 64). **FRONT** 1½–$2\frac{15}{16}$ in. (3.8–7.5 cm) L × 1½–2 in. (3.8–5.0 cm) W. Single proximal pad often shows in summer. **HIND** 1½–3½ in. (3.8–8.9 cm) L × $1\frac{5}{16}$–2¼ in. (3.4–5.7 cm) W. **TRACK PATTERNS** In deep snow, uses 2x bound or lope. Tracks are rarely more than 3–4 in. (7–10 cm) deep, even in loose snow. **2x BOUND & LOPE** Stride: 21–52 in. (53.3–132.1 cm). Trail width: $2\frac{3}{8}$–$4\frac{11}{16}$ in. (6.0–11.9 cm). Group length: 2¾–$6\frac{11}{16}$ in. (7–17 cm). **3x LOPE** Stride: 27–35 in. (68.5–89.0 cm). Trail width: 2½–3¾ in. (6.4–9.5 cm). Group length: $6\frac{1}{8}$–$14\frac{3}{8}$ in. (15.5–36.5 cm).

Right front (above) and hind tracks of an American marten. North Cascades, Washington.

Right hind foot of an American marten. The furred heel has made a light impression below the palm pad. North Cascades, Washington.

American marten bound

Marten 3x lope
(a hind foot has directly registered on a front in the middle of each group)

SIGNS

SCAT 1/4–5/8 in. (0.6–1.6 cm) D × 1 1/4–4 5/16 in. (3.2–11.0 cm) L. Twisty and ropey. Contents vary depending on diet but can include small mammal remains, feathers, and berries. Scat is often left prominently on rocks or logs on animal or human trails.

In winter, a marten will climb trees (probably searching for squirrels or mice) and then jump to the ground. When the snow is soft, it can sink more than 12 in. (30 cm) into the snow, leaving a full body print including four extra-deep spots, one for each foot. Photo by Amy Gulick/amygulick.com.

Small scat of a marten. Middle Rockies, Idaho.

This track pattern is the most common for American marten in loose snow—a lope with the hind feet landing directly on top of the fronts. Okanagan, Washington.

FISHER

Martes pennanti

The fisher is the larger cousin of the American marten, similar in its habits and biology. Fishers readily take to trees, though they probably travel from tree to tree less often than once believed. They rest in trees, especially conifers, but in winter they often rest on the ground or in the subnivean zone. Fishers are solitary except during breeding season. They are active year-round and both day and night (shifts seasonally).

SIZE Male length: 90–120 cm. Weight: 3.5–5.5 kg. Female length: 75–95cm. Weight: 2.0–2.5 kg. **FIELD MARKS** Larger and heavier than American marten. Dark fur with white spot on belly. Tapered, moderately bushy tail, and pointy face. **REPRODUCTION** Leading up to breeding in the spring, both sexes create scent stations in their home ranges with secretions from their anal glands, likely to inform other fishers of their breeding condition. As with other members of the Mustelidae, parturition induces estrus and breeding. An average of three young are born in the spring, often in a tree cavity. They disperse in the fall. **DIET** Studies in northern California revealed that fishers consume small mammals up to the size of a snowshoe hare, fruit and berries, truffles (*Rhizopogon* spp.), birds, eggs, amphibians, and carrion. Much has been written about the fisher's ability to kill porcupines, but it is unlikely that porcupines are an important part of the fisher's diet in our region. **HABITAT & DISTRIBUTION** Prefer low- to mid-elevation dense forests (with American marten occupying higher elevations). Associated with areas of less snowpack, as they are less well adapted for travel in deep snow than is the American marten. This habitat partitioning is apparently less distinct in the Klamath Mountains ecoregion than in other places where these two species coexist. As with the marten, forest floor structural diversity is important for hunting and thermoregulation in winter. Populations in the Oregon Cascades and Washington's Olympic Peninsula are from reintroduction efforts. Studies in the western United States show home ranges of females to be 1.5–12.3 square miles (4–32 km²) and males 7.7–32.8

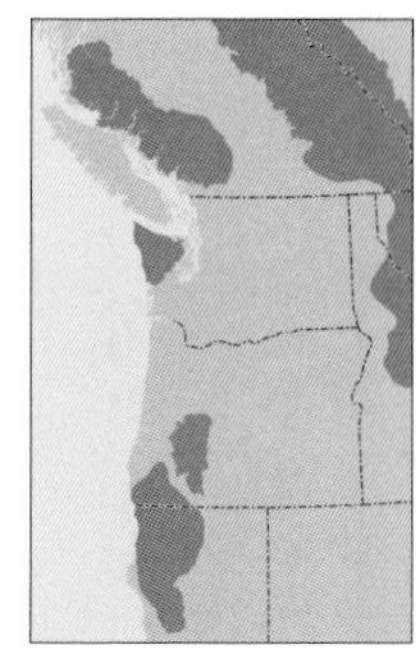

Fisher range

square miles (20–85 km^2). A male's home range overlaps with several females but not other males. (Females also rarely overlap with one another.) **SIMILAR SPECIES** Raccoon, river otter, American marten. **CONSERVATION ISSUES** Populations in our region have plummeted from habitat destruction and fur trapping. Fishers have a relatively short dispersal distance that limits their ability to repopulate an area once they have been extirpated. Several reintroduction attempts occurred in the Oregon Cascades and Blue Mountains in the late 20th century. Starting in 2008, fishers were translocated to the Olympic Peninsula from British Columbia, and plans call for the same in the Washington Cascades.

TRACKS

Typical mustelid structure. Tracks are easily confused with other more common species, including raccoons, river otters, and American marten. Fur appears around toes, metacarpal pads, and proximal pad. Fur is thicker in winter, obscuring pads in tracks more often than in summer months. Fronts and hinds are similar in size, with fronts slightly larger (page 66).

Versus raccoon. As with other mustelids, the fisher's toe pads generally do not connect to the palm in tracks. In raccoons, toes generally do connect to the palm. Raccoon feet are furless, which produces much clearer

All four feet of a female fisher in early winter; fur is thickest on the soles of the feet and details in tracks are obscured. Fronts are at bottom. Northwest Coast, Washington.

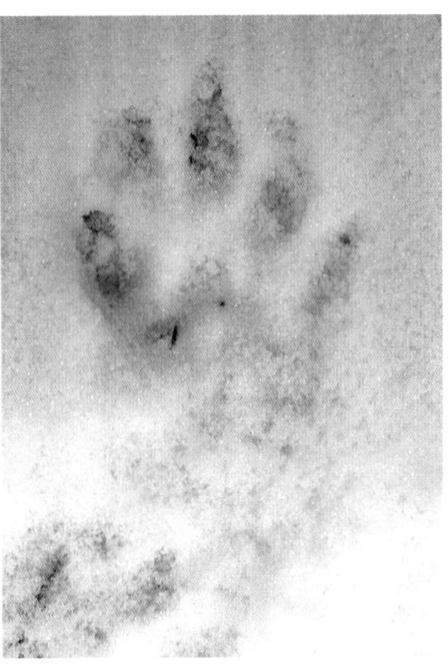

Left hind track of a female fisher. Northwest Coast, Washington.

Fisher lope and bound variations

tracks than those of fisher. The raccoon's palm pad is one entirely fussed pad and takes up a substantial portion of the overall track. In fisher, the palm is narrower and comprises four distinct subpads. The typical walking track pattern of a raccoon could be confused with a bounding track pattern used by fisher in deep snow, despite their being created by entirely different gaits.

Versus river otter. Fisher tracks can be similar in size to those of otters. In otters, the hind is distinctly wider and the foot is completely unfurred. In otter tracks, the toe and metacarpal pads appear clearer and are larger relative to the overall size of the track. Fishers rarely enter water, but most otter trails start and end at water. Unlike fishers, otters never climb trees.

Versus American marten. The majority of marten tracks and trails are smaller but are otherwise nearly identical to those of fishers. Some overlap exists in track and trail measurements between male marten and female fishers. Both species employ similar gaits, leave similar track patterns, readily

Common track pattern for fisher in shallow snow. One of the hind feet registers on one of the fronts, creating a pattern that appears to be three tracks. Photo by Dan Gardoqui.

Scats, urine, and tracks of a male fisher, showing scent-marking behavior associated with spring breeding behavior. Photo by Susan C. Morse.

climb trees, and spend lots of time exploring under logs and in the subnivean zone. Marten are lighter, and tracks in snow may appear to float where a fisher's tracks would sink more deeply.

FRONT 1⅝–3¹⁵⁄₁₆ in. (4.2–10.0 cm) L × 1⅝–3⁵⁄₁₆ in. (4.1–8.4 cm) W. Slightly larger than hind foot. Single proximal pad registers occasionally. **HIND** 2–4⁵⁄₁₆ in. (5–11 cm) L × 2–3⅛ in. (5.0–7.9 cm) W. Toe 1 may register lightly and is the smallest. **TRACK PATTERNS** Primarily lopes and bounds; walks for short distances. **2x LOPE** Stride: 18½–43⁵⁄₁₆ in. (47–110 cm). Trail width: 3⅜–7⁵⁄₁₆ in. (8.5–18.5 cm). Group length: 4¾–8⅞ in. (12.0–22.5 cm).

SIGNS

SCAT Generally less twisted than smaller mustelids. Averages larger than marten scats but similar in contents.

NORTHERN RIVER OTTER

Lontra canadensis

River otters are excellent swimmers and spend more time in the water than on land. The most social of all the mustelids, they have been observed in various sex and age class associations. They are more active at night in summer and during the day in winter. In coastal populations, tides influence their behavior.

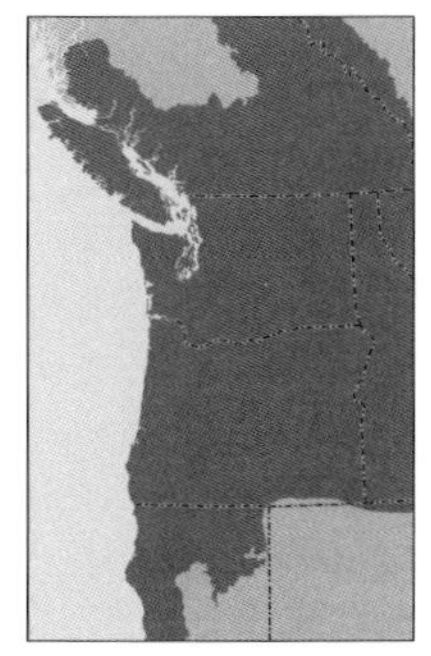
River otter range

SIZE Length: 0.9–1.1 m. Weight: 4.5–13.6 kg. Far less sexually dimorphic than other mustelids, but males are slightly larger. **FIELD MARKS** Sleek, tubular appearance. Long whiskers and a flattened tail. **REPRO-**

Northern river otter

Look carefully to see the trails of three otters that loped across a beach toward the waters of the Puget Sound, Washington.

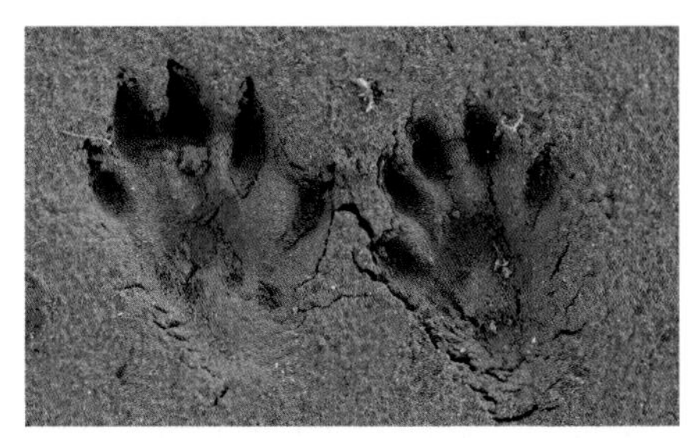

Left hind (at left) and right front tracks of a river otter. Columbia Plateau, Oregon.

Common track pattern of a loping otter. Columbia Plateau, Oregon.

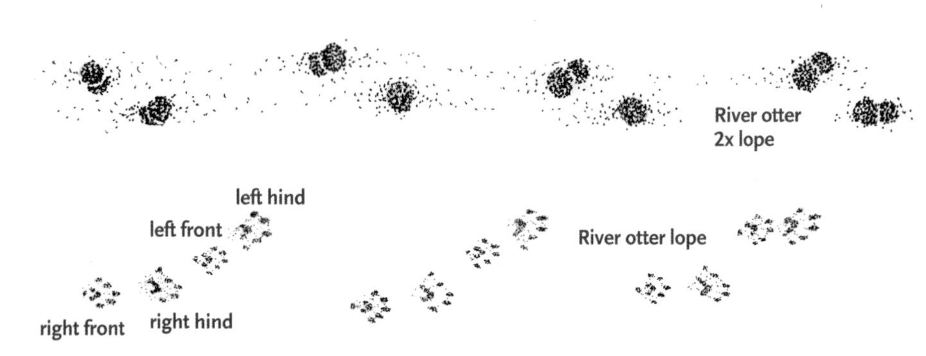

DUCTION Female breeds shortly after giving birth, around April; females that are not already pregnant breed earlier. Natal den is located in the abandoned burrow of another animal, in thick brush, or under the root wad of a fallen tree. Generally close to water but occasionally up to 1.5 miles (2.4 km) distant. Average litter is 2–3 kits. **DIET** Mostly fish (up to 80% of its diet); also crustaceans, amphibians, birds, and occasionally muskrats and fruit. **HABITAT & DISTRIBUTION** Rivers, lakes, ponds, wetlands, and coastal marine environments. Travels overland for considerable distances from one body of water to another. Found throughout the region in appropriate habitat, more abundantly west of the Cascades. Often found in association with beavers. **SIMILAR SPECIES** Fisher, raccoon.

TRACKS

An otter's feet are highly adapted to its aquatic lifestyle but still retain similarities to those of other mustelids. Webbing on both front and hind feet connects toes at about the middle of each toe pad but does not always register. The bottom of the foot is not furred. Toes on both feet appear oval or tear-drop shaped, either disconnected from palm pad or distinctly bulbous at tips, which helps distinguish otter tracks from those of raccoons. Claws often register on both front and hind feet, sometimes merging with the toe's imprint (page 68). **FRONT** 2 1⁄16–3 9⁄16 in. (5.2–9.0 cm) L × 1 13⁄16–3 in. (4.6–7.6 cm) W. Palm pad is larger compared to the overall track than in other mustelids. The single proximal pad sometimes registers. **HIND** 2 3⁄16–4 1⁄8 in. (5.6–10.4 cm) L × 2¼–3 7⁄8 in. (5.7–9.8 cm) W. Larger and wider than front. Toe 1 is set lower and more to the side than other four toes, creating a wide track. **TRACK PATTERNS** Most common terrestrial gait is a lope with the body turned slightly sideways to the direction of travel. Also walks for short distances, leaving direct or indirect register pattern. **LOPE** Stride: 17 5⁄16–41 5⁄16 in. (44–105 cm). Trail width: 4¾–10 in. (12.0–25.4 cm). Group length: 9 7⁄8–16 1⁄8 in. (25–42 cm).

SIGNS

SCAT ½–7⁄8 in. (1.2–2.2 cm) D × 2–9 1⁄16 in. (5–23 cm) L. Rarely in consolidated form.

Two otters ran and slid across the snow. North Cascades, Washington.

Otter scat. Puget Trough, Washington.

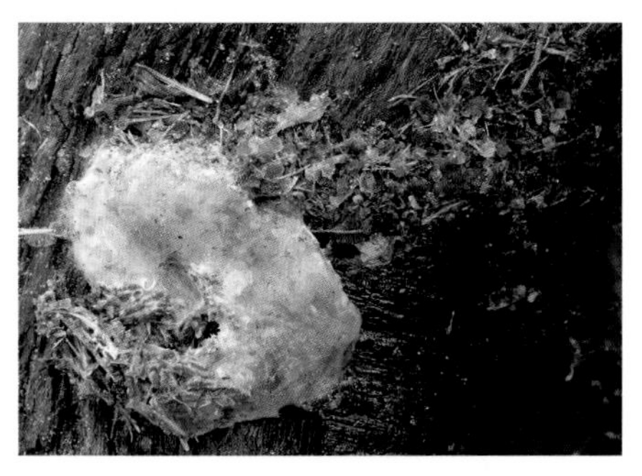

Otter scat containing fish bones and scales and covered with a gelatinous mucus secretion produced in the animal's intestines.

Usually contains fish scales, bones, and crustacean pieces. Deposited at water's edge, sometimes accompanied by a yellowish, green, or brown mucus secretion. **SCENT STATION** Created at prominent points around water, such as a small isthmus on a pond, the outlet to a lake, or close to a beaver bank burrow. Scats, gelatinous intestinal secretions, odorous secretions from anal glands, role marks, and twisted grass also found. Rolling helps clean fur. **SLIDES** Frequently slide on snow, downhill and on flat terrain. Slides may be used repetitively, or you may see tracks disappear for a few feet, replaced by a trough created by the animal's body. Slides are often associated with the otter's playful behavior, but they may also be an efficient means of travel for these streamlined critters. Wolverines also slide on snow on occasion.

WOLVERINE

Gulo gulo

Wolverines are the largest terrestrial mustelid. This wide-ranging species is adapted to deep snow and cold conditions. It scavenges on carcasses of animals killed by other predators or other causes such as avalanches, and its powerful jaws allow it to consume frozen carcasses and large bones. Individuals have large home ranges and travel a circuit through it during the course of about a week. In summer, they may seek out remnant snow patches for travel or play.

SIZE Length: 0.86–1.07 m. Weight: 6.6–16.2 kg. Males are larger. **FIELD MARKS** Wide body with short, powerful legs; short, rounded ears; and moderately bushy tail. Pale stripe along its side, from head to tail. Light-colored face mask. **REPRODUCTION** Low reproductive rate: females breed later in life than other similar-sized species, litter sizes are small, and population density is low. Delayed implantation. Young are born in subnivean cavities in talus fields or under snow-covered fallen logs in the spring. Polygamous males may visit young they sire from several females. While they are considered solitary, except for breeding behavior, growing evidence shows that this species may engage in more complex social interactions. **DIET** Carnivorous. Carrion is the primary winter food source. In our region, hoary marmots and ground squirrels are probably an important part of their summer diet. **HABITAT & DISTRIBUTION** Exact extent

Wolverine. Photo by Tom and Pat Lesson.

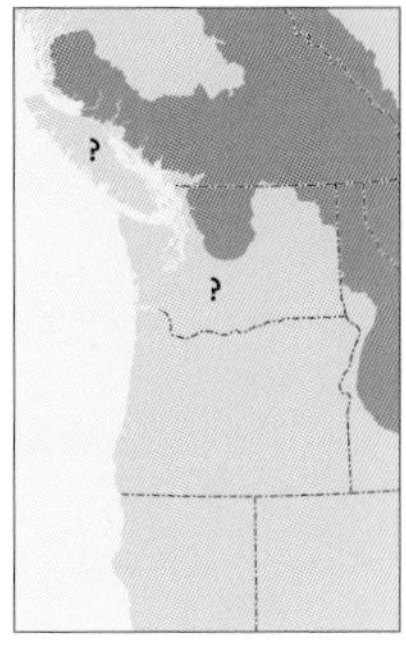

Wolverine range

of its range in our region is unclear. Mainly high-elevation alpine tundra and subalpine forest. Descends to lower elevations during foraging trips but rarely lingers. Vancouver Island subspecies (*Gulo gulo vancouverensis*) may be extinct. Large home ranges of 39–350 square miles (100–900 km^2). Dispersal distances of more than 500 miles (800 km) for juvenile males have been recorded; animals have been observed hundreds of miles from known breeding populations, sometimes in atypical habitat. **SIMILAR SPECIES** Canada lynx, gray wolf, black bear, domestic dogs, river otter, hoary marmot. **CONSERVATION ISSUES** Research has linked wolverine natal dens to late-winter snowpack, so global climate changes may affect this species in our region. Population status in most of the region is unknown.

TRACKS

Typical mustelid structure with adaptations for travel on snow. Tracks are large compared to the animal's size. Feet are furred (thicker in winter), which can diffuse track detail. Front and hind tracks similar in size and shape. Large claws may show in both front and hind tracks (page 69). **FRONT** $4\frac{5}{16}$–$5\frac{11}{16}$ in. (11.0–14.5 cm) L × $3\frac{9}{16}$–$4\frac{5}{16}$ in. (9–11 cm) W. Single proximal pad occasionally registers in tracks. **HIND** $3\frac{1}{16}$–7 in. (7.8–17.8 cm) L × $3\frac{5}{16}$–$5\frac{1}{2}$ in. (8.4–14.0 cm) W. Slightly narrower than front. Toe 1 may not register or may lightly register, leaving a track that may be confused with a that of a canine or feline. Heel is completely furred. **TRACK PATTERNS** Track pattern, plus overall track size and shape, are important for identifying this creature in the snow. Loping gait is most common for long-distance travel. Lope track pattern varies, with one or both of the hind feet landing near or directly on top of the front feet in deep snow. Lynx do not use this gait. Wolverines also walk with a direct register pattern, which could be mistaken for a lynx, but wolverines do not use this gait for great distances. Look for slides (similar to those of otters) when this animal is moving down steep snow slopes, another trail char-

Loping trail of a wolverine with hinds falling directly on top of fronts. Canadian Rockies, Montana.

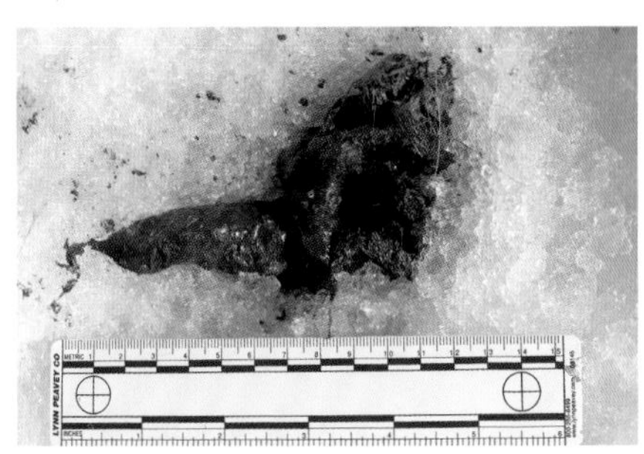

Wolverine scat. Canadian Rockies, Montana.

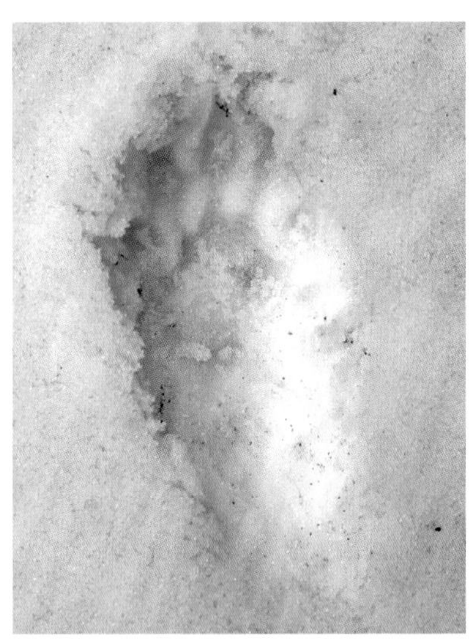

Right hind foot of a wolverine in snow. Canadian Rockies, Montana.

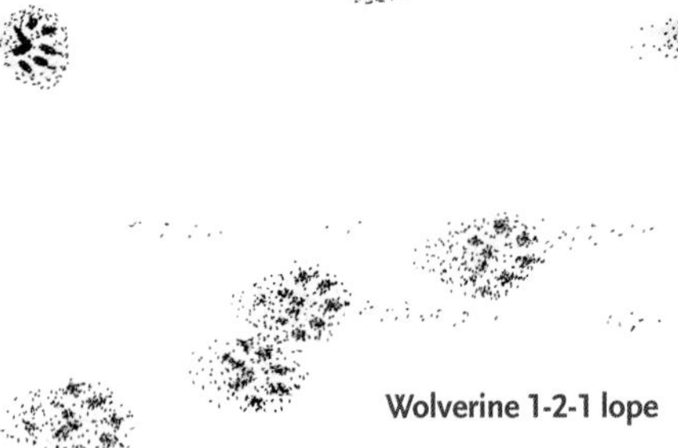

acteristic not shared with lynx. **2x BOUND & LOPE** Stride: 26¾–55 in. (68.0–139.7 cm). Trail width: 7–9 15⁄16 in. (17.8–24.5 cm). Group length: 9 11⁄16–26 in. (24.5–66.0 cm). **1-2-1 LOPE** Stride: 35–57⅞ in. (89–147 cm). Trail width: 7–10 in. (17.8–25.4 cm). Group length: 19 11⁄16–38½ in. (50.0–97.8 cm).

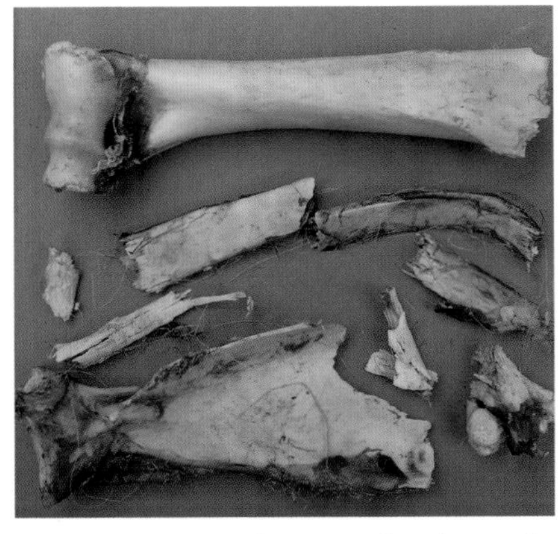

Mountain goat bone fragments found around a carcass consumed by one or more wolverines. Canadian Rockies, Montana.

SIGNS

SCAT 9⁄16–¾ in. (1.4–1.9 cm) D × 4¾ in. (12 cm) L. Pungent and varied depending on diet. Twisted scats filled with hair and bone fragments. When eating a richer diet, produces darker, more amorphous scats. **FEEDING SIGNS** In winter, excavates buried carcasses of mountain goats and other animals in avalanche debris, and makes large digs in the snow elsewhere to access potential food items likely detected by scent. Even the largest bones can be consumed, and bone fragments are common at feeding locations, similar to wolves. In summer, rolls rocks and tears up logs, similar to bears and skunks, presumably searching for insects. **DENS** Natal dens are located in snow, often in areas that accumulate disproportionately large snow loads and retain snowpack longer, such as leeward high-elevation locations or deep, cold-shaded valley floors. Natal dens are usually clean, containing no remnants from carcasses or dirt in the snow. Entrances in snow may be as small as 10 in. (25 cm) in diameter.

This fallen tree was surrounded by multiple wolverine trails, and a well-worn trail led under the log, indicating that this structure had been used repeatedly for a refuge or den. The photo was taken in mid-May; the winter snowpack was still more than 6 ft. (2 m) deep at this location. Canadian Rockies, Montana.

AMERICAN BADGER

Taxidea taxus

Badgers are efficient and prolific diggers, aided by long claws and shovel-like pigeon-toed hind feet. They are solitary, though more than one may be active in the same area. They may become less active for short periods during winter cold spells. Often rest in excavations they have dug while foraging.

SIZE Length: 52–87 cm. Weight: 3.6–11.4 kg. Males are larger. **FIELD MARKS** Squat, flattened appearance; white stripe from the nose to the shoulders; face black with white splotches. **REPRODUCTION** Breeds in June or July. Young are born in March or April, averaging three kits per litter. Natal burrow similar to other excavations but with significantly more tracks, scats, and other signs of habitation. **DIET** Burrowing mammals, such as ground squirrels and pocket gophers. Also voles, tree squirrels, and muskrat, and occasionally plant material and scavenged carrion. **HABITAT & DISTRIBUTION** Arid grassland and mountain environments. Most common in areas with abundant prey and loose soils. **SIMILAR SPECIES** Bobcat, coyote.

TRACKS

The badger's distinctive foot morphology reflects its digging lifestyle, though it still retains some features similar to those of other mustelids. Five toes on front and hind feet. Toe 1 is the smallest and lowest. The palm pad is relatively larger than that of other mustelids, and when toe 1 doesn't register clearly, tracks may be confused with those of a feline, though the pad's anterior edge is not bilobed as in felines. Claws are not always apparent in tracks (page 65). **FRONT** 1 15/16–2 9/16 in. (5.0–6.5 cm) L × 1 9/16–2 3/8 in. (3.9–6.1 cm) W. Larger than hind. Long claws (average 1¼ in., or 3.2 cm) often register; the center three register so far in front of the toes that you might miss them. Toes are oblong and taper at the leading edge; may curve slightly inward. Toe 3 is farthest forward (similar to a feline). Single proximal pad sometimes registers. **HIND**

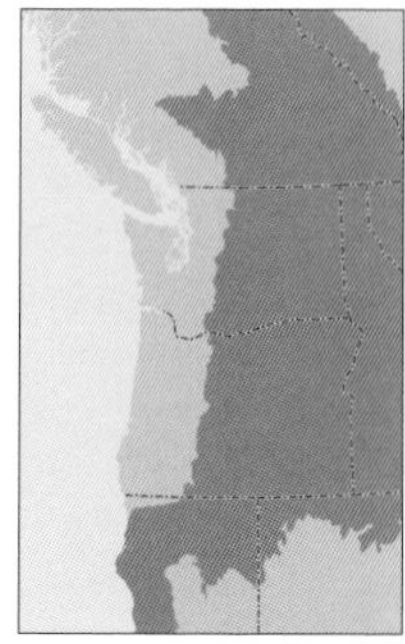

Badger range

1 11/16–2 5/8 in. (4.3–6.7 cm) L × 1 1/4–2 3/16 in. (3.2–5.5 cm) W. Similar to front but narrower. Toes may curve inward. Claws often register but are not as long as those on the front. Toe 1 often doesn't register or registers lightly. **TRACK PATTERNS** Direct register walk and trot. Tracks are pigeon-toed, angled in toward the trail's midline. **WALK** Stride: 11 1/4–21 1/4 in. (28.5–54.0 cm). Trail width: 3 1/8–7 5/16 in. (8.0–18.5 cm). **DIRECT REGISTER TROT** Stride: 20 1/2–24 3/16 in. (52.0–61.5 cm). Trail width: 2 3/4–4 15/16 in. (7.0–12.5 cm).

SIGNS

SCAT Twisty and filled with the fine hairs of small mammals. Found near excavations. Easily confused with coyote scat and is often

Right front (above) and hind tracks of a badger. Look closely to see the long claw marks on the front foot. The smaller tracks are from Columbian ground squirrels, an important prey species for badgers. Middle Rockies, Idaho.

A badger's left hind track directly registers atop its front track. Northern Great Basin, Oregon.

Badger walk

impossible to distinguish without accessory clues. **EXCAVATIONS** Large with obvious throw mounds at the mouth. Entrance holes to fresh digs measure 7½–10¼ in. (19–26 cm). Holes are often wider than high, mimicking the animal's squat appearance. Often found in the middle of a ground squirrel colony or in areas with lots of pocket gopher signs. Digs intended for foraging may not be very deep and are not necessarily revisited.

Badger scats. Columbia Plateau, Washington.

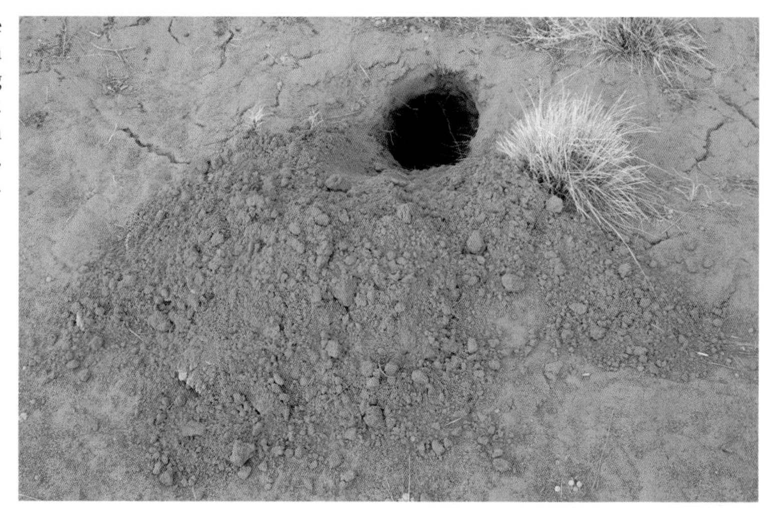

Fresh hole dug by a foraging badger. Northern Great Basin, Oregon.

SKUNKS

Family Mephitidae

Skunks are best known for their distinctive coloration and well-developed, muscle-encased, anal gland, from which they can propel their musk to about 20 ft. (7 m). The foul-smelling substance can cause temporarily blindness if it contacts the victim's eyes. Both species in our region are adept diggers and consume large quantities of insects. They are primarily nocturnal, solitary in their habits, and active year-round, though considerably less so during the winter.

Tracks and Signs

Plantigrade. Overall track shape is oval. Five toes on both front and hind feet do not splay; they are fused at the first joint, just below the toe pad. Claws on the front feet are longer than those on the hinds. Soles are hairless.

Track patterns. Both species use a lope that leaves a 1-2-1 track pattern. Also walk in direct register and overstep patterns.

Digs. Skunks dig out ground-nesting insects and tear up rotting logs looking for insects.

STRIPED SKUNK

Mephitis mephitis

The striped skunk is the better known and larger of our two native species, often seen in suburban and rural locations. It may harbor diseases, including rabies and distemper, which, while posing a health hazard to humans and other animals, also periodically decimate the population. (Rabies is occasionally documented in skunks in Northern California.) Spread of diseases is likely aided by the use of communal dens shared by several individuals during the winter.

SIZE Length: 57.5–80.0 cm. Weight: 1.2–6.3 kg. Males are larger. **FIELD MARKS** Distinctive black fur with prominent white stripe from head to tip of large, bushy tail. **REPRODUCTION** Polygamous. Breeding occurs in February and March; parturition in late May or June. Litters of 1–10 kits; more than half die within their first year. Natal dens are underground burrows excavated by the skunk or renovated existing burrows. Also uses human structures for dens. Females and young travel together until fall when young disperse. **DIET** Primarily insectivorous, but also a highly opportunistic scavenger when insects are not abundant. Also deer mice, voles, bird eggs, fruit, and carrion. **HABITAT & DISTRIBUTION** Diverse habitats, including brushy cover, meadows and fields, and forests. Highest abundance in our region is likely in agricultural areas and close to water in arid landscapes. Absent from higher elevations. Home range for males averages 1.9 square miles (5 km^2). Average for females is 1.5 square miles (3.8 km^2). **SIMILAR SPECIES** Spotted skunk.

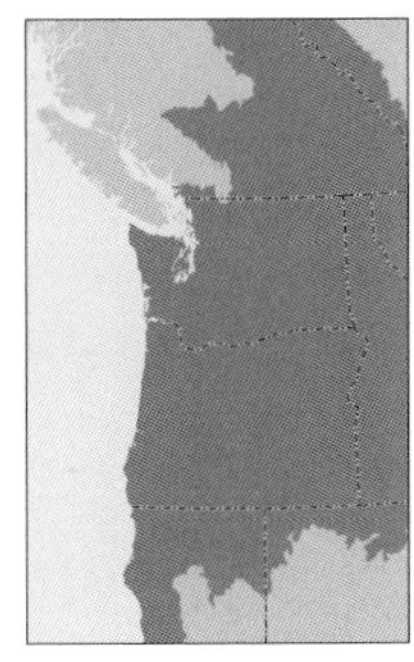

Striped skunk range

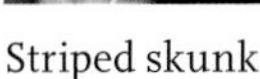

Striped skunk

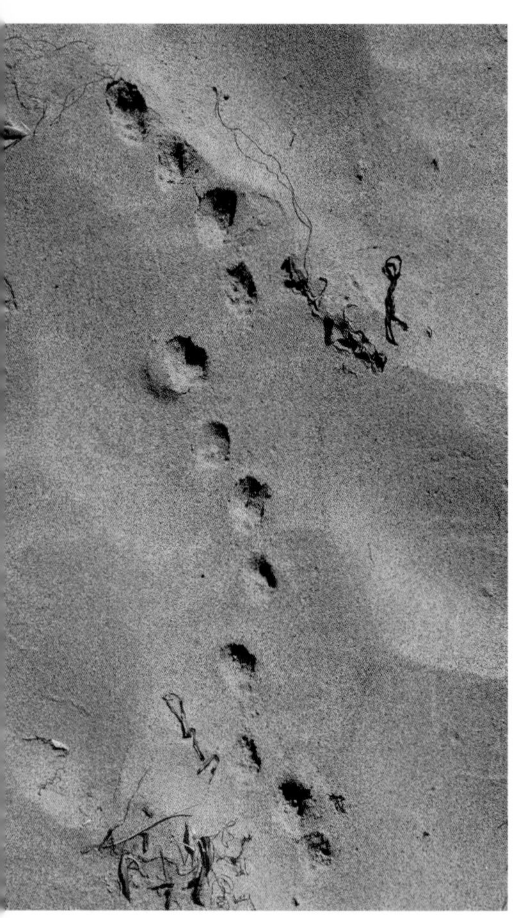

Typical track pattern of a slowly loping striped skunk. From bottom: right front, right hind, left front, left hind. North Coast California.

Right front (at right) and hind tracks. Willamette Valley, Oregon.

Left front (below) and hind tracks of a striped skunk. Willamette Valley, Oregon.

Striped skunk 1-2-1 lope

Striped skunk overstep walk

Striped skunk scats containing bees or wasps. North Coast California.

TRACKS

Striped skunks have relatively small feet for their body size (page 62). **FRONT** 7⁄8–1½ in. (2.2–3.8 cm) L × ¾–1 3⁄16 in. (1.9–3.0 cm) W. Long, curved claws (average ½ in., or 1.2 cm) are usually prominent in the track, registering well ahead of the toe pads, and are noticeably longer than hind claws in the track. The axis of each toe is parallel with the overall track. Palm pad is smooth and undivided. Heel sometimes registers and shows two distinct pads. **HIND** 15⁄16–1 13⁄16 in. (2.4–4.6 cm) L × 13⁄16–1¼ in. (2.0–3.1 cm) W. Flat palm and attached heel pad are prominent track features. Palm and heel make up about two-thirds of the track. Only one heel pad (unlike in the spotted skunk), which is clearly separated from the palm pad. **TRACK PATTERNS** Distinctive, though often confusing, track patterns. Walks with a wandering and inconsistent overstep pattern and commonly uses a lope at faster speeds. **WALK** Stride: 9¼–16 15⁄16 in. (23.5–43.0 cm). Trail width: 2 5⁄16–4 1⁄8 in. (5.8–10.5 cm). **1-2-1 LOPE** Stride: 11–21¼ in. (28–54 cm). Trail width: 1 15⁄16–4¾ in. (5–12 cm). Group length: 8 7⁄8–14¾ in. (22.5–37.5 cm).

SIGNS

SCAT 3⁄8–¾ in. (1.0–1.9 cm) D. Tubular and often fragile. Found at feeding and den locations. Contents often include insects. **FEEDING SIGNS** Digs in sod for insects, rips open rotting logs for larvae (similar to bears but on a smaller scale), and excavates ground nests of bees and wasps.

WESTERN SPOTTED SKUNK

Spilogale gracilis

The western spotted skunk is smaller and quicker than the striped skunk, an able climber, and an adept hunter. In areas where their range overlaps with that of mountain beavers, spotted skunks often use mountain beavers' burrows.

SIZE Length: 32–58 cm. Weight: 0.2–0.9 kg. Males are larger. **FIELD MARKS** Black with white spots that vary, with a prominent white mark between the eyes. Short, unstriped tail. **REPRODUCTION** Breeding occurs in summer, followed by a prolonged period of delayed implantation. Young are born the following April or May. Litters of 2–6 young travel with the mother at about 8 weeks and disperse in the fall. **DIET** More vertebrate prey than striped skunk, plus large amount of insects. **HABITAT & DISTRIBUTION** Mature forests, coastal areas, and brushy and rocky areas. **SIMILAR SPECIES** Striped skunk, long-tailed weasel.

TRACKS

Tracks are smaller than those of striped skunks and similar in size to long-tailed weasel tracks. Front claws are longer than hind claws, but the difference is not as great as in the striped skunk (page 61). **FRONT** 7⁄8–1 3⁄8 in. (2.3–3.5 cm) L × 5⁄8–1 1⁄8 in. (1.6–

Spotted skunk. Photo by Anthony Mercieca.

Spotted skunk range

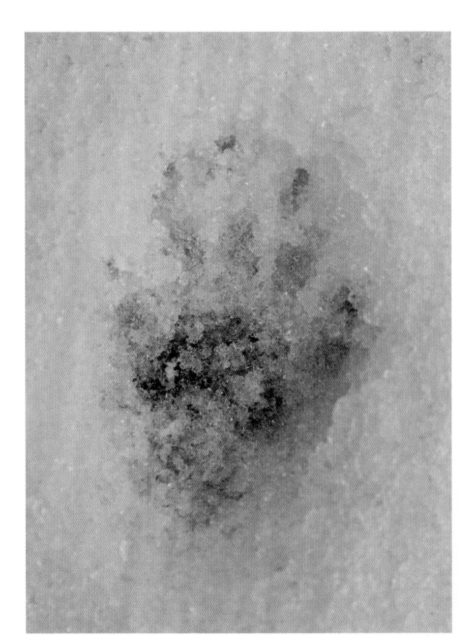

Right front track of a spotted skunk. Northwest Coast, Washington.

Left hind track of a spotted skunk. The divided heel and palm pad are made of distinct subpads, two key identifying features of this species. Northwest Coast, Washington.

2.8 cm) W. Unlike the striped skunk, palm comprises four distinct subpads that are distinguishable in clear tracks. The divided heel may show in tracks, with two pads. **HIND** 13⁄16–1¼ in. (2.1–3.1 cm) L × 13⁄16–1 in. (2.0–2.6 cm) W. Palm partially divided into four distinct subpads, and heel divided into two pads; both features distinguish the track from that of a striped skunk. **TRACK PATTERNS** Uses a lope for most travel and exploration, leaving a 1-2-1 pattern similar to that of a striped skunk. Also uses direct register and overstep walk. In deep snow, uses a bound with a short stride. **DIRECT REGISTER WALK** Stride: 8 11⁄16–10¾ in. (22.0–27.3 cm). Trail width: 2–2¾ in. (5–7 cm). **LOPE & BOUND** Stride: 11⅜–15 9⁄16 in. (29.0–39.5 cm). Trail width: 1⅞–2 15⁄16 in. (4.8–7.5 cm). Group length: 4 5⁄16–7 9⁄16 in. (11.0–19.2 cm).

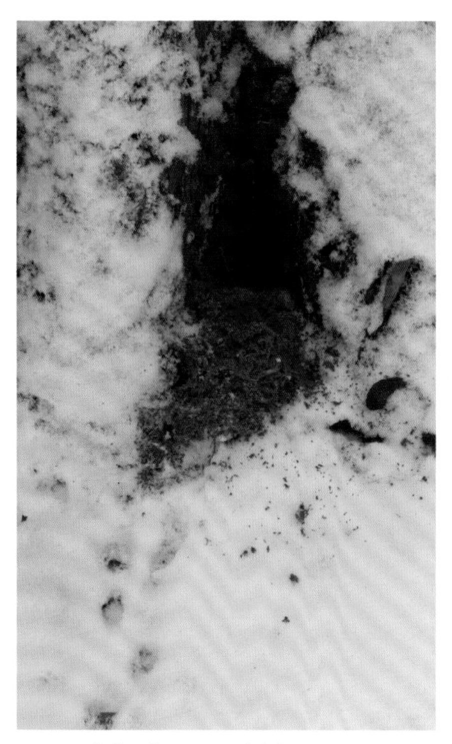

A spotted skunk excavated debris at the base of a large conifer tree. Northwest Coast, Washington.

SIGNS

SCAT Literature reports about ¼ in. (0.6 cm) D. Tubular or irregular in shape, similar to scats of striped skunk. **FEEDING SIGNS** Similar to striped skunks, creates small digs while foraging for insects and other subterranean treats.

At left is the trail of a loping spotted skunk. The larger tracks at right are from a raccoon. Northwest Coast, Washington.

HOOFED MAMMALS Order Artiodactyla

All members of this order are herbivores with a four-chambered stomach; all ruminate, regurgitate, rechew, and reingest food to enable complete digestion of large quantities of plant material. All have a keen sense of smell and hearing. Their field of view is wide but their vision is limited to detection of movement and gross shapes. Because their eyes are on the sides of their head, they have limited binocular vision directly in front. Three families of hoofed mammals are native to the Northwest.

Tracks and Signs

The hoofed, unguligrade feet of these animals leave a distinctive simplified track that, in clear prints, is rarely confused with any other group. In all species, toes 3 and 4 are enlarged and are the primary weight-bearing portion of the foot. Front tracks are larger than hinds. Distinguishing between various species can be difficult. Pay attention to the cleaves' overall shape, amount of splay between cleaves, the subunguis size and shape within the track (when visible), track patterns, and habitat to help you distinguish among similar species.

Dewclaws. All but one species, the pronghorn, have two additional toes on each foot called dewclaws. Dewclaws register only when the animal is running or moving in deep substrate (except woodland caribou, whose dewclaws regularly register on the front foot). On the front feet, dewclaws are larger and closer to toes 3 and 4 than on the hind feet, and they register oriented perpendicular to the direction of travel. On the hinds, dewclaws are thinner and farther from the main toes. Hind dewclaws usually register oriented parallel to the direction of travel.

Track patterns. Most common gait is a walk. All species have long legs and correspondingly long strides. At faster speeds, different species prefer to move in different gaits, leaving distinctive track patterns that can aid in identification.

Scat. Ungulates produce large amounts of scat because they consume large quantities of indigestible plant material. Scats are distinctive in some species and overlap with one another in others. In all species, a dry diet produces distinct oblong pellets, while a moist spring diet can produce larger amorphous patties. Tubes and clumps of partially formed pellets are also common.

Browse. All members of this order lack upper incisors and leave a distinct pattern on plants upon which they feed. Cut vegetation (grasses, forbs, and woody browse) often shows a ragged appearance, as it was more or less ripped rather then neatly sliced

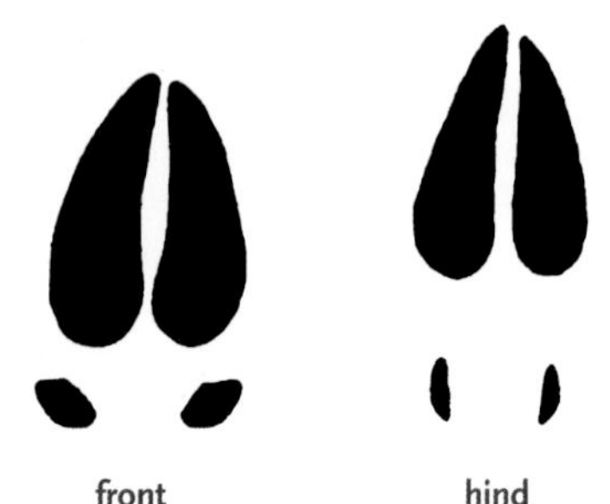

(as is the case for rodents and lagomorphs with their sharp incisors). Heavy browsing by ungulates can leave shrubs and saplings with a distinctive stunted form and can create a browse line on trees when all the leaves and buds within reach have been consumed. Several ungulates use their lower incisors to scrape bark off trees.

Determining Sex Through Tracks and Signs

Numerous theories exist concerning how to determine the sex of ungulates using their tracks—far too many to list. Many have limited or no practical merits, but several reliable clues can be helpful. More specific information can be found in relevant species accounts. Except for the last section on marking behavior, the information presented should be used as guidelines and not rules.

Tracks

For all species, mature male tracks are distinctly larger than those of any other group, and tracks at the upper end of size ranges are likely from males. As with many other species, subadult tracks of both sexes are closer in size to those of mature females. Mature male elk tracks show a greater difference in size between front and hind feet than mature females and subadults, particularly in the width (Halfpenny 1999). This is consistent with data I have collected of known-sex elk prints and similar to reports on African hoofed mammals (Liebenberg 1990b). Hypothetically, this size difference occurs because of the greater weight and more massive structure carried over the front legs—the large neck, chest, head, and antlers or horns of males. Anecdotal observations (mine and those of others) indicate that these differences are present in other Northwest cervids and bighorn sheep, though in some species, differences in track length may be more pronounced than differences in width (Mark Elbroch; Brian McConnell; personal communication). Measure the tracks' minimum outline when comparing relative track sizes, as substrate can exaggerate or obscure subtle differences.

Social Structure

You can also identify male and female tracks by understanding the species's social structure. Many hoofed mammals travel in groups that include females, the year's young, and subadults, whose tracks are distinctly smaller. During much of the year, mature males do not regularly travel with females and young. Tracks of a group of animals that include full-sized animals and subadults are likely made by females and young of either sex. Conversely, at certain times of the year, males travel in groups called "bachelor herds." Tracks of a group of full-sized animals that do not include subadult tracks are likely to be from males. Finally, a set of large tracks from an animal apparently traveling alone are more likely from a male. (Moose are an exception to this, as both sexes are solitary, except a mother and calf.)

Scent-Marking and Breeding Behaviors

Males of many of these species engage in scent-marking and breeding behaviors that leave distinctive signs, including marks on trees, and wallows or scrapes on the ground.

Hoofed Mammal Families

CERVIDS

Family Cervidae

The cervids include deer, elk, moose, and caribou. All cervids have antlers, bony structures that grow anew each year for the breeding season and are later shed. Typically, only males grow antlers, except in caribou, where both sexes have them, though female antlers are significantly smaller. The functions of antlers are related to breeding behaviors such as scent-marking and sparring with other males. Shed antlers are occasionally found in the field.

MULE DEER
BLACK-TAILED DEER

Mule deer, *Odocoileus hemionus hemionus*
Black-tailed deer, *O. hemionus columbianus*

Mule deer and black-tailed deer are each a subspecies of *Odocoileus hemionus* and together are the most common hoofed mammal in our region. In the Cascades, animals that appear to be hybrids of the two subspecies are found. Social groupings change seasonally within age and sex classes. Several females and their young often band together throughout the year. In spring and summer, several males may travel together in a bachelor herd. The largest groups of animals are usually found in winter, where animals congregate at preferred feeding locations.

SIZE Length: 1.3–1.7 m. Male weight: 40–120 kg. Female weight: 30–80 kg. **FIELD MARKS** Large, pointed ears. The black-tailed deer's tail is black. The mule deer's tail is white, tipped in black. Antlers branch evenly. **REPRODUCTION** Breeding occurs in late fall. Males produce pheromones that are released through scent-marking behaviors that induce estrus in females. Breeding males mate with numerous females. Spotted fawns (1–3) are born in late spring. Juveniles travel with their mother for at least a year, with males dispersing upon reaching sexual maturity. Yearling does may continue to travel with their mothers indefinitely. **DIET** Shifts seasonally. Variety of grasses, forbs, and woody plants. In arid locations, bitterbrush is a key winter browse species. **HABITAT & DISTRIBUTION** Black-tailed deer—Cascades west to the coast. Mule deer—From eastern slopes of the Cascades east, migrating out of the high mountains to lower elevations for the winter. In the northeastern part of our region, mule deer may be declining because of increasing numbers of white-tailed deer. **SIMILAR SPECIES** White-tailed deer, bighorn sheep, mountain goat.

TRACKS

Front and hind tracks are heart shaped, with the heart's tip pointing in the direction of travel. Toes spread and dewclaws show in tracks in deep substrate or when the animal is running. The interior hoof wall is concave and the outer wall is convex. The pad takes up a large portion of the track's interior. The subunguis is a small strip inside the hoof wall in the anterior of the print. Tracks of black-tailed deer average slightly smaller than those of mule deer. In areas where these deer overlap with white-tailed deer, I have found it impossible to differentiate the species through tracks alone (page 74). **FRONT** 2 1⁄16–3 1⁄8 in. (5.2–8.0 cm) L × 1 9⁄16–2 5⁄16 in. (3.9–5.8 cm) W. Cleaves tend to splay more often than in the hind track. **HIND** 1 13⁄16–2 7⁄8 in. (4.6–7.3 cm) L × 1 5⁄16–2 5⁄16 in. (3.4–5.9 cm) W. Smaller than front.

Female mule deer

TRACK PATTERNS Deer have a long stride for their foot and body sizes. Trots when moving fast but not seriously alarmed. When fleeing, uses a pronk, an energy-intensive and powerful gait that involves pushing off and landing simultaneously with all four feet. White-tailed deer rarely use this gait. **DIRECT REGISTER WALK** Stride: 30 1/8–48 in. (76.5–121.9 cm). Trail width: 3 5/16–10 7/16 in. (8.5–26.5 cm). **DIRECT REGISTER TROT** Stride: 44 1/8–65 in. (112–165 cm). Trail width: 3 1/8–5 1/8 in. (8–13 cm). **OVERSTEP TROT** Stride: 85¼–114 3/16 in. (216.5–290.0 cm). Trail width: 7½–15 3/8 in. (19–39 cm). **PRONK** Stride: 73 5/8–255 7/8 in. (187–650 cm). Trail width: 7½–18 in. (19.0–45.7 cm). Group length: 24–35 in. (61.0–88.9 cm).

SIGNS

SCAT ¼–7/16 in. (0.6–1.1 cm) D × ¼–¾ in. (0.7–1.9 cm) L. Indistinguishable from scats of white-tailed deer. Usually a pellet about

Male black-tailed deer

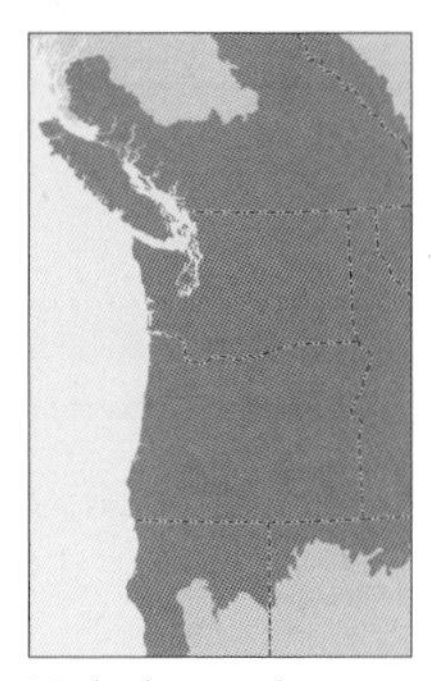

Mule deer and black-tailed deer range

Front (above) and hind tracks of a mule deer. East Cascades, Washington.

This willow shrub has been browsed repeatedly by deer, giving it a stunted appearance. Puget Trough, Washington.

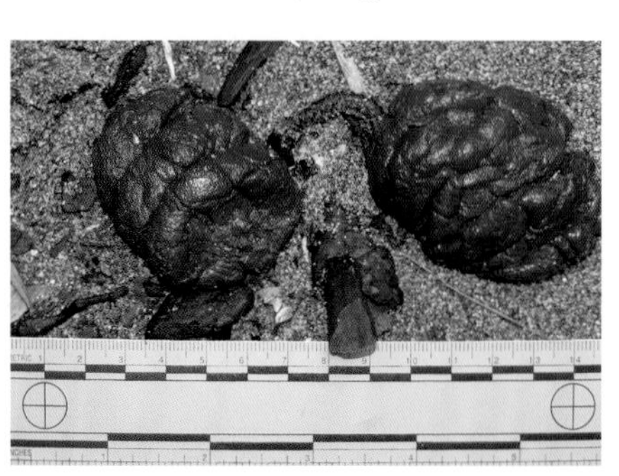
Black-tailed deer summer scat. Puget Trough, Washington.

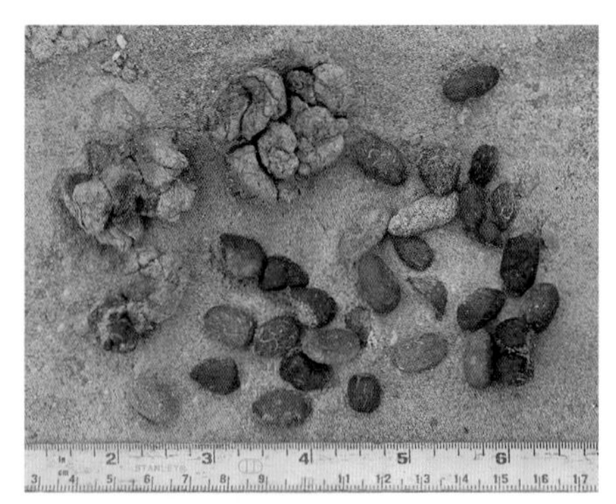
Several variations of mule deer scat. Columbia Plateau, Washington.

the size and shape of a jelly bean. When diet contains a lot of moist, green vegetation, scat may be more amorphous in shape. **ANTLER RUBS** Males make conspicuous scars on saplings prior to the breeding season. Bucks use their antlers to shred tree bark and then rub scent glands located on their heads and necks on the trees to advertise their presence and readiness to breed. Antler rubs can be found on shrubs, thin saplings, and sometimes large-diameter trees. For mule deer, the act of rubbing and the noise it makes serves as an auditory advertisement to challenge other males. During its antler rubbing, it stops to listen for a response from another buck, which it may seek out to challenge. This appears to be less important for black-tailed and white-tailed deer. Rubs are

This ponderosa pine sapling died after a mule deer buck rubbed its antlers on its bark and broke off branches. Columbia Plateau, Oregon.

6–48 in. (15–120 cm) in height, with most about 18–36 in. (45–91.4 cm). **BROWSE ON WOODY VEGETATION** In areas used heavily by deer, especially winter range, preferred browse species take on a distinct stunted form caused by years of repetitive feeding on the terminal buds.

Antler rub of a black-tailed deer on a black cottonwood. The bark's shredded appearance at the top of the rub and the linear scratch marks are from the antler tines. Most antler rubs occur on smaller trees. The work of a beaver appears on the right trunk's lower portion. A deer removed the bark on the upper portion. Puget Trough, Washington.

WHITE-TAILED DEER

Odocoileus virginianus

Odocoileus virginianus is the most abundant hoofed mammal in North America. Its range is limited but expanding in our region east of the Cascades. West of the Cascades, a subspecies, the Columbia white-tailed deer (*O. virginianus leucurus*), has a limited distribution along the Columbia River and in a small part of southwestern Oregon.

SIZE Length: 0.8–2.4 m. Weight 22–137 kg. Males are larger. **FIELD MARKS** Large tail fringed in white and completely white on the underside, displayed when the animal runs and flags its tail. On mature males, the antler tines sprout from a single main beam. **REPRODUCTION** Strategy, behavior, and timing are similar to those of mule deer. **DIET** Grasses and shrubs, plus fruit from abandoned orchards where available. **HABITAT & DISTRIBUTION** Range is expanding. In areas where it overlaps with *Odocoileus hemionus, O. virginianus* tends to occupy river bottoms and agricultural fields and edges. **SIMILAR SPECIES** Mule and black-tailed deer, bighorn sheep, mountain goat. **CONSERVATION ISSUES** West of the Cascades, the Columbia white-tailed deer is endangered. Conversely, east of the Cascades, its range expansion may be detrimentally impacting mule deer and endangered woodland caribou populations.

TRACKS

In our region, track shape and size overlap completely with those of mule deer and black-tailed deer. My data does not include Columbia white-tailed deer. **FRONT** 2 11/16–3 1/16 in. (6.8–7.8 cm) L × 1 3/4–2 3/8 in. (4.4–6.0 cm) W. **HIND** 2 1/16–3 in. (5.3–7.6 cm) L × 1 5/8–2 5/16 in. (4.2–5.8 cm) W. **TRACK PATTERNS** When fleeing, usually gallops as

White-tailed deer

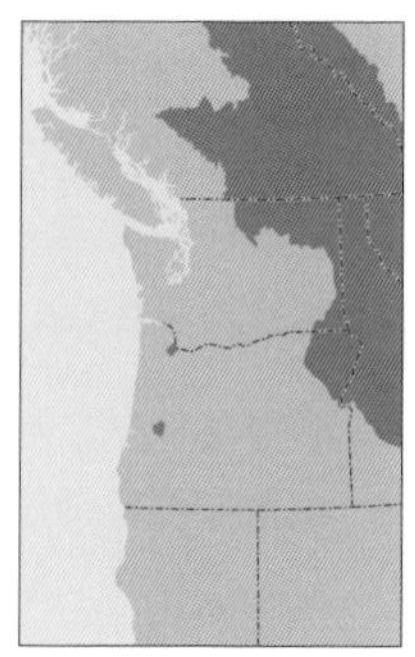

White-tailed deer range

opposed to pronking (as commonly used by mule deer and black-tailed deer). Measurement ranges for walks and trots fall entirely within measurement ranges for black-tailed deer and mule deer. **DIRECT REGISTER WALK** Stride: 39 3⁄8–46 11⁄16 in. (100.1–118.5 cm). Trail width: 4 5⁄16–7 1⁄16 in. (10.9–18.0 cm).

SIGNS

Scats, feeding signs, and antler rubs are indistinguishable from those of mule deer and black-tailed deer. **SCRAPES** During the breeding season, males create scrapes on the ground with their front hooves and then urinate in them. This scent-marking behavior is common in white-tailed deer but not observed in mule deer and black-tailed deer. Scrapes are 12–36 in. (30.5–91.4 cm) in diameter and reach bare soil; they are usually made under a low branch of a tree that the buck licks, chews, or rubs to deposit scent.

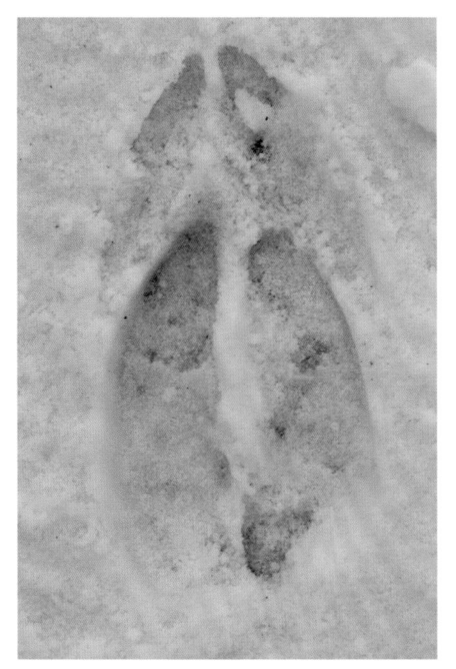

Hind over front track of a female white-tailed deer. Okanagan, Washington.

ELK

Rocky Mountain elk, *Cervus elaphus nelsoni*
Roosevelt elk, *C. elaphus roosevelti*

Elk are large herd animals, and their presence in a landscape is often apparent. In areas where elk have recently visited, their scent is often noticeable. Current populations in our region include both native animals and transplants from other parts of North America. It is believed that these large animals evolved for life in open country but have adapted to forested and mountainous environments.

SIZE Length: 2.1–2.6 m. Male weight: 178–497 kg. Female weight: 171–292 kg. **FIELD MARKS** Tawny brown with dark mane and off-white rump patch. Roosevelt elk are slightly larger and darker and have more "webbing" between their antler tines. **REPRODUCTION** A complex social structure regulates breeding activity. Mature females and juveniles form herds with a lead cow. In the fall, large bulls battle for possession of such groups and breed with all the females in the group while attempting to prohibit other bulls from breeding. The dominant bull commonly changes several times throughout the course of the breeding season, which lasts from early September to mid-November (peaking in early October). After the breeding season, bulls often become solitary for the winter, sometimes forming bachelor herds in the spring and summer. Females give birth to one calf (sometimes two) in late spring or early summer. Female offspring often stay with their mother's herd indefinitely. Male offspring disperse at about 1½ years of age, when they are driven off by their mothers as rut begins in the fall. **DIET** Herbivorous. During spring and summer, primarily grazes on grasses and forbs rather than browsing on woody plants. Browsing increases during the fall and winter. **HABITAT & DISTRIBUTION** Forest and mountain environments throughout our region. Roosevelt elk—Coastal forests and mountains. Rocky Mountain elk—Cascades and east. Elk use meadows and grasslands with access to cover for protection from predators and cold weather. Beds can be found in many settings, from open fields to dense forests on steep slopes. They migrate from upland to lowland sites for the winter, and then return to higher elevations in the spring. **SIMILAR SPECIES** Moose, caribou.

TRACKS

Elk tracks are significantly larger than deer tracks, with the exception of tracks of young calves. Overall shape is a squat oval. Tips of toes are rounded. Front half to two-thirds of the track has a domed appearance because of the large, concave subunguis that makes up most of the cleave. A relatively small pad appears in the posterior portion of each cleave (page 75). **FRONT** 3–4⅝ in. (7.6–11.7 cm) L × 2¾–3 15/16 in. (7–10 cm) W. Larger than hind. **HIND** 3–4½ in. (7.6–11.5 cm) L × 2⅝–3½ in. (6.6–8.9 cm) W. **TRACK PATTERNS** Commonly walks (direct or indirect register). Trotting is the most common faster gait (direct register or overstep). **WALK** Stride: 42⅛–71½ in. (107.0–181.8 cm). Trail width: 5 5/16–13 9/16 in. (13.5–34.5 cm). **TROT** Stride: 67½–123¼ in. (171.5–313.0 cm). Trail width: 7½–14 3/16 in. (19–36 cm).

SIGNS

SCAT ⅜–⅞ in. (1.0–2.3 cm) D × ⅝–1¼ in. (1.6–3.1 cm) L. Larger than deer pellets, smaller than moose. Varies in size and shape seasonally. A diet of moist vegetation

Rocky Mountain elk

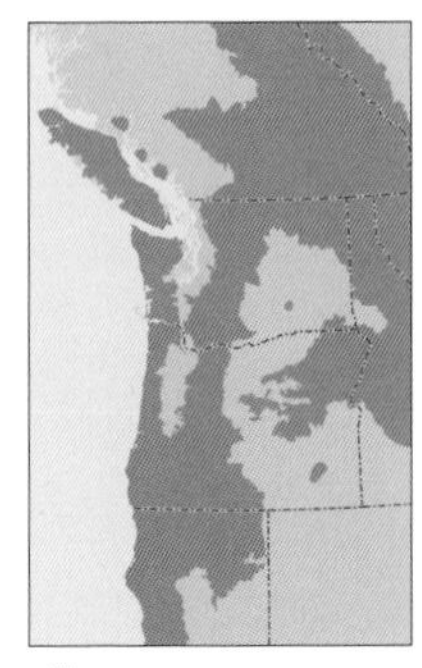

Elk range

(typical in spring) results in more amorphous-shaped patties. Typically large, asymmetric pellets, often with a small nipple on one end and either rounded or dimpled on the other end. Drier forage produces oblong pellets convex on both ends. **TRAILS** In heavily used areas, herds create well-worn trails that are also used by other wildlife species. **ANTLER RUBS** Similar to other cervids, bull elk rub their antlers on trees, disturbing the bark and breaking branches, and then rub their bodies on the tree to impart their scent. Elk generally use larger trees than used by deer and disturbance is higher on the trunk. Saplings and shrubs may also be severely damaged. Rub marks range in height, 16–72 in. (40–183 cm), with most activity occurring at 24–60 in. (61–152 cm). Long, dark hairs from the bull's shoulder are often found sticking to the tree along with dirt or mud from a wallow in which the bull has coated itself. The bark is usually shredded at the scar's top and bottom. **WALLOWS**

Left front (top) and hind tracks of a Rocky Mountain elk. West Cascades, Washington.

Winter scats of a Roosevelt elk. Northwest Coast, Washington.

Summer scats of a Roosevelt elk. Northwest Coast, Washington.

Elk walking

Elk antler rubs on red alder trees. West Cascades, Washington.

Old incisor marks on the trunk of a quaking aspen. The short paired scrapes all at the same angle help differentiate these from antler rub scars. East Cascades, Washington.

These red osier dogwood (*Cornus stolonifera*) branches have been stunted by repeated browsing by elk. East Cascades, Washington.

An elk wallow pit in a mountainside spring

Males create large, muddy or dusty depressions by scraping the ground and rolling in the area after urinating in the location and on themselves. This behavior begins about a month prior to the rut and is thought to increase their scent and attractiveness to females. **INCISOR MARKS** Male and female elk scrape tree bark with their lower incisors (they have no upper incisors) for feeding on the nutritious cambium layer and for marking territory. A dominant female may scrape bark off a tree to mark the presence of her herd in the vicinity. Bulls also use their incisors to mark trees in conjunction with antler rubs. **BROWSE** In heavily used winter range, browse species take on a stunted form from repetitive feeding. **BEDS** For defense, bedding areas are often sited where approaching predators can be detected through scent, sound, or sight—perhaps in the middle of a large meadow or high up on a steep, forested slope with a difficult approach from above, where the elk can detect a predator that might be following its trail. In winter, elk often seek lower elevations and south-facing hillsides with less snowpack and strong sun exposure. Summer beds are often in dense timber on north-facing slopes, a relatively cool location. Bull elk scratch the ground with their antlers while lying on their sides in a bed, leaving a distinctive sign.

MOOSE

Alces alces

Moose are the largest members of the Cervidae family in our region. However, the subspecies *Alces alces shirasi* in our region is smaller than moose in more northern areas. These mainly solitary creatures are well adapted for life in cold environments. Their large size and insulating winter pelage help them conserve heat. Their long legs allow them travel in deep snow in winter and in marshes and aquatic environments in summer, where they feed and stay cool. They are strong swimmers. Moose are unable to survive without shade or access to water for immersion in temperatures greater than 81°F (27°C).

SIZE Length: 2.4–3.1 m. Weight: 275–500 kg. Males are larger. **FIELD MARKS** Gray-brown pelage. Long, rounded nose and long legs. Male has a flap of skin, or dewlap, hanging from its chin. Antlers are large and palmate (tines flattened and greatly connected). **REPRODUCTION** Breeding occurs in September and October. Bulls create scent posts by rubbing trees and creating large wallows, which attract cows and induce estrus. Calves (usually one, occasionally two) are born in the late spring or early summer. Calves stay with the mother through their first year and sometimes longer. **DIET** Browses aquatic vegetation when available in summer and stems, buds, and bark of woody vegetation, including willows and quaking aspens, in winter. Their long legs prevent them from reaching the ground to feed without stooping on their front legs. **HABITAT & DISTRIBUTION** Summer range includes riparian and aquatic habitats. Winter range is selected for access to forage and ease of movement in snow and is often more upland than in summer. Also uses recent clearcuts and recovering burns, as this habitat often provides a substantial amount of forage. Range expanding in our region, down the Cascades and into the Blue Mountains of northeastern Oregon. **SIMILAR SPECIES** Elk, deer, caribou.

TRACKS

Moose tracks (except the youngest calves) are larger than those of deer but share their

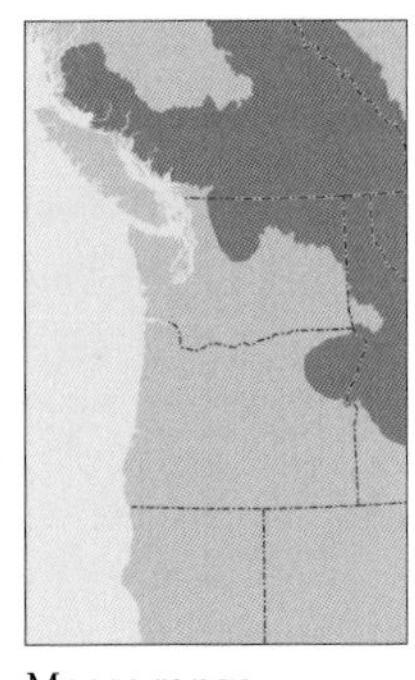

Moose range

heart-shaped appearance. Also larger than most elk tracks. In larger moose tracks (most pronounced in males), the outer hoof wall is flat for about three-quarters of the track's length, and it then angles inward to a point, creating a relatively blunted appearance. In tracks of smaller moose, a more even curve appears on the cleave's outside edge, giving tracks a pointy, more deerlike appearance. A small space appears between the tips of each cleave, even when toes are not splayed. The lower half of the medial edge of each cleave is convex, and the upper section is concave. As in deer, the subunguis is small, with the pad filling most of the cleave's interior. As a result, the track floor is relatively flat, with only a small strip of raised subunguis apparent in clear tracks, another feature that distinguishes them from elk tracks. Toes spread and dewclaws show in deep substrate or when the animal is running. Hind foot is smaller than front foot (page 75). **FRONT** 3½–6½ in. (8.9–16.5 cm) L × 3 3⁄16–4½ in (8.1–11.5 cm) W. **HIND** 3 7⁄16–6 1⁄8 in. (8.8–15.5 cm) L × 2 9⁄16–4 5⁄16 in. (6.5–11.0 cm) W. **TRACK PATTERNS** When

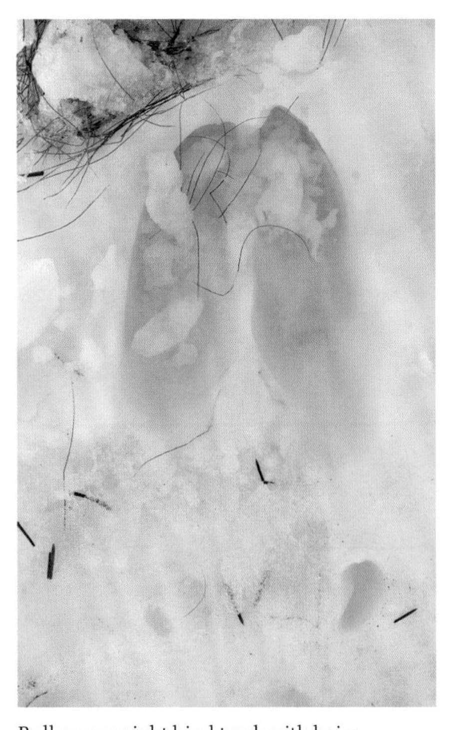

Bull moose right hind track with hairs. Canadian Rockies, Montana.

Right hind track of a smaller moose. Canadian Rockies, British Columbia.

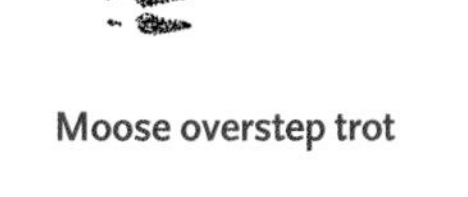

Moose overstep trot

Typical winter moose scats. Okanagan, Washington.

These willow branch tips were browsed by a moose. Canadian Rockies, Washington.

Cambium feeding sign from a moose on an aspen trunk parallel to the ground. If the trunk were upright, the incisor grooves would be parallel with the trunk. Canadian Rockies, Alberta.

running, uses a distinctive trotting gait with the hinds overstepping and falling to the outside of fronts. **DIRECT REGISTER WALK** Stride: $53\frac{15}{16}$–$80\frac{5}{16}$ in. (136.7–204.0 cm). Trail width: $6\frac{5}{16}$–20 in. (16.0–50.8 cm). **DIRECT REGISTER TROT** Stride: $87\frac{3}{8}$–$98\frac{7}{16}$ (222–250 cm). Trail width: $6\frac{1}{2}$–$12\frac{9}{16}$ in. (16.5–32.0 cm). **OVERSTEP (STRADDLE) TROT** Stride: $98\frac{13}{16}$–$141\frac{3}{4}$ in. (251–360 cm). Trail width: $11\frac{13}{16}$–$22\frac{7}{16}$ in. (30–57 cm).

SIGNS

SCAT $\frac{1}{2}$–1 in. (1.3–2.5 cm) D × $\frac{13}{16}$–$1\frac{5}{8}$ in. (2.1–4.2 cm) L. Most pellets are larger than those of other hoofed mammals in our region. Often rounded on both ends and fairly uniform. As with other hoofed mammals, spring scats are often amorphous patties. **BROWSE** Branches up to $\frac{3}{8}$ in. (1.0 cm) D. In summer, browse marks can be 15 ft. (4.6 m) from the ground, made while the moose fed atop the winter snowpack. Also breaks upright branches to access buds at their tips. **CAMBIUM FEEDING** Use their incisors to scrape cambium off aspens in winter. **ANTLER RUBS & WALLOWS** Males create antler rubs and wallows similar to those of elk. Rubs are slightly higher on trees, with marks up to 8 ft. (2.4 m).

WOODLAND CARIBOU

Rangifer tarandus caribou

The woodland caribou is one of the most imperiled mammals in our region. Our caribou have a specialized migration pattern; they move back and forth between high and low elevations several times in a year, in part to separate spatially from low-elevation ungulates and predators in the winter. Their large, round feet are well adapted for travel in the snow. Their dewclaws, which often register in tracks, increase the feet's surface area for flotation.

SIZE Length: 1.4–2.1 m. Weight: 63–153 kg (avg. 110 kg males, 81 kg females). Males are larger. **FIELD MARKS** Creamy white hair on neck, underbelly, and rump. Males' antlers are larger than those of females. **REPRODUCTION** Travels in small herds throughout the year. Breeding occurs in October, and cows give birth to a single calf in June. A high mortality rate in calves during the first year of life is due to predation. **DIET** Most important part of their winter diet is arboreal lichen, as other food sources are buried under deep snowpack. Forest stands must be at least 80–150 years old to support lichens in quantities required in winter. During summer, includes forbs, sedges, grasses, and lichen. **HABITAT & DISTRIBUTION** Mountainous terrain with deep winter snowpack. Summer range is mainly alpine and subalpine. In early winter, animals migrate to lower elevation late-successional forest. In late winter, once the snowpack has consolidated, they migrate to higher elevations. Small transborder herds located in the Selkirk Mountains of British Columbia, Idaho, and Washington. **SIMILAR SPECIES** Elk, moose. **CONSERVATION ISSUES** Listed as endangered in the United States and as threatened in Canada, their population is declining not only along the international border but across their entire range in southern British Columbia and Alberta. Large-scale habitat destruction from timber operations in its critical winter habitat has altered the predator–prey dynamics of this species with wolves and mountain lions.

TRACKS

Tracks are distinctive among hoofed mammals. Cleaves are rounded in the front and back of the foot. The center of each foot is

Woodland caribou bull. Photo by John E. Marriott.

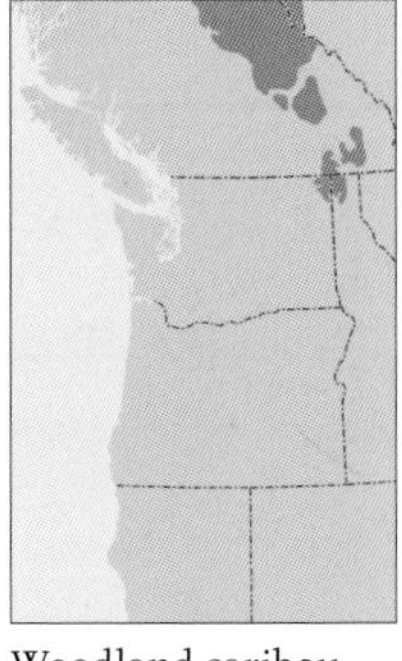

Woodland caribou range

concave, leaving a distinctive raised, round dome in the center of the track. The dome in elk tracks is smaller and more forward in the track (page 76). **FRONT** 4 1/16–5 1/2 in. (10.3–14.0 cm) L × 4 5/16–5 3/16 in. (11.0–13.1 cm) W. Larger than hind. Dewclaws are crescent shaped and usually register in tracks. **HIND** 3 7/16–5 1/2 in. (8.8–14.0 cm) L × 3 9/16–5 1/8 in. (9–13 cm) W. Slightly narrower than front. Dewclaws show less often in hinds than in fronts. **TRACK PATTERNS** Similar to other ungulates, walks and trots

Front track of a woodland caribou; dewclaws have clearly registered. Canadian Rockies, British Columbia.

Variety of scats from woodland caribou. Canadian Rockies, British Columbia.

Antler rub of a bull caribou on a subalpine fir. My hand is pointing out the highest scrapes on the tree from the animal's antlers. Similar to rubs made by moose and elk. Canadian Rockies, British Columbia.

in a direct register pattern. **DIRECT REGISTER WALK** Stride: 45 11/16–77 15/16 in. (116–198 cm). Trail width: 6½–14 in. (16.5–35.5 cm).

SIGNS

SCAT Deer-sized, irregular, dimpled pellets, at times clumped together into tubes or patties. Their winter diet—almost exclusively black tree lichens (*Bryoria* spp.)—produces black, tarlike pellets. Caribou urine creates an orange stain in the snow. **TRAILS** Create and use well-developed trails through their forested habitat, which connect feeding and bedding areas (page 45). **WALLOWS & RUBS** Bulls create rubs and wallows similar to those produced by elk and moose.

BOVIDS

Family Bovidae

The bovids include bighorn sheep and mountain goats. Both sexes have unbranched horns that, unlike antlers, are permanent structures and are not shed yearly. Males' horns are larger than those of females. The American bison (*Bos bison*), whose historic range included the Pacific Northwest, has been extirpated regionally.

MOUNTAIN GOAT

Oreamnos americanus

An icon of the Northwest's mountains, mountain goats are exceptionally well adapted to winter and steep mountain terrain. Their thick, white pelage protects them from the elements and provides camouflage in the snow. Their stocky, muscular build makes them able climbers. Their legs are shorter and placed closer to one another compared to those of other hoofed mammals—both adaptations for climbing and movement on steep terrain, narrow ledges, and tight spaces. While mountaineering in the Cascades, I have watched them negotiate places I dared not go myself. These social animals travel in small bands of males, females, and juveniles.

Mountain goat

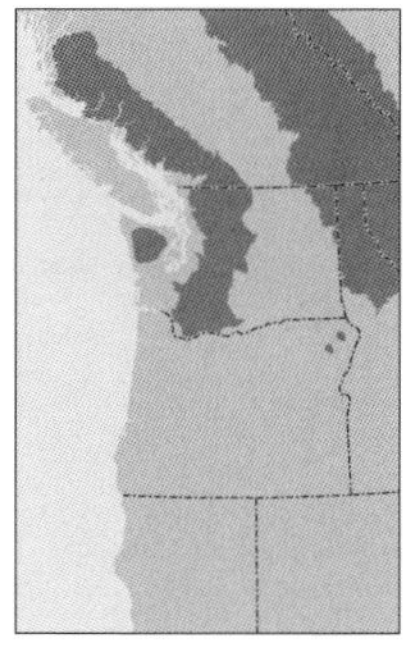

Mountain goat range

Typical mountain goat scat. North Cascades, Washington.

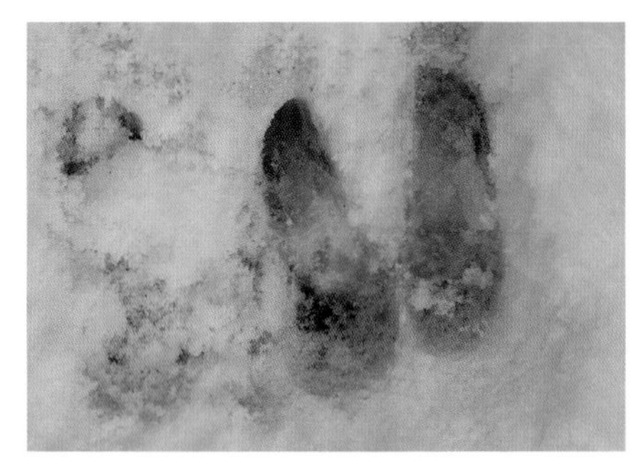

Mountain goat hind track on top of front. West Cascades, Washington.

The oval depression in the foreground is a typical mountain goat bed, situated on a high ridgeline with a commanding view of its surroundings. North Cascades, Washington.

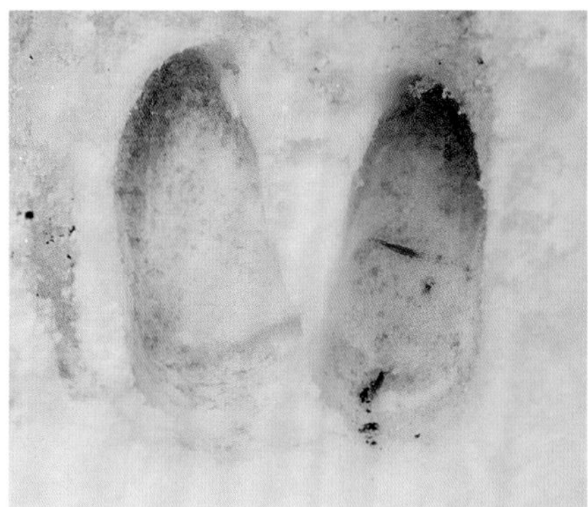

Mountain goat right hind track. West Cascades, Washington.

A clump of mountain goat hair is caught on a branch of a stunted mountain hemlock in the Olympic Mountains. Northwest Coast, Washington.

SIZE Length: 1.2–1.8 m. Weight: 46–126 kg. Males are larger. **FIELD MARKS** Creamy white coat in summer and winter, stocky build. Relatively small horns (male horns are slightly larger than those of females and more curved; female horns are straighter with a distinctive hook at the tips). **REPRODUCTION** Breeding occurs later than for other members of this order in our region, often not until late November. Precocial young (usually one, occasionally two or three) are born in late May. The female selects a cliff ledge on which to give birth. Kids stay close to their protective mothers for the entire summer and into the fall. Males may travel alone or on the periphery of several females with their young throughout the year. **DIET** In summer, primarily grasses, with sedges and forbs. Winter diet includes lichens and mosses taken from trees and rocks. **HABITAT & DISTRIBUTION** Native to the Washington Cascades and Canadian Rockies, but introduced to the Olympic and Blue Mountains. Scattered populations found along the crest of the Washington Cascades in rocky alpine habitats. In winter, they descend in elevation to forested cliff bands. **SIMILAR SPECIES** Mule deer, big horn sheep, elk.

TRACKS

Similar in size to deer tracks. Tips of cleaves are more blunt than those of deer or bighorn sheep. Point of each cleave falls along its centerline. Individual cleaves are wide, and tips or entire cleaves often spread apart when the animal walks, giving the track a boxy appearance at times, almost as wide as long. The pad is large and convex, taking up the entire inside of each cleave, leaving little topography in the track's interior, even when it registers clearly in fine substrate (page 74). **FRONT** 2¼–2 15⁄16 in. (5.7–7.5 cm) L × 2–2 15⁄16 in. (5.0–7.5 cm) W. Larger than hinds. **HIND** 2 1⁄16–2 15⁄16 in. (5.3–7.5 cm) L × 1 9⁄16–2 9⁄16 in. (4.0–6.5 cm) W. **TRACK PATTERNS** Walks (direct or slightly indirect register). For faster movement, usually bounds over broken terrain. **WALK** Stride: 23 5⁄8–44 1⁄16 in. (60–112 cm). Trail width: 4½–9 5⁄8 in. (11.5–24.5 cm).

SIGNS

SCAT ¼–7⁄16 in. (0.7–1.1 cm) D × 7⁄16–9⁄16 in. (1.1–1.4 cm) L. Similar to deer scat. Cylindrical, to multifaceted and dimpled (most common). **SHED HAIR** In the spring, goats shed their thick winter coats, leaving large clumps of white hair on branches of trees and shrubs around treeline. **BEDS** Often located on exposed ridge tops and cliff ledges, occasionally in clumps of stunted ridgeline trees. Oval shape, similar to other ungulate beds. **RUBS** Goats of both sexes rub trees with their horns, depositing scent. During the breeding season, males create rut pits in which they urinate and roll.

BIGHORN SHEEP

Ovis canadensis

Bighorn sheep are found in steep, rocky terrain, which they use to evade predators. During most of the year, rams travel and congregate in loose herds separate from females and juveniles. Dominance among males is determined by their horn size and their sparring ability, which involves charging and head-on collisions.

SIZE Length 1.6–1.9 m. Male weight: 75–135 kg. Female weight: 48–85 kg. **FIELD MARKS** Deer-sized. Brown with light rump patch.

Bighorn sheep

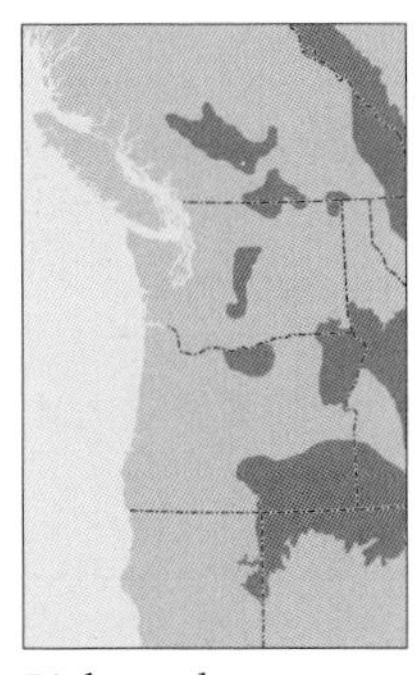

Bighorn sheep range

Males have large, heavilycurvedhorns. Females have smaller spiked horns. **REPRODUCTION** Breeding occurs in the fall and lambs are born in the spring. Exact time varies with location and latitude (earlier in the south). Females give birth to a single lamb (sometimes two), often in a protected location along a cliff ledge. Males leave their mothers and join a group of rams after their second year. **DIET** Herbivorous: primarily grasses, but also forbs and shrubs. **HABITAT & DISTRIBUTION** Open, arid, and cliffy mountainsides or canyons. Scattered populations along the eastern flank of the Cascades and mountains to the east. **SIMILAR SPECIES** Mule deer, mountain goat. **CONSERVATION ISSUES** Bighorn sheep were extirpated from most of our region in the first half of the 20th century. Subsequent reintroductions of the original native subspecies, *Ovis canadensis canadensis* and *O. canadensis californiana*, have been largely successful.

TRACKS

Tracks are heart shaped and can be difficult to distinguish from those of mule deer. Tips are less pointy than those of deer but more pointy than a mountain goat. The outside edge of the hoof wall is flat rather than curved as in deer. The subunguis is large, leaving a raised dome in the central portion of the track. The hoof wall and pad register as a thick crescent shape in the posterior portion of the track (smaller in hind

Front (above) and hind tracks of a female bighorn sheep. East Cascades, Washington.

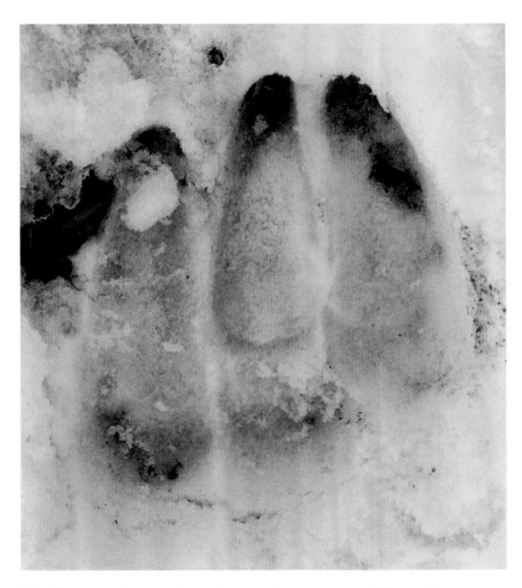

Bighorn sheep hind track on top of front. Columbia Plateau, Washington.

Scats of a bighorn sheep. Columbia Plateau, Oregon.

track) (page 73). **FRONT** 2 7/16–3 5/16 in. (6.2–8.5 cm) L × 1 15/16–2 5/8 in. (4.9–6.7 cm) W. Larger than hind. Even when toes are not splayed, space appears between the tips of each cleave. **HIND** 2¼–3¼ in. (5.7–8.2 cm) L × 1 11/16–2½ in. (4.3–5.9 cm) W. Tips of toes slightly narrower and pointier than fronts. More difficult to distinguish from deer than fronts. **TRACK PATTERNS** Direct register walk and trot. Lopes and gallops when fleeing across open terrain. They usually stay close to the refuge of cliffs, where they can deftly navigate rocky, loose terrain. **WALK** Stride: 33–48 in. (83.8–122.0 cm). Trail width: 4¼–9 in. (10.8–22.9 cm).

SIGNS

SCAT ¼–5/8 in. (0.7–1.6 cm) D × 5/16–13/16 in. (0.8–2.1 cm) L. Similar to that of deer, mountain goats, and pronghorns. Indistinguishable in places where they overlap with these species. **BEDS** Prefer areas around cliffs and ledges, often between cliff bands. Bedding areas are used repeatedly, leaving dusty, oval depressions with abundant scats.

PRONGHORN

Family Antilocarpridae

The pronghorn is distinct in having a forked horn, the outer sheath of which is shed annually. Its foot structure is also distinct in that it lacks dewclaws (toes 2 and 5) on both front and hind feet.

PRONGHORN

Antilocapra americana

Pronghorns occupy flat, open country, where they can use their excellent vision and exceptional speed to detect and evade predators. They have been recorded running at speeds of 55 mph (88.5 kph), with speeds of 45 mph (73.4 kph) commonly recorded. In the spring, when their forage has a high moisture content, they don't need to drink, but in late summer they tend to linger within 4 miles (6.4 km) of a water source. Dominant males defend a territory associated with a herd of mature females and juveniles from the beginning of spring through the end of breeding season. Other males form bachelor herds that are not welcome within a dominant male's territory.

SIZE Length: 1.3–1.5 m. Weight: 41–59 kg. Males are larger. **FIELD MARKS** Distinctive facial markings: dark patch on top of the nose and light patches under the head. Horns are dark and short, shed annually. Males' horns are longer and pronged, females usually not pronged. **REPRODUCTION** Breeding occurs in late September or early October. Most breeding is between dominant males that control a territory and the females in the herd associated with the territory. Members of a bachelor herd may occasionally breed. **DIET** Prefers to graze on grasses and forbs. Late summer and winter, consumes mostly woody browse, especially sagebrush. **HABITAT & DISTRIBUTION** Arid, flat, open country. Southern portions of the Columbia Plateau ecoregion. **SIMILAR SPECIES** Bighorn sheep, mule deer.

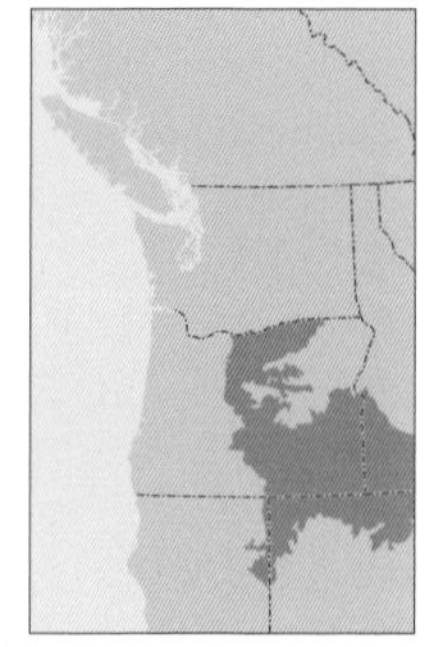

Pronghorn range

TRACKS

Tracks are heart shaped and often difficult to distinguish from those of deer, especially the hinds. Pronghorns are the only hoofed mammal in our region that do not have dewclaws. In clear tracks, the subunguis creates a raised area in the central portion (smaller than in bighorn sheep). As with bighorn, the hoof wall and pad create a hooked crescent shape in the posterior of each cleave. The outside edge of each cleave is usually either concave or flat at about one-third of the way up the track (convex in deer) (page 73). **FRONT** 2 3⁄8–3 9⁄16 in. (6–9 cm) L × 1¾–2 9⁄16 in. (4.5–6.5 cm) W. Tips of each cleave are long and tapered, forming well-defined points. Posterior portion of cleaves are bulbous. Larger than hinds. **HIND** 2 3⁄16–3 3⁄8 in. (5.5–8.5 cm) L × 1 3⁄8–2¼ in. (3.5–5.7 cm) W. Smaller with a slightly less elongated appearance than fronts. Track posterior may appear bulbous as in the front, but often less distinctly so. **TRACK PATTERNS** Lopes and gallops when fleeing. Their great strides can make it difficult to determine the exact pattern of their footfalls. **DIRECT REGISTER WALK** Stride: 30 5⁄16–45 11⁄16 in. (77–116 cm). Trail width: 4 5⁄16–10 5⁄8 in. (11–27 cm).

SIGNS

SCAT 3⁄16–9⁄16 in. (0.5–1.5 cm) D × 3⁄8–13⁄16 in. (1–2 cm) L. Similar to other similar-sized ungulates. Sometimes produces distinctive pellets that are more slender, elongated, and pointy at the tips than I have seen in deer, bighorn sheep, or mountain goats. Tubes or patties of clumped pellets produced when pronghorns are feeding on moist vegetation. **TERRITORIAL MARKING** Starting in late winter, dominant males mark and defend a breeding territory from other males and protect females that graze there from harassment by other males. Marking behavior consists of pawing the ground with the forelimbs and urinating and defecating on the scrapes. They also rub glands below their eyes on clipped or broken vegetation at these locations.

Left front track beneath left hind. Northern Great Basin, Oregon.

Three variations of pronghorn scats. Northern Great Basin, Oregon.

PART THREE
Other Wildlife

Pileated woodpecker at a nest cavity. Klamath Mountains, Oregon.

BIRDS Class Aves

A bird's hidden life is often revealed in its tracks and signs. Some species, such as the secretive common snipe, are rarely observed, but by knowing how to identify their tracks and signs, you can learn a great deal about their habitat and foraging behaviors. In some places, highly terrestrial species, such as quail, can be the most prolific track makers. Other species leave signs that can be confused with those of mammals. Some birds, such as woodpeckers, leave distinctive and conspicuous signs that last for years.

BIRD TRACKS

Bird Foot Morphology

Feet and tracks can tell you much about a bird's life history. Like the feet of mammals, birds' feet have evolved to accommodate their diverse habitats and behaviors. All birds in our region are digitigrade. As their forelimbs have become highly specialized for flight, their hind limbs and feet must perform many functions other than walking, such as feeding, swimming, preening, and perching.

In general, taxonomically related birds that share similar behavioral characteristics create similar tracks. In my experience, a fairly consistent relationship exists between the ratio of track size to the bird's body size within a related group of birds (ducks, for instance). You can use this ratio to help sort out species not specifically mentioned. This criteria is less useful for small perching birds, as many similar sized species do not regularly leave tracks on the ground. I have limited my discussion of bird foot morphology to birds that commonly leave tracks. The foot structures of several species (such as swifts) that typically do not land on the ground are not mentioned.

Toes. The toe size, shape, and arrangement make up the basic shape of bird tracks. In clear tracks, toes may show distinct pads divided by small, constricted ridges. Most birds have four toes. Toe 1 is called the hallux and points backward, except when it is absent, as is the case for some shorebirds. In perching birds, toe 1 is well developed and may be longer than the other toes. In some birds that are highly terrestrial, the hallux is reduced and elevated and therefore may not register as clearly as the other three toes, if it registers at all. In most species, toes 2–4 point forward, a foot structure called anisodactyl. In some birds in our region (woodpeckers and owls), toe 4 registers backward in a zygodactyl arrangement, leaving an X- or K-shaped track.

Metatarsal pad. Toes radiate from the end of the metatarsal bone, whose tip is covered by a pad that may or may not register. In tracks, the presence or absence of this pad can sometimes serve as a useful clue.

Claws. Claw sizes and shapes vary depending on the type of bird. Claws are often the most distinct part of a track and can be difficult to distinguish from the rest of the toe. Because of this (and unlike in the mammals section), all measurements of bird tracks include the claws.

Webbing. Webbing appears between two or more toes of many birds. As in mammals, webbing can be proximal, mesial, or distal. Ducks and geese are examples of birds with distal webbing, while wading birds, such as spotted sandpipers, can have proximal webbing between toes 3 and 4. Few birds in our region have mesial webbing. A few birds associated with water have lobed toes—rather than a web that connects toes, a discrete flap appears around the basic structure of each toe, increasing the surface area for swimming and flotation on mud and soft substrates.

Bird Gaits and Track Patterns

With only two feet, birds leave simpler track patterns than quadrupeds, and their patterns are more intuitively understood by humans, since we are also bipeds. Similar to quadruped tracks, bird track patterns are in two main groups that represent distinct forms of locomotion.

Track patterns of one: Walking and running. When walking or running, a bird moves each foot independently and for an equal

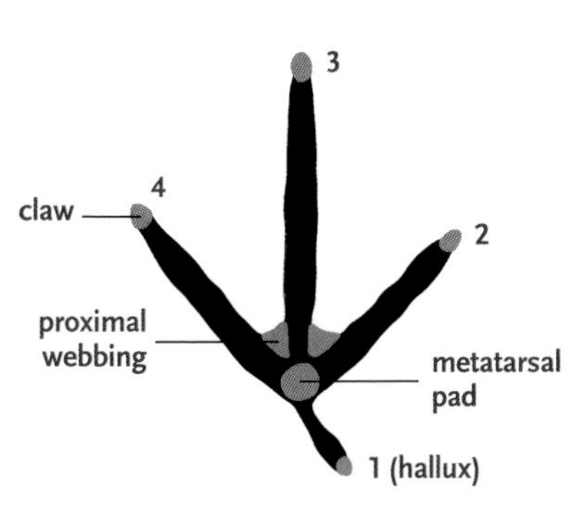

Shore and game bird left track

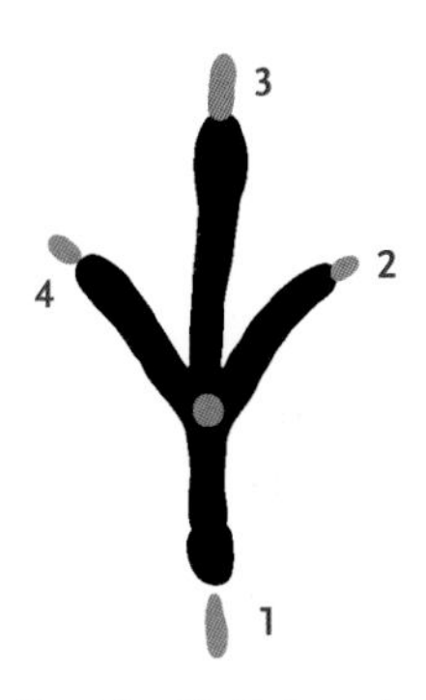

Perching bird left track

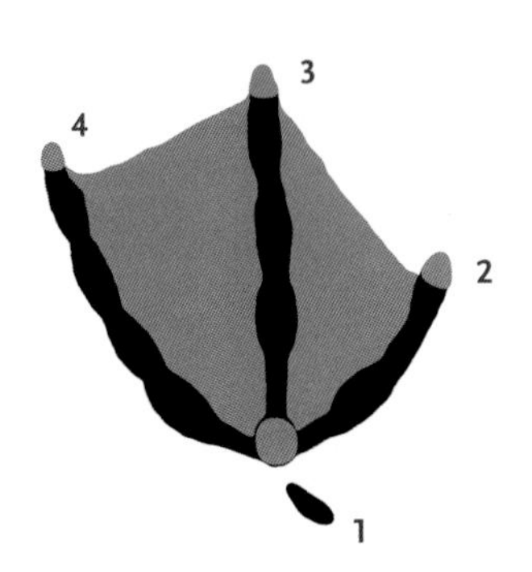

Distally webbed left track (ducks, geese, gulls)

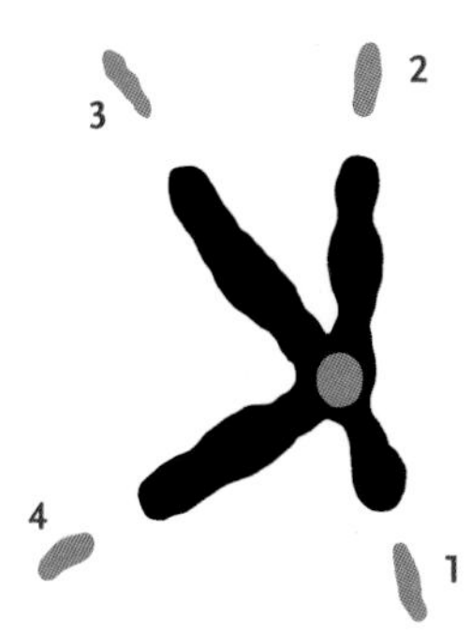

Zygodactyl left track (owls and woodpeckers)

length of time. In a run, the bird may become airborne briefly between each footfall. In both walks and runs, the distance between each footfall is roughly equal, leaving a regular alternating pattern on the ground, defined by individual feet from one side of the body and then the other. In a run, the bird's stride is longer and the trail width is narrower than in a walking gait for the same bird. Walking and running are commonly used by birds that spend a lot of time on the ground, such as geese and quail.

Track patterns of two: Hopping. In a hop, a bird pushes off with both feet at the same time, briefly becoming airborne before landing with both feet simultaneously. Hopping produces a paired pattern, with the two feet registering side by side. The stride increases as the gait becomes faster, but the trail width may not change substantially.

Track patterns of two: Skipping. A skip is distinguished from a hop by the fact that feet move independently between airborne phases. This gait leaves tracks that register one after the other, followed by a larger space produced by the airborne phase. This gait is often used by perching birds that are comfortable moving on the ground.

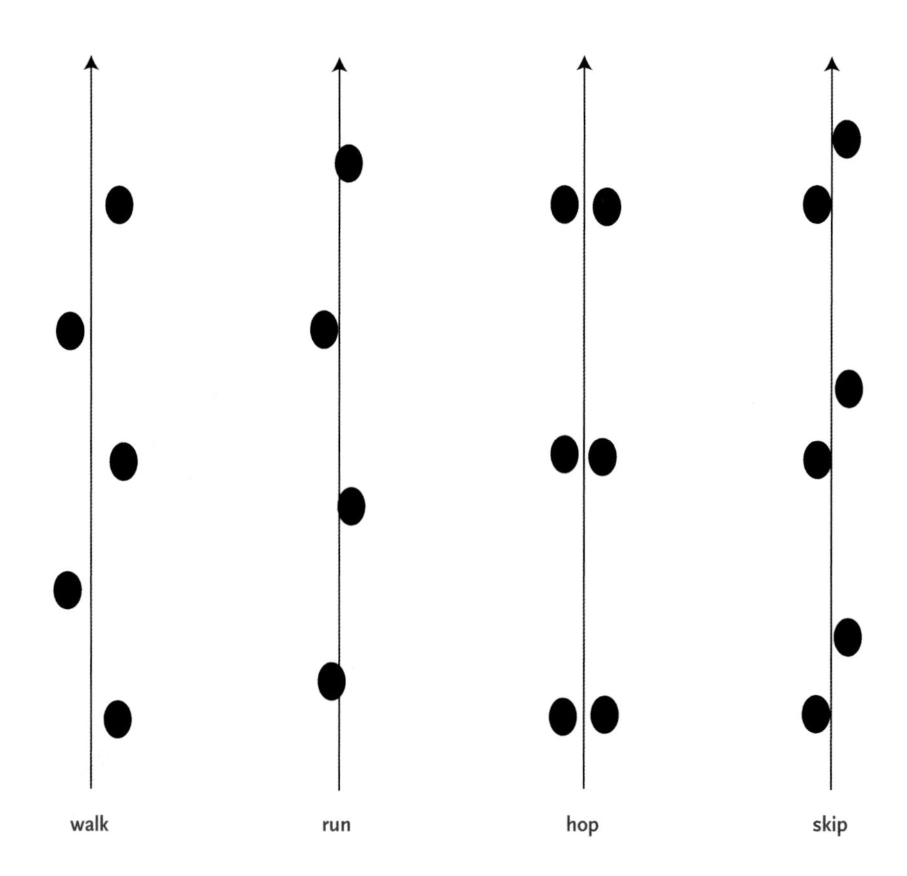

Measurements

Individual tracks. Length and width measurements include claws because they almost always register, and in many tracks it is impossible to distinguish the end of the toe pad from the claw. For perching bird tracks, zygodactyl tracks, and pelican and cormorant tracks (where the hallux shows consistently and clearly), length is measured from the end of the claw on toe 1 to the end of the claw on toe 3. In birds with a reduced hallux, this toe is not included in length measurements. Width is measured at the track's widest portion, perpendicular to the length.

Track pattern measurements. Stride is measured from one foot placement to the next place that exact foot registers. Trail width is measured across the widest portion of a trail for an animal moving in a straight line.

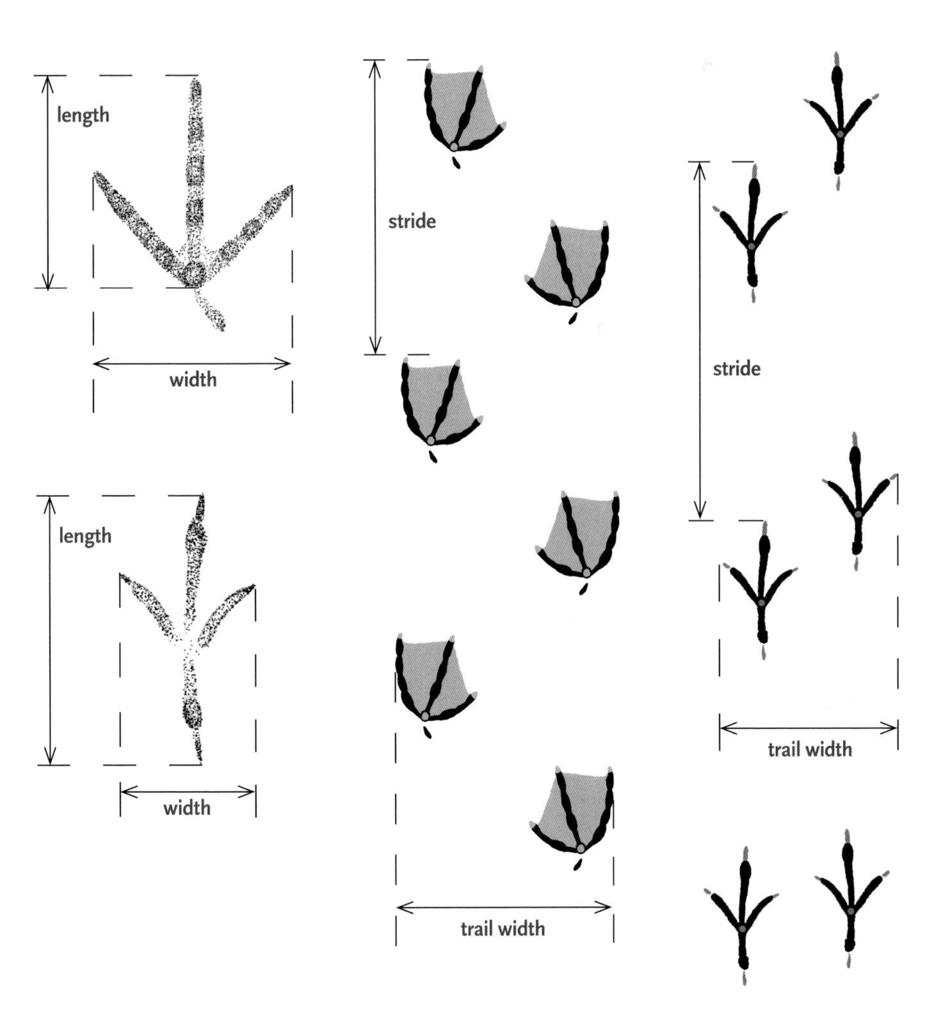

KEY TO BIRD TRACKS

This key shows life-sized or exactly scaled left tracks.

Tracks with Distal Webbing
Half Life Size

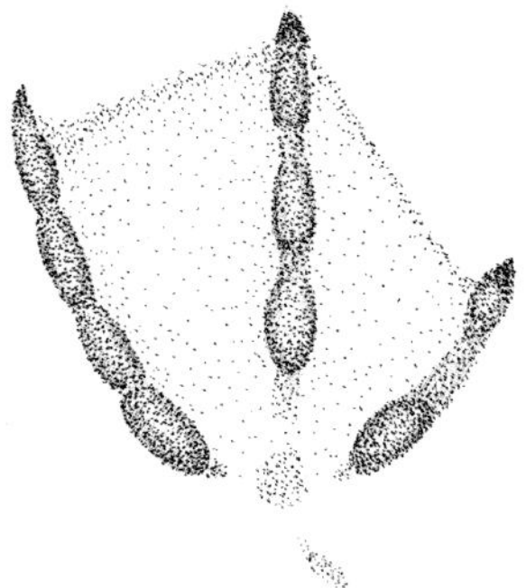

Gulls (*Larus* spp.), page 312

Caspian tern (*Sterna caspia*), page 312

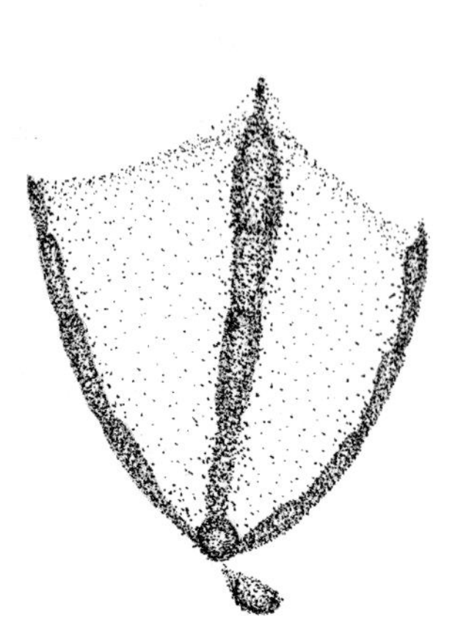

Common merganser (*Mergus merganser*), page 312; ducks, geese, and swans, page 311

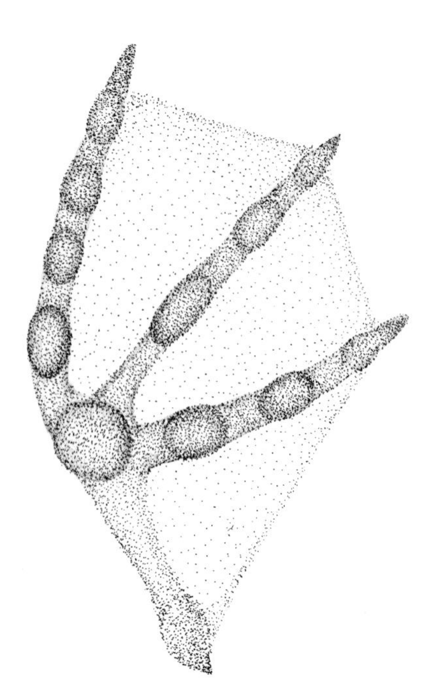

Brown pelican (*Pelecanus occidentalis*), page 312

Shorebird and Game Bird Tracks
Life Size

Killdeer (*Charadrius vociferus*) and other plovers, page 314

Spotted sandpiper (*Actitis macularia*) and other sandpipers, page 314

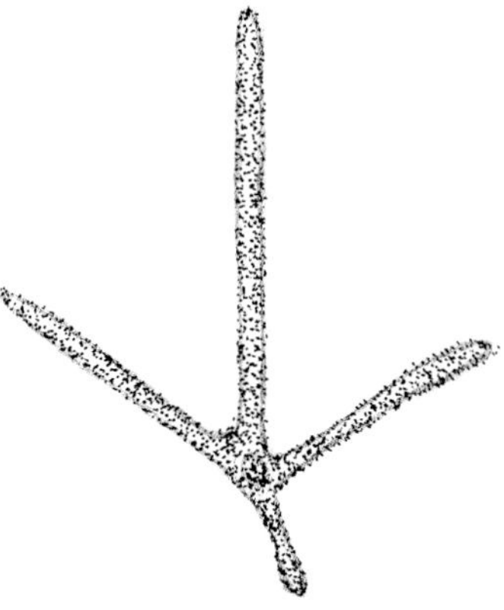

Wilson's snipe (*Gallinago delicata*), page 314

California quail (*Callipepla californica*), page 316

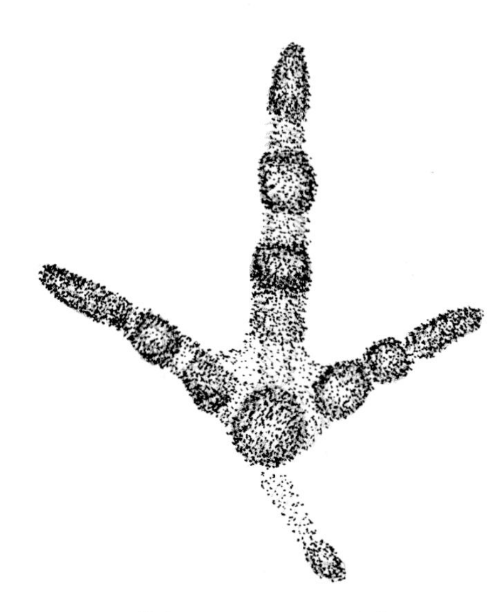

Ruffed grouse (*Bonasa umbellus*) and other grouse, page 317

Shorebird and Game Bird Tracks
Half Life Size

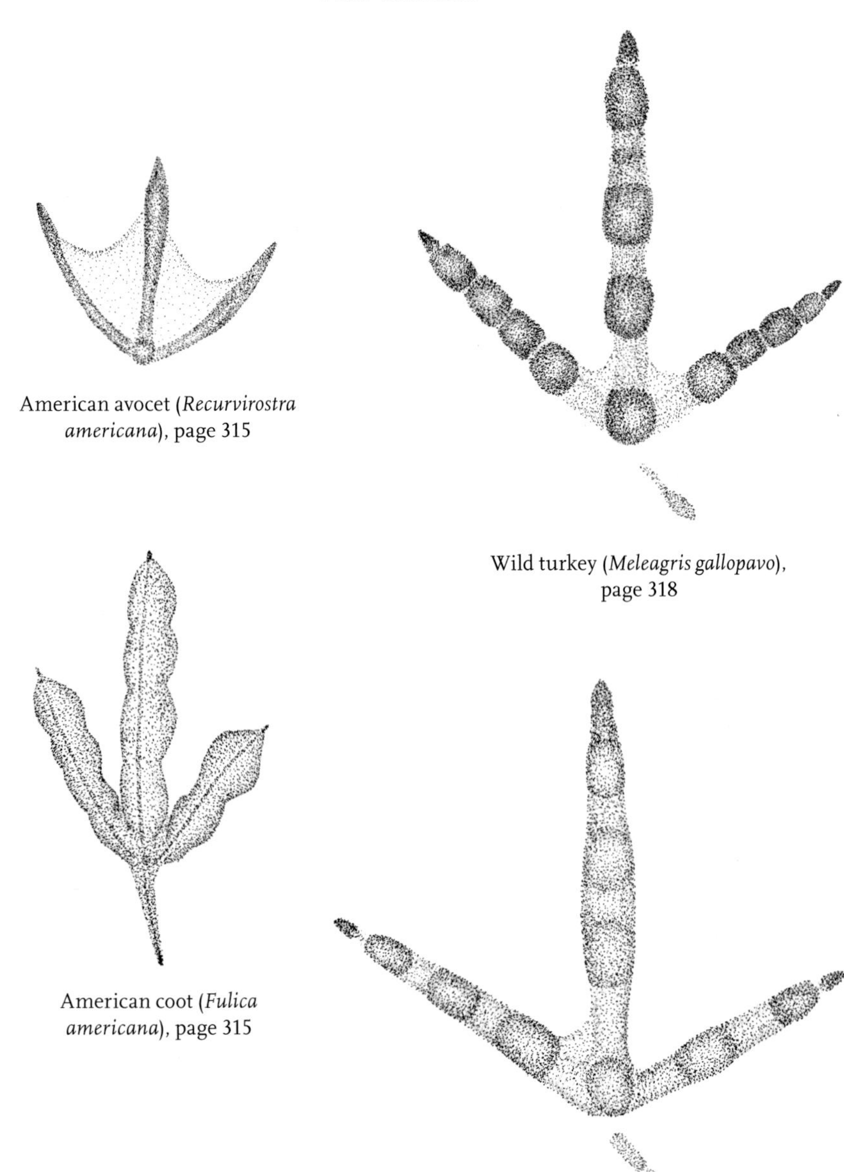

American avocet (*Recurvirostra americana*), page 315

Wild turkey (*Meleagris gallopavo*), page 318

American coot (*Fulica americana*), page 315

Sandhill crane (*Grus canadensis*), page 316

Perching Bird Tracks
Life Size

Winter wren (*Troglodytes troglodytes*), page 318

Dark-eyed junco (*Junco hyemalis*), page 318

Song sparrow (*Melospiza melodia*), page 318

Mourning dove (*Zenaida macroura*), page 320

Red-winged blackbird (*Agelaius phoeniceus*), page 318

American robin (*Turdus migratorius*), page 319

European starling (*Sturnus vulgaris*), page 319

Perching Bird Tracks
Life Size

Steller's jay (*Cyanocitta stelleri*), page 320

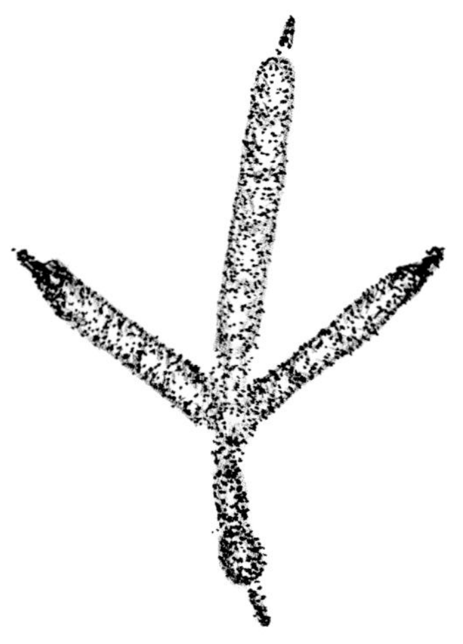

Rock dove (*Columba livea*), page 320

Black-billed magpie (*Pica pica*), page 320

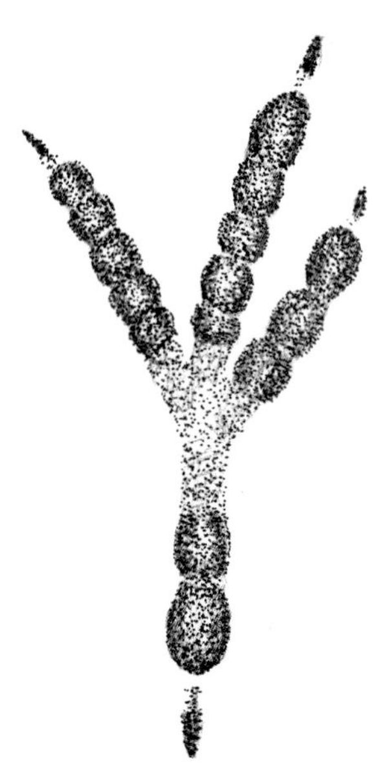

American crow (*Corvus brachyrhynchos*), page 320

Perching Bird Tracks
Half Life Size

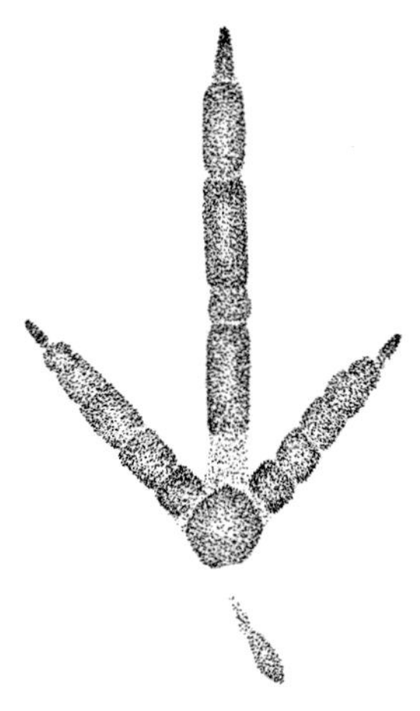

Turkey vulture (*Cathartes aura*), page 322

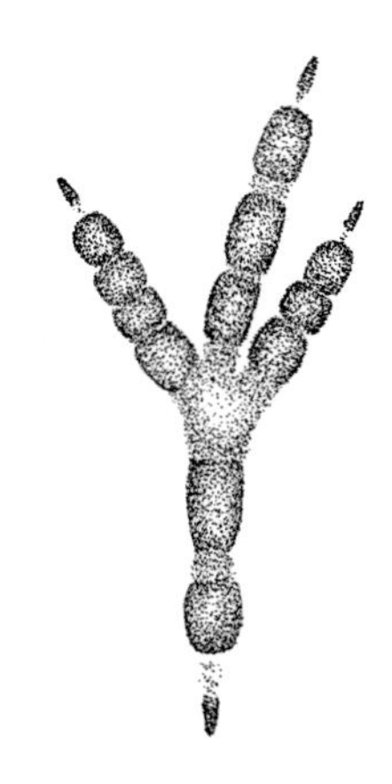

Common raven (*Corvus corax*), page 320

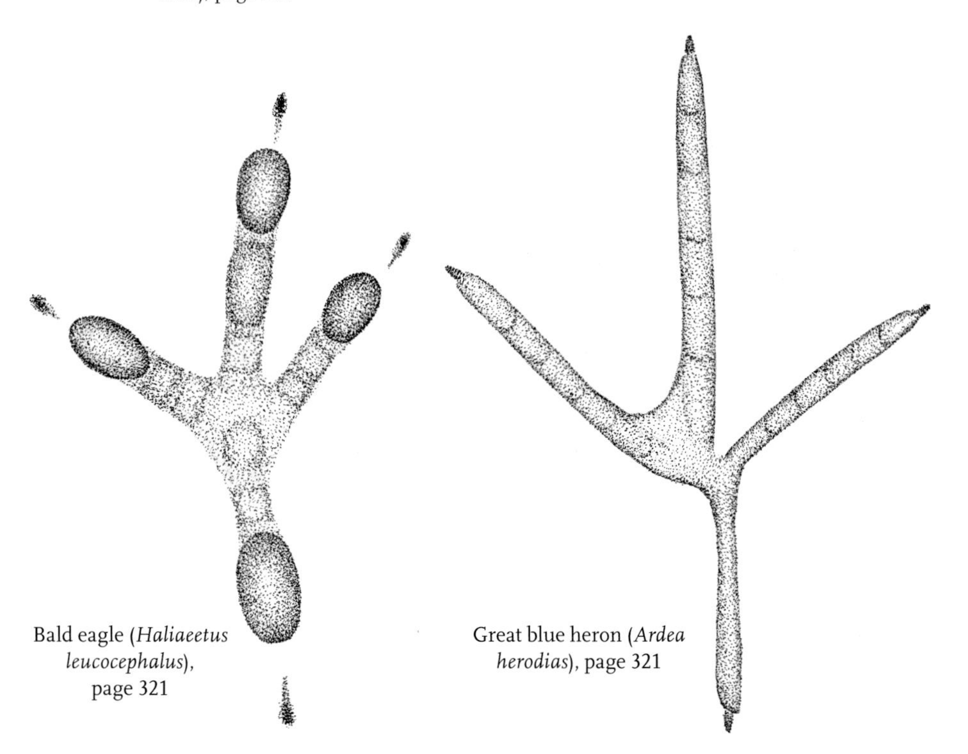

Bald eagle (*Haliaeetus leucocephalus*), page 321

Great blue heron (*Ardea herodias*), page 321

Zygodactyl Tracks
Life Size

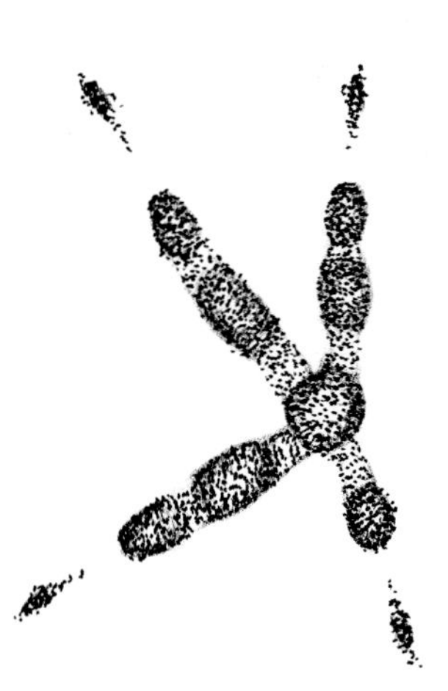

Burrowing owl (*Athene cunicularia*), page 322

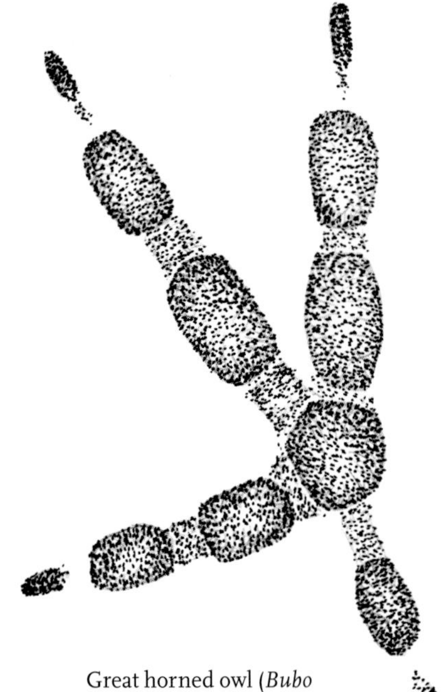

Great horned owl (*Bubo virginianus*), page 322

Northern flicker (*Colaptes auratus*), page 323

BIRDS WITH DISTAL WEBBING

Ducks, Geese, Swans, Gulls, Terns, Pelicans, and Cormorants

Because of their webbed feet, these birds are strong swimmers, and their tracks are often found close to the water's edge. Several species, however, feed in fields and other areas far from water. This group also includes birds with a greatly reduced hallux that is not webbed (ducks, geese, and gulls) and birds with a more fully developed hallux that is distally webbed, as are toes 2–4 (cormorants and pelicans).

DUCKS, GEESE, AND SWANS

Family Anatidae

TRACKS Toe 1 is reduced and curves inward. Toes 2 and 4 curve inward strongly toward toe 3 (more so than in gulls). Webbing connects the tips of the toes, and claws extend beyond toes. The metatarsal pad often registers in tracks, as does the hallux. Track size correlates to the bird's size (page 304). **TRACK PATTERNS** Walks and runs. Short stride. Feet may turn slightly inward.

CANADA GOOSE

Branta canadensis

In Canada goose tracks, the metatarsal pad and central toe often register deeper than the rest of the foot. Tracks are larger and broader than those of ducks.

TRACKS 2 15⁄16–4½ in. (7.5–11.5 cm) L × 3 3⁄16–4 5⁄16 in. (8–11 cm) W. **STRIDE** Walk: 13¾–19 7⁄8 in. (35.0–50.5 cm).

MALLARD

Anas platyrhynchos

Mallard tracks are indistinguishable from other dabbling duck tracks of similar size.

TRACKS 1 7⁄8–2½ in. (4.8–6.4 cm) L × 1 7⁄8–2 7⁄16 in. (4.8–6.2 cm) W. **STRIDE** Walk: 7 1⁄16–10¼ in. (18–26 cm).

Left track of a Canada goose. The metatarsal pad and central toe have registered deeply. Columbia Plateau, Washington.

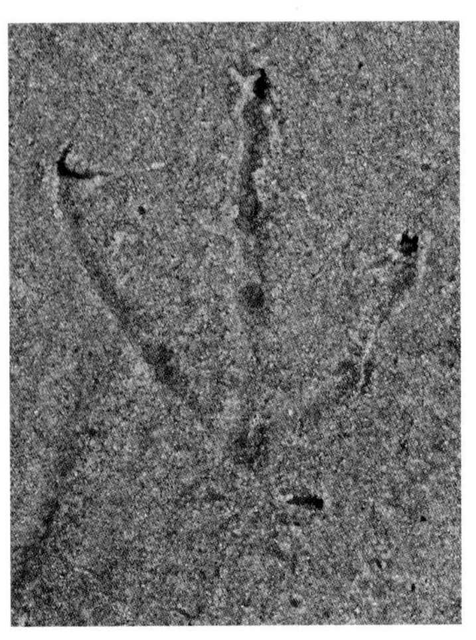

Left track of a mallard. Northwest Coast, Oregon.

Male wood duck

COMMON MERGANSER

Mergus merganser

Diving ducks such as common mergansers have a lobed hallux that increases the foot's surface area and can register more broadly in tracks than do those of dabbling ducks such as mallards. They also have shorter legs and therefore a shorter stride than similar-sized dabbling ducks.

TRACKS 2⅝–2 15⁄16 in. (6.7–7.5 cm) L × 2½–3 in. (6.3–7.6 cm) W (page 304). **STRIDE** Walk: 6 5⁄16–8½ in. (16.0–21.5 cm).

GULLS AND TERNS

Family Laridae

TRACKS Toe 1 is reduced and elevated. Toes 2 and 4 curve inward, but remain divergent at their tips. Claws extend beyond the line of webbing. The metatarsal pad and hallux often do not register. **TRACK PATTERNS** Mainly walks and runs.

GULLS

Larus spp.

TRACKS Distal webbing connects the tips of toes 2–4. Track size varies depending on the species's size, but tracks are otherwise indistinguishable between species (page 304).

TERNS

Sterna spp.

TRACKS Webbing connects distally to toes 2 and 4 and mesially to toe 3. The Caspian tern (*Sterna caspia*) is the largest tern in our region, but even its tracks are smaller than those of most gull species (page 304).

PELICANS AND CORMORANTS

Families Pelecanidae and Phalacrocoracidae

TRACKS Distal webbing is present between all the toes in the track, including the hallux, which is almost as long as the other three toes and curves inward toward the trail's midline. These birds occasionally land and leave tracks on the ground, but they prefer to land on pilings and rocks surrounded by water.

BROWN PELICAN

Pelecanus occidentalis

TRACKS 5 11⁄16–6⅞ in. (14.5–17.5 cm) L × 3⅞–4½ in. (9.8–11.5 cm) W (page 304). **STRIDE** Walk: 12 3⁄16–17 15⁄16 in. (31.0–45.5 cm).

SHORESBIRDS

Plovers, Sandpipers, Avocets, Coots, and Cranes

Most shorebirds spend a lot of time at water's edge, often in a location with excellent substrate for holding tracks. In coastal and wetland locations with large numbers of shorebirds, a good bird guide with detailed range maps and relative sizes of birds can help refine your estimation of the potential maker of tracks you have found. The tracks of these birds closely resemble tracks left by game birds.

Tracks. Toe 1 is either greatly reduced and elevated, or absent. Toes 2–4 are slender. Toe 2 is smaller than toe 4. The metatarsal pad may or may not register depending on the species. In several species, proximal webbing is present between toes 3 and 4.

PLOVERS

Family Charadriidae

TRACKS Hallux is entirely absent. Toes are asymmetrically oriented, with toes 3 and 4 forming a smaller angle than toes 2 and 3. Proximal webbing may register between toes 3 and 4. Often the metatarsal pad does not register or registers lightly. Various species of plovers inhabit coastal and estuarine locations; tracks are indistinguishable between species (page 305). **TRACK PATTERNS** Walks and runs with a narrow trail width. Feet tend to cant inward.

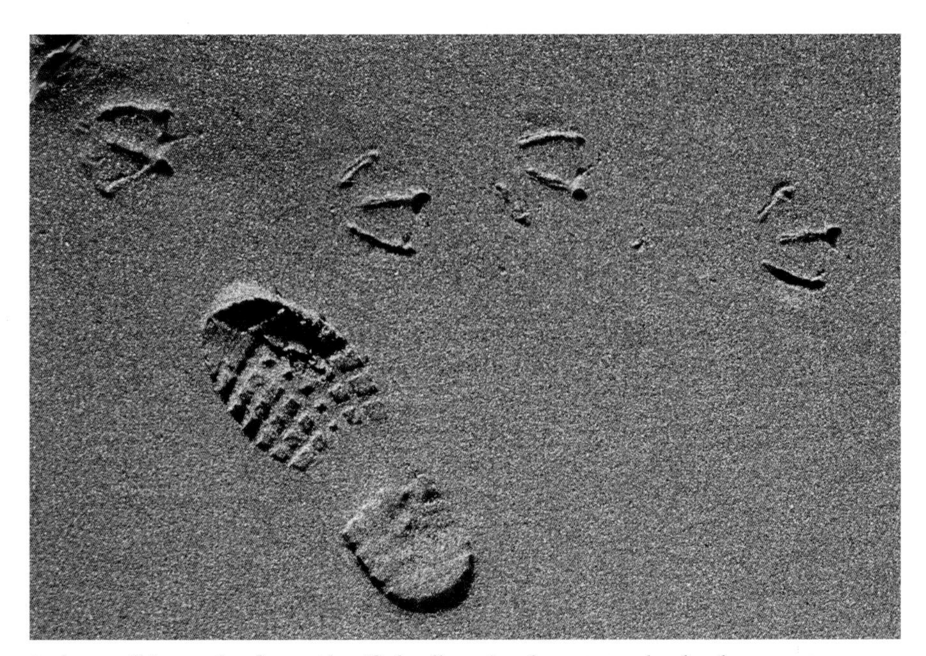

In these walking tracks of an unidentified gull species, the metatarsal pad and hallux have not registered, which is typical. Northwest Coast, Washington.

Killdeer tracks. Columbia Plateau, Washington.

KILLDEER

Charadrius vociferus

Most widespread species of plover across inland locations in the Northwest.

TRACKS 1–$1\frac{5}{16}$ in. (2.6–3.3 cm) L × $1\frac{1}{16}$–$1\frac{3}{8}$ in. (2.7–3.5 cm) W (page 305). **STRIDE** Walk/run: $2\frac{11}{16}$–$7\frac{11}{16}$ in. (6.8–19.5 cm)

SANDPIPERS

Family Scolopacidae

TRACKS Toes 2–4 are more symmetrical than plover toes. When the elevated hallux registers, it is an excellent clue to distinguish sandpiper tracks from those of plovers. One small species, the sanderling (*Calidris alba*), lacks a hallux. Proximal webbing is present between toes 3 and 4. **TRACK PATTERNS** Walks and runs.

Spotted sandpiper

SPOTTED SANDPIPER

Actitis macularia

The most common sandpiper in our region inhabits the shores of streams and lakes and is similar in size to the killdeer. It often forages at the water's edge, and its bill probings provide occasional signs.

TRACKS $1\frac{1}{16}$–$1\frac{5}{16}$ in. (2.7–3.3 cm) L × 1–$1\frac{3}{8}$ in. (2.5–3.5 cm) W (page 305). **STRIDE** Walk/run: $2\frac{3}{8}$–$7\frac{5}{16}$ in. (6.0–18.5 cm)

WILSON'S SNIPE

Gallinago delicata

Snipes create deep, thin holes in the mud with their long, thin bills, which are often apparent even when tracks are not (page 327).

TRACKS 1 3/8–2 1/16 in. (3.5–5.2 cm) L × 1 5/8–2 in. (4.1–5.1 cm) W. Exceptionally thin toes. The hallux often registers. Look for tracks in tall grass and brush in and along marshes (page 305). **STRIDE** Walk/run: 3 3/8–8¼ in. (8.6–21.0 cm). Walking stride is small relative to the size of individual tracks.

GREATER YELLOWLEGS

Tringa melanoleuca

TRACKS During the fall, winter, and spring, tracks are commonly encountered along rivers and streams west of the Cascades and east to the Columbia River in central Washington. Larger overall than the spotted sandpiper, its toes are broader at the tips, and the metatarsal pad and hallux show less commonly. **STRIDE** Stride of this long-legged bird is considerably longer than that of the spotted sandpiper.

Spotted sandpiper tracks; the hallux has registered and the angle between toes are relatively even.

LONG-BILLED CURLEW

Numenius americanus

TRACKS 1 7/8–2 1/8 in. (4.8–5.4 cm) L × 1¾–2 11/16 in. (4.5–6.7 cm) W. Slightly larger than greater yellowlegs tracks, with proximal webbing between toes 2 and 3 as well as 3 and 4. **STRIDE** Walk: 11–14 9/16 in. (28–37 cm).

AVOCETS

Family Recurvirostridae

AMERICAN AVOCET

Recurvirostra americana

These long-legged shorebirds frequent the edges of marshes and ponds in our region during the summer months.

TRACKS Distinctive in having mesial webbing between toes 2, 3, and 4. Tracks similar in size to those of curlews (page 306).

COOTS

Family Rallidae

AMERICAN COOT

Fulica americana

Ubiquitous in our region year-round, coots often feed in flocks on the edges of urban ponds along with Canada geese and mallards.

TRACKS 3 1/8–3 3/8 in. (8.0–8.5 cm) L × 2 3/16–2¾ in. (5.5–7.0 cm) W. All four toes are lobed rather than webbed, a characteristic shared with several other rails and grebes. Tracks are large for the bird's size (page 306). **TRACK PATTERNS** Walks and runs.

CRANES

Family Gruidae

SANDHILL CRANE

Grus canadensis

A regular visitor during migration to many parts of our region, the sandhill crane also breeds in parts of the Columbia Plateau ecoregion and in marshy mountain meadows. Frequents wet meadows and open fields.

TRACKS $3\frac{15}{16}$–$5\frac{5}{16}$ in. (10.0–13.5 cm) L × $5\frac{1}{16}$–$6\frac{5}{16}$ in. (12.8–16.0 cm) W. Similar in size to wild turkey tracks. Proximal webbing present between toes 3 and 4. Hallux and metatarsal pad show less consistently, and toes are relatively thinner, with less distinct individual toe pads than in turkeys (page 306). **TRACK PATTERNS** Walks and runs. A long-legged bird, its stride is longer than that of the wild turkey. **STRIDE** Walk: $9\frac{7}{16}$–$29\frac{1}{2}$ in. (24–75 cm).

Left track of a sandhill crane; the hallux has registered lightly. Middle Rockies, Idaho.

GROUND BIRDS WITH A REDUCED HALLUX

Quail, Grouse, Ptarmigan, Pheasant, and Turkey

Also referred to as game birds, many species were introduced into our region for hunting and have become well established.

Tracks. Similar in overall shape to those of shorebirds, but toes are more bulbous and robust. The hallux is elevated and does not always register. On some species, proximal webbing is present between toes 2, 3, and 4. The metatarsal pad is often pronounced.

Track patterns. Walks, runs when alarmed.

QUAIL

Family Odontophoridae

CALIFORNIA QUAIL

Callipepla californica

Quail are native in parts of our region and introduced in others (Columbia Plateau). Two similar-looking species, the chuckar (*Alectoris chukar*) and gray partridge (*Perdix perdix*), have also been widely introduced in the Columbia Plateau. These species are slightly larger than quail. Chukars occupy cliffs and canyons. Partridge primarily inhabit the brushy edges of agricultural areas, a habitat often shared with quail.

TRACKS $1\frac{3}{8}$–2 in. (3.5–5.1 cm) L × $1\frac{3}{8}$–$1\frac{7}{8}$ in. (3.5–4.8 cm) W. Toes are thin and straight. The metatarsal pad often registers clearly. The hallux often registers, angled medially but occasionally toward the track midline (page 305). **STRIDE** Walk: $3\frac{9}{16}$–$7\frac{1}{16}$ in. (9–18 cm). Run: $8\frac{11}{16}$–$19\frac{5}{16}$ in. (22–49 cm).

GROUSE AND PTARMIGAN

Family Phasianidae

At high elevations, grouse and ptarmigan fly into deep, loose snowbanks and create a small chamber in the snow to weather a storm. At the end of the storm, they explode out of the drift, leaving behind a neat little chamber, often with several scats within.

TRACKS Small amount of proximal webbing between toes 2 and 3 as well as 3 and 4. Hallux and metatarsal pad often register. In winter, grouse grow a fringe of horny material around their toes, which aids in flotation on snow. The white-tailed ptarmigan (*Lagopus leucura*) grows feathers around its toes, which increases flotation on snow and provides insulation.

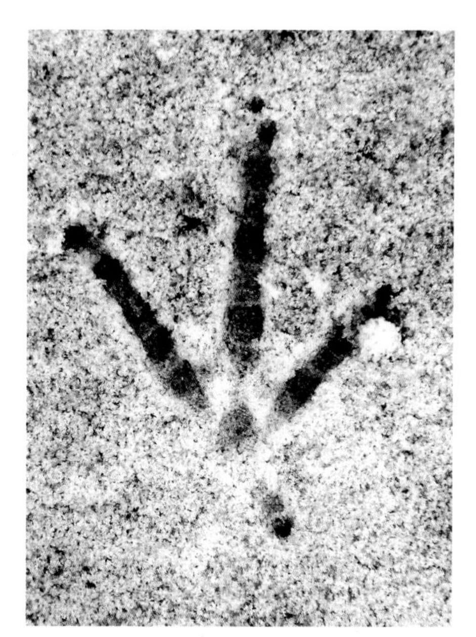

Left track of a California quail. Columbia Plateau, Washington.

RUFFED GROUSE

Bonasa umbellus

TRACKS 1 7/8–2 7/16 in. (4.7–6.2 cm) L × 1 7/8–2 5/16 in. (4.8–5.9 cm) W (page 305). **STRIDE** Walk: 5 1/2–12 3/16 in. (14–31 cm).

BLUE GROUSE

Dendragapus obscurus

TRACKS 2 5/8–3 1/4 in. (6.7–8.3 cm) L × 2 3/8–2 15/16 in. (6.0–7.5 cm) W. **STRIDE** Similar to that of ruffed grouse.

PHEASANT

Family Phasianidae

RING-NECKED PHEASANT

Phasianus colchicus

Pheasants were introduced widely throughout our region, especially in agricultural areas.

White-tailed ptarmigan in summer plumage. North Cascades, Washington.

TRACKS 2–3 1/16 in. (5.1–7.8 cm) L × 2 3/8–3 1/8 in. (6–8 cm) W. Tracks and stride are significantly larger than those of grouse, while pheasant toes are relatively thinner. Smaller than wild turkey tracks. **STRIDE** Walk and run: 14 3/16–36 1/4 in. (36–92 cm).

TURKEY

Family Phasianidae

WILD TURKEY

Meleagris gallopavo

Native to the eastern part of the continent, the wild turkey was introduced into many forested parts of our region east of the Cascades. Males are larger than females.

TRACKS Larger than pheasant tracks. Similar in size to tracks of sandhill crane, but toes are more bulbous and the hallux shows more often (page 306).

SMALL PERCHING BIRDS

Songbirds, Doves, and Corvids

Because many similar-sized perching songbirds spend little time on the ground, many tracks are difficult, if not impossible, to identify. Several groups do have distinctive characteristics that can be consistently identified in tracks, especially those that spend significant time on the ground.

Tracks. Three toes forward and one long toe straight back.

Track patterns. Perching birds that spend little time on the ground generally hop. More terrestrial species walk, run, skip, or hop.

WRENS

Family Troglodytidae

WINTER WREN

Troglodytes troglodytes

TRACKS 1 1/8 in. (2.8 cm) L × 7/16 in. (1.1 cm) W. Toes are long and thin. Toe 1 is proportionally longer than that of similar-sized birds (page 307). Tracks often found under logs or at water's edge in forested brush areas. Usually hops.

SPARROWS AND THEIR KIN

Family Embrizidae

DARK-EYED JUNCO

Junco hyemalis

TRACKS 1 3/8 in. (3.5 cm) L × 3/4 in. (1.9 cm) W. Hallux may appear shorter than in sparrows or other similar-sized perching birds. Hallux and toe 3 often register in a straight line rather than curved as in many sparrow tracks (page 307). Usually hops.

SONG SPARROW

Melospiza melodia

TRACKS 1 1/2 in. (3.8 cm) L × 9/16 in. (1.4 cm) W. Hallux and toe 3 often form an inward curve (page 307). Hops and skips.

BLACKBIRDS AND MEADOWLARKS

Family Icteridae

RED-WINGED BLACKBIRD

Agelaius phoeniceus

TRACKS 1 7/8 in. (4.8 cm) L × 11/16 in. (1.7 cm) W. Base of toe 3 is slightly offset toward toe

4, where it attaches to the track; same in starling tracks. Toe 4 often spreads less than in other similar-sized bird tracks, creating a smaller angle between toes 3 and 4, a characteristic shared with other blackbirds (page 307). Walks and hops.

American robin tracks are some of the most common perching bird tracks found in our region.

STARLINGS

Family Sturnidae

EUROPEAN STARLING

Sturnus vulgaris

TRACKS 2 in. (5.1 cm) L × 1 in. (2.5 cm) W. Similar in size to robin tracks. Toe 3 often curves inward at tip and is offset at its base toward toe 4 (page 307). This species is often found in large flocks; look for a large number of tracks in an area. Commonly walks.)

THRUSHES

Family Turdidae

AMERICAN ROBIN

Turdus migratorius

TRACKS 1⅞–2¹⁄₁₆ in. (4.7–5.2 cm) L × ⅞–1¹⁄₁₆ in. (2.2–2.7 cm) W. Toes 2 and 4 curve outward. Pad at the base of hallux is well developed and often registers prominently (page 307). One of the most common perching bird tracks to find in our region. Usually runs.

American robin track. West Cascades, Washington.

DOVES AND PIGEONS

Family Columbidae

TRACKS Outer toes spread widely, and if the hallux does not show clearly, tracks can resemble game bird tracks.

MOURNING DOVE

Zenaida macroura

TRACKS 1 5/8 in. (4.1 cm) L × 1 1/4 in. (3.1 cm) W. Outer toes straight or slightly curved inward (page 307). Walks with a short stride.

ROCK DOVE

Columba livea

TRACKS 2 5/16 in. (5.9 cm) L × 1 11/16 in. (4.3 cm) W. Larger than mourning dove tracks. Outer toes more consistently straight than those of mourning doves (page 308). Look for tracks of the larger band-tailed pigeon (*Columba fasciata*) at the edge of mud puddles in forested locations. Walks.

CORVIDS

Family Corvidae

TRACKS This conspicuous group of birds are all prolific track makers. All species in our region share an easily recognized diagnostic feature: toe 3 is aligned with toe 2, creating a smaller angle between these two toes than between toes 3 and 4.

STELLER'S JAY

Cyanocitta stelleri

TRACKS 1 5/8–2 3/8 in. (4.1–6.0 cm) L × 1/2–7/8 in. (1.2–2.2 cm) W. Tracks are narrower than those of other similar-sized perching birds (page 308). Hops. Western scrub jay (*Aphelocoma californica*) and gray jay tracks (*Perisoreus canadensis*) are similar.

BLACK-BILLED MAGPIE

Pica pica

TRACKS 2 5/16–2 15/16 in. (5.9–7.5 cm) L × 1–1 3/8 in. (2.6–3.5 cm) W. Typical corvid structure with robust toes for the size of the tracks (page 308). Comfortable on the ground. Walks, runs, and skips. **STRIDE** Walk: 7 1/16–8 11/16 in. (18–22 cm).

AMERICAN CROW

Corvus brachyrhynchos

TRACKS 2 11/16–3 1/4 in. (6.8–8.2 cm) L × 1 1/8–1 5/8 in. (2.9–4.1 cm) W. Typical corvid structure. Larger than magpies and smaller than ravens (page 308). Walks, hops, and skips. **STRIDE** Walk: 8 1/4–16 1/8 in. (21–41 cm).

COMMON RAVEN

Corvus corax

TRACKS 3 11/16–4 15/16 in. (9.4–12.5 cm) L × 1 3/4–2 1/16 in. (4.5–5.2 cm) W. Typical corvid structure. Tracks are large and toes are robust (page 309). **STRIDE** Walk: 10 5/8–17 11/16 in. (27–45 cm).

LARGE PERCHING BIRDS

Herons, Egrets, and Bitterns; Eagles and Hawks; and Vultures

Though taxonomically diverse and distinct from songbirds, these birds share a similar foot structure that enables them to perch on tree limbs. Herons and egrets leave abundant tracks that are often confused with those of eagles and hawks.

Tracks. Resemble typical shape of songbirds but larger. The hallux is similar in length to the other three toes and is almost always oriented straight back in tracks. Typically walks.

HERONS, EGRETS, AND BITTERNS

Family Ardeidae

TRACKS Herons and their kin have slender, even-width toes and proximal webbing between toes 3 and 4, similar to many shorebirds. Toe 3 is not aligned with toe 1, giving tracks a distinctive offset appearance where the toes converge. Size is the largest distinguishing feature among members of this family, and distinguishing similar-sized species such as great egrets and great blue herons is not always possible. Even though the great egret often has a narrower track with more slender toes, these features can be distorted by substrate. **TRACK PATTERNS** Walks. Great blue herons and egrets are long-legged and comfortable walking, commonly leaving long strings of tracks with a longer stride than that of eagles.

GREAT BLUE HERON

Ardea herodias

TRACKS 6 1⁄16–8 7⁄16 in. (15.4–21.5 cm) L × 4 5⁄16–5 7⁄8 in. (11–15 cm) W (page 309). **STRIDE** Walk: 19 11⁄16–34 1⁄4 in. (50–87 cm).

EAGLES AND HAWKS

Family Accipitridae

TRACKS Eagles and hawks have more robust toes and bulbous pads than herons and egrets. Their long, sharp, curved claws register prominently. Unlike herons, no proximal webbing is present, and all the toes converge at the metatarsal pad, which may or may not register (page 309). **TRACK PATTERNS** Walking stride is shorter relative to the track size compared to herons. Raptors spend less time on the ground. Long strings of tracks are not common, and tracks on the ground are often associated with feeding signs such as animal remains.

BALD EAGLE

Haliaeetus leucocephalus

TRACKS 6 5⁄16–7 3⁄16 in. (16.0–18.2 cm) L × 3 3⁄4–4 5⁄8 in. (9.5–11.7 cm) W. Tracks are fairly common along sandbars of salmon-

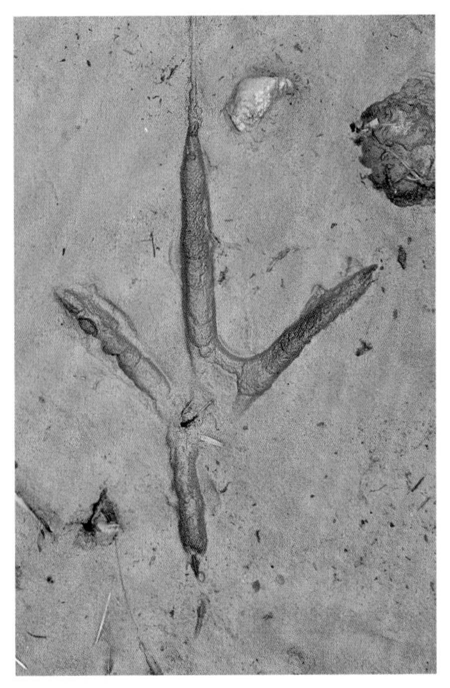

Great blue heron track.
Puget Trough, Washington.

Red-tailed hawk

Bald eagle track. Puget Trough, Washington.

bearing rivers (page 309). **STRIDE** Walk: 15 3⁄8–21¼ in. (39–54 cm).

VULTURES

Family Cathartidae

TURKEY VULTURE

Cathartes aura

TRACKS 3 5⁄16–4¾ in. (8.5–12.0 cm) L × 2 3⁄8–3¾ in. (6.0–9.5 cm) W. Hallux is narrower than the other toes and often registers lightly (page 309).

BIRDS WITH ZYGODACTYL TRACKS

Owls and Woodpeckers

Although owls and woodpeckers are evolutionarily and behaviorally quite different, they share a similar foot structure, with two toes facing forward and two facing back.

OWLS

Families Tytonidae (Barn Owl) and Strigidae (All Others)

TRACKS Toes 1 and 4 point backward, and toes 2 and 3 point forward, giving tracks an X- or K-shape. Toes 1 and 2 often create a

Large owl tracks in mud.
Okanagan, Washington.

straight line on the medial side of the track. Toe 4 is flexible and can change where it registers from track to track, more so than in other birds' tracks (page 310). The burrowing owl (*Athene cunicularia*) is the only highly terrestrial species in our region and often runs down its prey.

WOODPECKERS

Family Picidae

With the vast and varied forested landscapes in our region, woodpecker signs are abundant and diverse.

TRACKS Toes 1 and 4 face backward and 2 and 3 face forward. One species, the northern flicker, feeds on the ground and tracks are common. Pileated woodpeckers land on the ground to ingest small stones to aid in digestion. Hairy woodpeckers occasionally spend time on the ground, sometimes hunting for ants in a fashion similar to a northern flicker, but often in association with feeding on fallen trees.

NORTHERN FLICKER

Colaptes auratus

TRACKS 2⅛–2¾ in. (5.4–7.0 cm) L × ½–⅞ in. (1.2–2.2 cm) W. Narrow tracks, often in a hopping pattern. Toes are thin (page 310). Pileated woodpecker tracks are similar in shape but larger.

BIRD SIGNS

Excretions

Birds excrete two forms of waste. Cough pellets are made of the indigestible remains of animals and insects a bird has ingested and then regurgitated. Droppings are excreted through the cloaca and are a mix of feces and urine.

Cough Pellets

Many omnivorous and all carnivorous birds produce cough pellets. They are most well known in owls, however, because their pellets often contain complete skulls, mandibles, and other bones of prey species. Dissecting them can yield a wealth of information about the bird's diet as well as the types of small mammals that inhabit the vicinity.

Eagle and hawk pellets can contain hair, feathers, or insect remains depending on species and diet. Fewer bones are found in their pellets, compared to owl pellets, because of the stronger digestive acids in their stomachs. Other birds, including ravens, crows, gulls, herons, and cormorants, also produce pellets that can be found at common roosting locations. Many pellets are fragile and disintegrate easily.

Droppings

In droppings, urine often appears as a white paste or crust on part of the defecation, a characteristic shared with some reptiles.

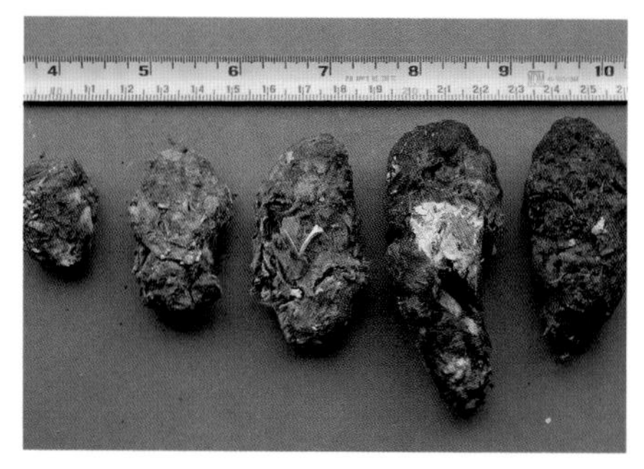

Great-horned owl (*Bubo virginianus*) pellets. Columbia Plateau, Washington.

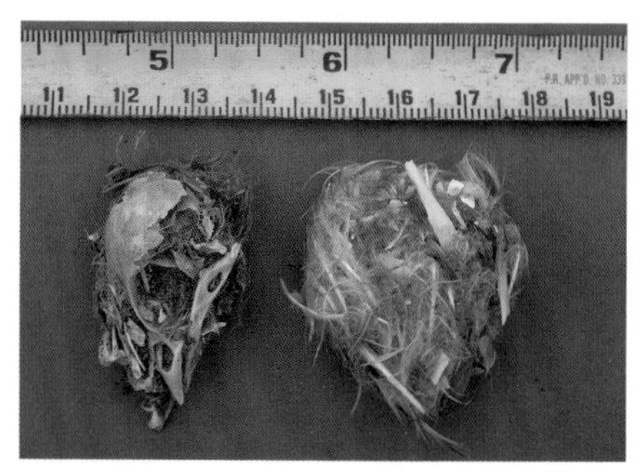

Short-eared owl pellets containing feathers and bird skull. Northwest Coast, Oregon.

Raven pellet. Northwest Coast, Oregon.

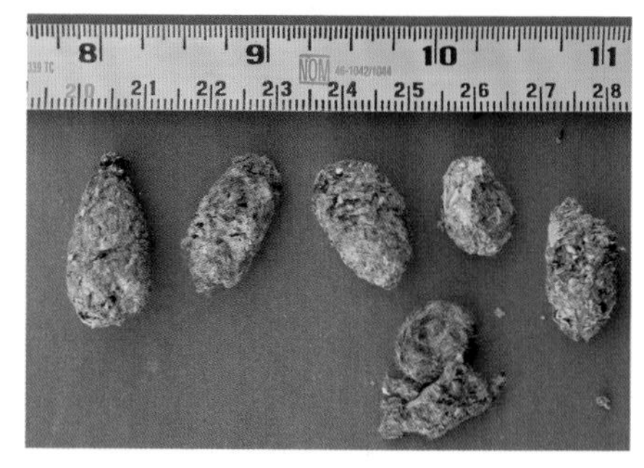

American kestrel (*Falco sparverius*) pellets containing insect remains, fur, and bones of small mammals. Columbia Plateau, Washington.

Dove droppings have a distinctive appearance; shown here are mourning dove droppings. Columbia Plateau, Washington.

Typical northern flicker droppings. Puget Trough, Washington.

Woodpeckers and other insect-eating birds. Woodpecker droppings consist mainly of insect exoskeletons and can be found at the base of trees with other feeding signs or around ant colonies where flickers have also been feeding. These droppings are usually white-capped and uneven in shape. The northern flicker produces slightly different droppings—smooth-sided, curved cylinders, usually with a uniform white coating. Flycatcher droppings found at favored perches often include insect parts and are smaller than woodpecker droppings.

Waterfowl and ground bird droppings with vegetation and fibrous remains. Droppings of many game birds and waterfowl can be confused with mammal scats. These bird droppings usually have a cap of white uric acid on one end of the excretion, which differentiates them from mammal scats. Mallards and Canada geese produce tubular to amorphous droppings that contain vegetation or the remains of invertebrates.

Grouse droppings are usually C-shaped, sometimes capped in white, and often made up of compressed vegetation. Abundant droppings are found at well-used roosts. Grouse produce two types of scat: a soft, amorphous cecal excretion and the more common coarse pellets. Sage grouse (*Centrocercus urophasianus*), listed as threatened in the state of Washington, leave pellets with distinctively rounded ends, which are found in small piles in their preferred habitat of open sagebrush country. Grouse droppings average ¼ in. (0.7 cm) in diameter and are occasionally as large as 13⁄16 in. (2.1 cm).

Pheasant droppings are relatively large for a bird and might be confused with wild turkey where the two overlap. Scats have a white cap on one end and appear to be a twisted conglomeration of thinner tubes.

Spruce grouse droppings. Canadian Rockies, Montana.

Canada goose droppings average ½ in. (1.3 cm) in diameter.

Distinctive pellets of a sage grouse. Columbia Plateau, Oregon.

Dropping of a female wild turkey. East Cascades, Washington.

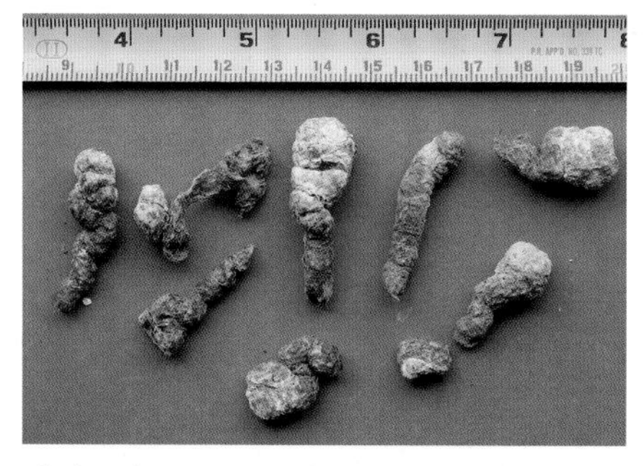

Chukar droppings. Columbia Plateau, Oregon.

Male and female wild turkeys leave droppings that can be distinguished from one another because of differences in the intestinal orientation of each sex. Male droppings are often elongated and J-shaped; female droppings are often twisted or clumped.

Chukar leave distinctive droppings that are often found under overhangs and at the base of rock formations in the Columbia Plateau.

Heron, merganser, and other bird droppings with a fishy smell. Piscivorous (feeding on fishes) birds produce whitish-gray, amorphous scats with a fishy smell. Quantity and location of deposits vary based on species. Heron and egret droppings can often be found along with their tracks at the base of a common roost.

Raptors. Raptors produce large, whitish droppings. When defecating from perches, hawks and eagles usually lean forward and project their scat out and away, leaving an elongated spray behind them. When owls defecate from perches, whitewash is excreted straight down. Turkey vultures excrete on their legs and feet, which deters mites and other parasites that otherwise affect these birds, since they spend their time prancing around on rotting carcasses.

Feeding Signs

Birds exploit diverse food resources in many ways. Some bird feeding signs can be confused with mammal signs (such as the work of bird-eating raptors), while others are quite distinctive (such as the work of woodpeckers).

Probing Holes in the Ground

Many shorebird species probe the mud and sand at the water's edge in search of invertebrates, a sign that is often as conspicuous (or more so) as the birds' tracks once you know what you are looking at. The size, depth, and location of these probings depend on the bill's shape and the animal's feeding behavior. Probably the most common probing signs to find inland are made by the Wilson's snipe—long, thin holes in the mud near a body of water. Shallower, wider holes in mud or grass thatch are created by European starlings. The northern flicker creates holes where it has been feeding on ants and other insects, often near ant mounds. The holes are wider at the top than at the base (unlike those of thin-billed

Holes in grass thatch made by the bills of a flock of starlings probing for insect larvae. Puget Trough, Washington.

Holes created by the bill of a northern flicker searching for ants (knife is about 3½ in., or 8.9 cm, long). East Cascades, Oregon.

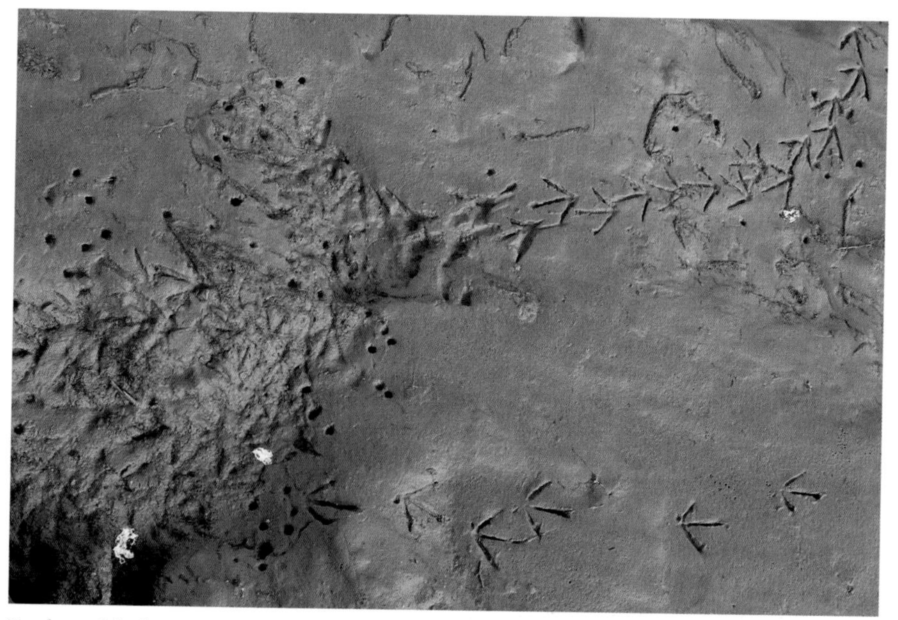

Tracks and feeding signs of a Wilson's snipe. Columbia Plateau, Washington.

shorebirds). Scats are also commonly found at these sites.

Disturbed Forest Litter

Like domestic chickens that forage in the yard, grouse disturb the ground with their feet to find invertebrates and seeds to eat, leaving behind patches of ruffled leaf litter. On a smaller scale, songbirds, including towhees and some sparrows, also turn over leaf litter with their feet.

Feeding Signs on Trees: Woodpeckers

Our region is graced with a variety of woodpeckers, many of whom have distinctive feeding behaviors and subsequent signs on trees.

Extensive feeding signs of a pileated woodpecker on a dead western red cedar tree. West Cascades, Washington.

PILEATED WOODPECKER

Dryocopus pileatus

This species creates the most conspicuous and largest holes made by woodpeckers in our region. Excavations are commonly 3 in. (7.5 cm) wide or larger and have distinctive rectangular appearances that, along with their size, identify its work. Holes are constructed to access carpenter ants in dead trees or in live trees with a rotten center.

Rotting logs often show signs of woodpecker excavation. I disturbed the hairy woodpecker that made these holes. West Cascades, Washington.

HAIRY WOODPECKER

Picoides villosus

Hairy woodpecker holes are smaller than those made by pileated woodpeckers; they are oval shaped and tapered toward the bottom. Similar holes are made by sapsuckers and flickers. Downy woodpeckers (*Picoides pubescens*) make such excavations only in soft, rotten wood.

SAPSUCKERS

Sphyrapicus spp.

Sapsuckers drill small holes through the bark of a live tree, out of which sap flows. These wells often appear in neat rows at the same height on several trees. Sapsuckers return to feed on the sap and the insects attracted to it. Red-breasted (*Sphyrapicus ruber*) and red-naped (*S. nuchalis*) sapsuckers often produce wells in linear patterns. Over time, a Williamson's sapsucker (*S. thyroideus*) creates a consistent grid pattern as it adds row after row of holes.

A Williamson's sapsucker tends his wells. Photo by Paul Bannick.

BLACK-BACKED AND THREE-TOED WOODPECKERS

Black-backed woodpecker, *Picoides arcticus*
Three-toed woodpecker, *P. tridactylus*

These two woodpecker species specialize in

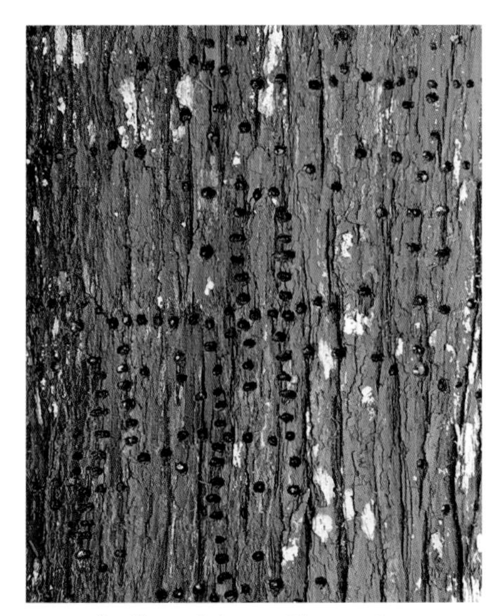

Rows of wells made by a red-breasted sapsucker on a western red cedar. Puget Trough, Washington.

Bark was removed from a slide alder by a red-naped sapsucker. The rectangular shapes were made by a sapsucker expanding previous wells that merged with neighboring holes. North Cascades, Washington.

Woodpeckers flaked off the outer bark from these Engelmann spruce to access spruce bark beetles. Such signs are easily spotted in mid-elevation and subalpine forests. In the Cascades, this feeding behavior is typical for three-toed woodpeckers. North Cascades, Washington.

A black-backed woodpecker removes a beetle larvae from a burned conifer. The lighter spots on the bark show where the woodpecker has flaked off the burned outer layer in search of larvae. Photo by Paul Bannick.

burned forests, where their feeding signs can be conspicuous. In their hunt for bark and wood-boring insects, they flake off sections of blackened bark or charred wood on standing trees, revealing patches of unburned trunk. Black-backed woodpeckers feed on wood-boring beetles in tree cambium and sapwood. Three-toed woodpeckers focus on bark beetles (family Scolytidae) that are found only in the cambium layer. Three-toed woodpeckers are also common in subalpine forests infested by bark beetles, especially spruce beetles on Engelmann spruce trees.

ACORN WOODPECKER

Melanerpes formicivorus

As its name implies, this woodpecker feeds primarily on acorns. It is found in oak woodlands in the southwestern section of our region. It constructs and maintains massive granaries by drilling hundreds of holes into a single tree and then collecting and storing acorns there. These granaries are used for generations, and their owners defend them against marauding squirrels and other birds.

This duck was killed and eaten by a sharp-shinned hawk (*Accipiter striatus*). Columbia Plateau, Oregon.

Carnivorous Feeding Signs

Herons consume their prey in one piece, leaving no evidence. Hawks and owls swallow small prey whole, and they consume larger prey in parts, so that pieces of the carcass may remain. An owl often consumes the head of a small mammal before consuming the body. Birds of prey consume the breast meat of birds of pigeon size or larger, leaving the carcass behind. Or the carcass may be carried to a perch, where the raptor will consume it. Several distinctive signs indicate that a bird of prey has fed on another bird: for example, feathers are plucked and may show creases on the quills made by the raptor's bill.

This vole was probably killed, beheaded, and then abandoned by an owl. Puget Trough, Washington.

Nests

Bird nests come in many shapes and sizes, ranging from simple scrapes on the ground, to elaborate woven structures suspended from tree branches, to cavities excavated in a tree trunk.

Nests in Tree Cavities

All of our region's woodpecker species excavate nest cavities in trees. Many excavations are never used to rear young, as males may create several cavities while courting females. These unused cavities are an important refuge location for a variety of animals, and for this reason, woodpeckers play an important ecological function. Pileated woodpeckers create the largest cavities, which are often oval in shape. The birds are powerful enough to excavate cavities in dead and living trees. Nest cavity entrances

Marsh wren (*Cistothorus palustris*) nests are completely enclosed and attached to vegetation in their marshy habitat. Columbia Plateau, Washington.

average 3½ in. (8.9 cm) high and 3¼ in. (8.3 cm) wide. Black-backed woodpeckers remove the bark around their entrance holes, which gives their cavities a distinctive appearance. Downy woodpeckers' bills are not as strong as other larger species; they create smaller cavities in softer, rotten wood.

Common snipe nests are located in wetlands, concealed in high grass and vegetation. Canadian Rockies, Montana.

Cliff swallows (*Petrochelidon pyrrhonota*) create gourd-shaped mud nests along cliffs—or, as in this case, under bridges. Cliff swallows are colonial. Middle Rockies, Idaho.

Nest of a ruffed grouse or blue grouse in the brush at the base of a tree. East Cascades, Washington.

Northern flicker nest cavities, such as this one in a western juniper, provide nest sites for other birds that cannot excavate cavities themselves. Northern Great Basin, Oregon.

Only pileated woodpeckers excavate holes of this size in standing trees. Such cavities are used by many other birds and mammals after being abandoned by their creators. West Cascades, Washington.

REPTILES AND AMPHIBIANS
Classes Amphibia and Reptilia

TRACKS

The tracks of some reptiles and amphibians can be confused with those of mammals and birds.

Gaits and Track Patterns

Four-legged reptiles and amphibians use fewer gaits than mammals use. Unlike most mammals, whose legs extend from under their bodies, reptile and amphibian limbs are oriented to the sides of their bodies. As a result, their bodies are close to the ground, and body or tail drag is common in tracks of many species.

At slow speeds, all groups use a slow-moving gait that resembles a mammalian walk or trot. The diagonally opposed front and hind feet move in synchronicity as in a trot or in succession as in a mammalian walk. From tracks, it is often not possible to determine whether the hind feet were moving synchronously or independently.

Limb motion may be accompanied by a lateral undulatory motion in the spine that

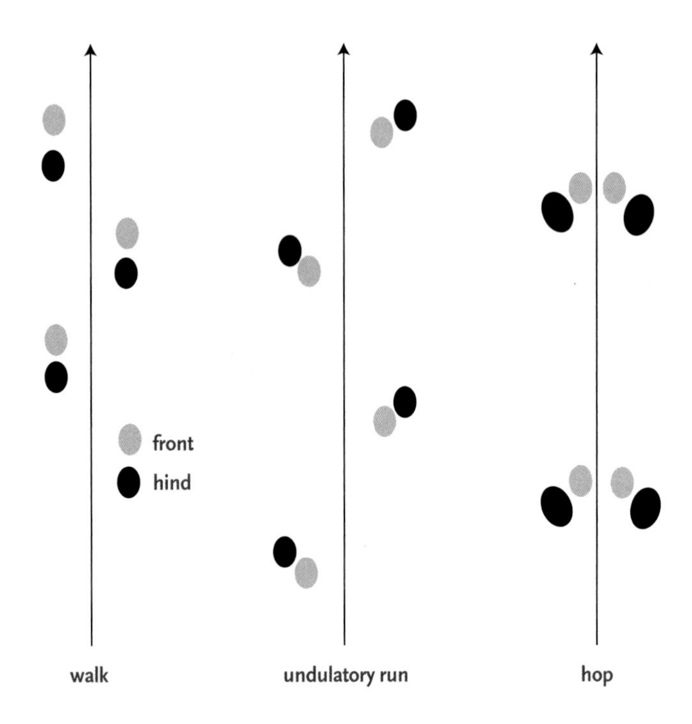

enables the hind feet to move farther forward, often causing hind tracks to register farther from the centerline of the trail than the front tracks. This motion becomes much more prominent at faster speeds, when some lizards can obtain an airborne phase as in a mammalian trot. As in mammals, in these instances the hind track can register far ahead of the same side front foot. The Great Basin collared lizard (*Crotaphytus bicinctores*) actually runs on only its hind legs at faster speeds.

Frogs and toads can use a totally different gait—a hop. The animal launches itself off all four feet at once, driven by its long, powerful hind legs, and lands on all four feet some distance away. This gait allows for a substantial airborne phase. Most frogs hop for almost all locomotion, though they occasionally walk. Toads hop less frequently, preferring the slow-moving walking gait.

Frogs and Toads

Frogs have four clawless toes on their front feet and five on the hinds. The soles of the feet are smooth. Hind feet are significantly larger than fronts and have mesial or distal webbing. The front feet tend to register ahead of the hinds and face inward toward each other. Usually, only the toes of the hind feet register in tracks.

Distinguishing among frog species is usually impossible from tracks alone. The Pacific tree frog (*Pseudacris regilla*), our smallest frog, creates distinctive tracks, with the tip of each toe showing an enlarged circular pad that registers in clear tracks. It sometimes walks. The largest species in our

Frog hop. Upper tracks are the fronts. Trail widths for bullfrogs are about 4 in. (10.2 cm). For Pacific tree frogs, trail widths are about 1 in. (2.5 cm).

Cascades frog (*Rana cascadae*). West Cascades, Washington.

Typical track pattern of a walking toad. The front foot registers ahead of the hind foot and the toes are turned inward. Often only the tips of the hind toes register, appearing similar to tracks of a mole's front feet. (Trail width 1 3/16–3 3/8 in., 3.0–8.5 cm.)

region is the bullfrog (*Rana catesbeiana*), an aggressive, introduced species found in many low-elevation wetlands. Jumping mice tracks can be mistaken for the hopping trails of a small frog.

Toads are distinguished from frogs primarily by their warty skin and large poison glands behind their eyes. Toad tracks are similar to those of frogs; however, toads have relatively shorter toes, mesial webbing is usually less apparent, and tubercles (knobby protuberances) appear on the bottom of the feet and may register in tracks. Toads are slower moving than frogs and tend to walk more than hop, and this is usually the best distinguishing feature in tracks. The walking trail of a toad could easily be confused with that of a mole. Western toads (*Bufo boreas*) are the most widespread species in our region and they range in size.

A slow-moving rough-skinned newt (*Taricha granulosa*)

Salamanders

Salamanders have four clawless toes on their front feet and five on their hinds. (Some exceptions to this are found in a few species with one or more greatly reduced toes.) Toes are straight and thinner at the tips than at the base. Hinds are slightly larger than the fronts. Trails often show an understep, with both front and hind feet registering in line with the direction of travel (rather than angled inward, as with toads and frogs). Tail drag is common. Size and habitat can help

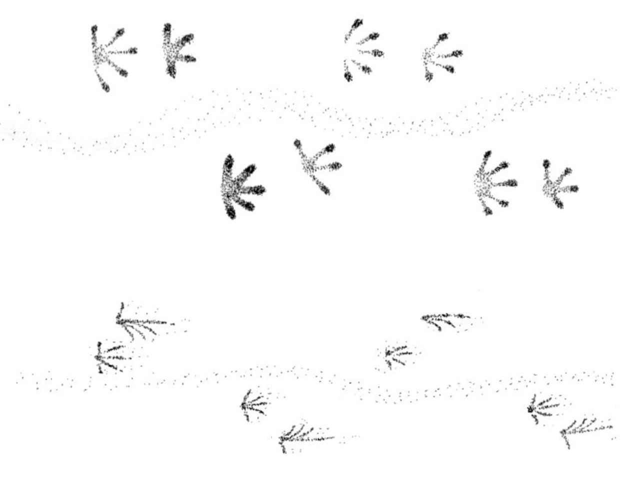

Common track pattern of a walking salamander. The smaller front foot registers ahead of the hind foot on the same side of the body. The toes in both front and hind tracks are oriented forward. Tail drag is common.

In this track pattern of a western fence lizard (*Sceloporus occidentalis*), the hind track registers ahead of and to the outside of the front track from the same side of the body. (Shown here about half life size.)

you interpret what species might have left tracks.

Lizards

Lizards have five long, slender, and often curved toes on their front and hind feet, with a thin, sharp claw at the end of each toe. Both feet are longer than wide. Hind feet are unique in that the outermost toe attaches to the foot lower than the rest and angles out at about 90 degrees. The inner and outer toes of the front feet can register perpendicular to the track's axis or closer to parallel. Tail drag is common. In arid environments where lizards are often encountered, you will often find little detail in individual tracks, leaving only the track pattern to help you identify the maker.

Turtles

Turtles have five clawed toes on their front and hind feet. Often only four toes register in hind tracks. Clear trails show an understep pattern, with the hind foot oriented parallel to the direction of travel and the front foot angled in slightly. Drag marks from the underside of the turtle's shell, the plastron, are common.

Snakes

Snakes in our region use two forms of locomotion. The slower rectilinear movement pattern leaves an almost straight trail as

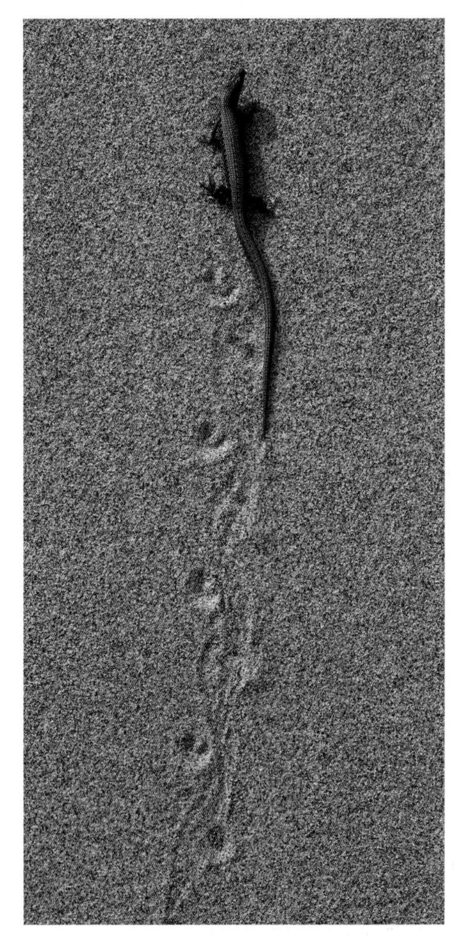

A northern alligator lizard (*Elgaria coerulea*) creates tracks from a fast gait, with a characteristic pattern that includes an undulating tail drag. Northwest Coast, Oregon.

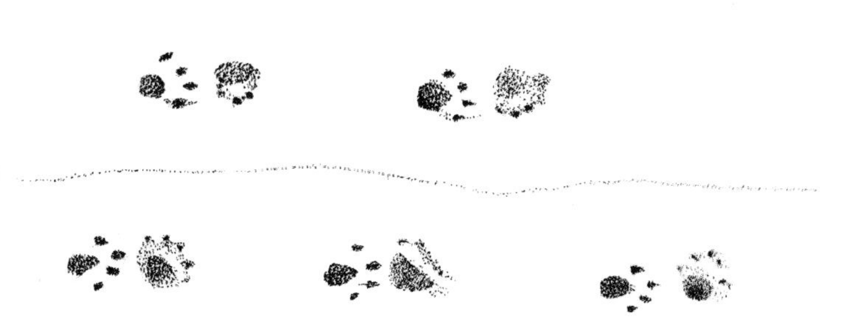

Turtle tracks. Paired tracks are a front ahead of a hind. (Shown here about half life size.)

the snake uses its belly muscles to move different portions of its body forward. For faster speeds, snakes use an undulatory pattern, making lateral waves of movement that propel them forward. Determining the direction of a snake's travel from its tracks can be difficult. The trail's sinuousness increases if the snake is going uphill, giving it more purchase. Downhill trails can become almost straight, as in rectilinear movement.

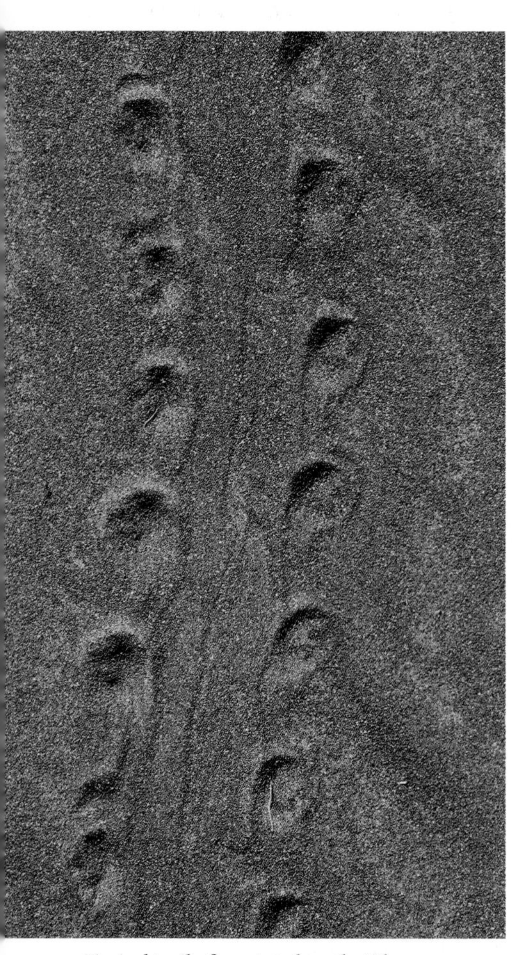

Typical trail of a painted turtle (*Chrysemys picta*), showing drag marks from the plastron. The trail width is about 5 3⁄4 in. (14.6 cm). Columbia Plateau, Washington.

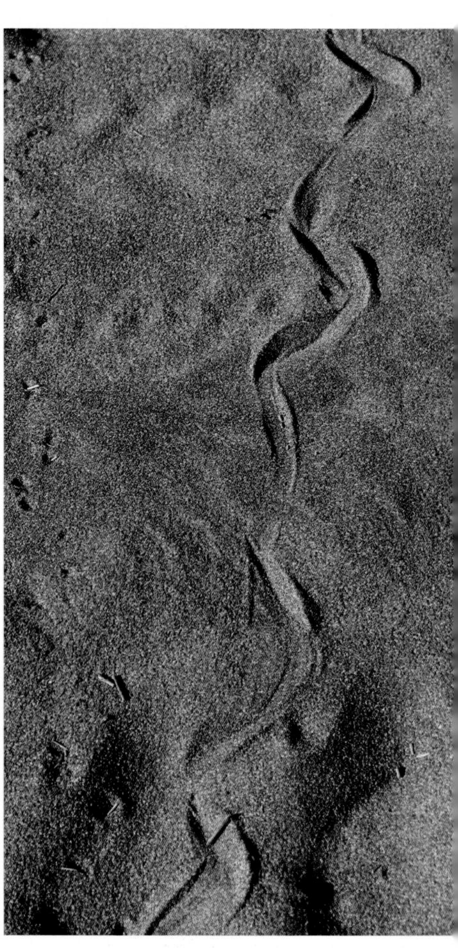

Trail of an unknown snake species. Columbia Plateau, Washington.

SIGNS

Aquatic Amphibian Eggs

Egg masses of many aquatic amphibians are conspicuous in the spring and are distinctive for several species. Amphibian eggs have a dark interior (the actual ovum), surrounded by a transparent sphere of a gelatinous substance and are marble-sized or smaller, including the jelly sphere. Eggs may be laid in strings, in clusters, or singly. Depending on the species, they are deposited terrestrially or in water, with certain species showing narrow preferences for the specific habitats they use. Amphibians lay eggs in the spring, and some species hatch within days. The most common and conspicuous amphibian examples are covered here. Reptile eggs are rarely found in the field.

Toad Eggs

Western toads (*Bufo boreas*) lay eggs in long and sometimes looping strands in shallow, slow-moving, or stagnant water. Eggs are surrounded by a double layer of jelly (visible with backlighting).

Woodhouse's toad (*Bufo woodhousii*) eggs are similar in appearance to those of the western toad, but with only one layer of jelly. They inhabit a limited range in southeastern Washington and eastern Oregon.

The Great Basin spadefoot (*Spea intermontana*) lays its eggs in loose, amorphous clusters in shallow water. Unlike all other amphibian egg clusters in our region, individual eggs can be easily removed from the cluster. They inhabit broad range across most of the eastern portion of our region.

Frog Eggs

All our native frogs lay their eggs in clumped clusters (as do some salamanders).

Pacific tree frog (*Pseudacris regilla*) egg masses are small, up to about 1.5 in. (4 cm) in diameter. Eggs are laid at the bottom or attached to submerged material in shallow water.

Red-legged frogs (*Rana aurora*) produce large egg masses (up to the size of a cantaloupe), which are laid in cool water and attached to submerged vegetation or wood debris, or deposited at the bottom under vegetation. They may eventually float to the surface.

Cascades frogs (*Rana cascadae*) produce orange-sized egg masses that are deposited in shallow water on vegetation that may be flooded during the spring snowmelt in mountain meadows of the Oregon and Washington Cascades and Olympic mountains.

Oregon spotted frogs (*Rana pretiosa*) and Columbia spotted frogs (*R. luteiventris*) produce similar sized masses that are deposited at the bottom of shallow ponds and slow-moving streams in British Columbia and mountainous regions from the Cascades east. Spotted frog eggs appear more closely packed than those of Cascades frogs, which have more jelly between individual ova.

The bullfrog (*Rana catesbeiana*) and the green frog (*R. clamitans*), two introduced species, lay eggs in broad sheets that at first float and eventually settle in shallow, still water. These frogs are destructive to native amphibians and other native components of their aquatic habitats.

Aquatic Salamander Eggs

Salamanders lay eggs singly or in masses, depending on species.

Single eggs. The long-toed salamander (*Ambystoma macrodactylum*) deposits eggs singly or in masses. Single eggs (including jelly) are usually larger than 3⁄8 in. (1 cm).

The tiger salamander (*Ambystoma tigrinum*) is found east of the Cascades and lays eggs singly, though often close together, attached to submerged vegetation. Eggs are smaller than ⅜ in. (1 cm).

Rough-skinned newts (*Taricha granulosa*) lay eggs singly, but they are rarely discovered because they are hidden under submerged vegetation in wooded ponds where the newts breed. The jelly layer is thinner than the interior ovum, the opposite of other salamanders.

Eggs in masses. The northwestern salamander (*Ambystoma gracile*) lays unique egg masses that are surrounded by an additional layer of jelly, giving the orange- to grapefruit-sized mass a smooth, compact appearance. Masses are attached to underwater branches or vegetation.

Long-toed salamander (*Ambystoma macrodactylum*) egg masses are smaller than an orange, have no additional jelly layer, and are often attached to a branch or vegetation in shallow water. Egg masses appear more widely spaced because of jelly spheres that are larger than similar-sized masses of Pacific tree frogs, with whom they share breeding habitat.

Reptile and Amphibian Scats

Lizard scats are small, compact pellets or tubes that contain insect exoskeletons, sometimes with a white tip on one end as with birds.

Toad scats (¼–½ in. [0.6–1.2 cm] D × ⅞–1⅝ in. [2.1–4.2 cm] L) are commonly found, also contain insect exoskeletons, and are often large enough to be mistaken for skunk scats.

Northwest salamander egg mass found in a shallow pond. West Cascades, Washington.

This western toad scat contains insect exoskeletons. East Cascades, Washington.

INSECTS AND OTHER INVERTEBRATES

Invertebrate life histories differ greatly from those of mammals and are often confusing and nonintuitive for us humans to understand. The number and variety of invertebrate signs in the Northwest could fill an entire field guide. Some signs, such as slug scat, are common in parts of our region. Other signs, such as the ground holes constructed by some spiders, could be confused with signs left by small animals.

Banana slug feeding on a mushroom (*Russula* spp.). Northwest Coast, Washington.

Covered here are the tracks and signs of common insects or those of significant interest from an environmental or conservation perspective. I give special attention to bark beetles, because their signs are ubiquitous in the diverse forests of our region, they are associated with the activities of other wildlife (especially woodpeckers), and their ecological role in dry forests east of the Cascade Crest is immense and evolving because of global climate change and human forest management practices.

TRACKS AND TRACK PATTERNS

DARKLING BEETLES

Order Coleoptera
Family Tenebrionidae

Darkling beetles are prolific track makers in the arid landscapes east of the Cascades. No matter their size, they leave a distinctive pattern of three that at times creates an arrow shape, pointing approximately in the direction of travel. Size varies depending on species.

Darkling beetle track pattern. Direction of travel is left to right.

CATERPILLARS AND OTHER INSECT LARVAE

Order Lepidoptera

Caterpillars are the larval forms of moths and butterflies. Their tracks are defined by leglike structures (prolegs) on the posterior segments of their bodies. Their true legs are connected to their thoracic segments farther forward in the body, but the marks they leave behind are covered over by those of the prolegs. Several other tube-shaped insect larvae also have leglike structures that can leave similar tracks.

STONEFLIES

Order Plecoptera

Stoneflies are aquatic insects well known to fly fishermen. The forked tails of larvae leave distinctive drag marks. Shed exoskeletons are commonly found along stream banks.

GRASSHOPPERS, CRICKETS, AND KATYDIDS

Order Orthoptera

The long hind legs of grasshoppers and crickets are apparent in their trails, and this

Caterpillar track pattern (life size). Direction of travel is left to right.

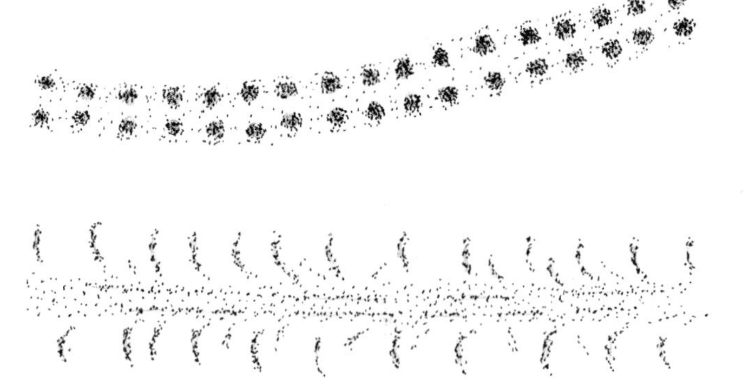

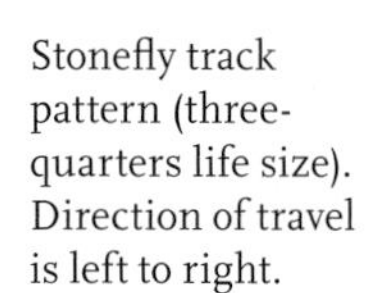

Stonefly track pattern (three-quarters life size). Direction of travel is left to right.

Millipede track pattern (life size). Direction of travel is left to right.

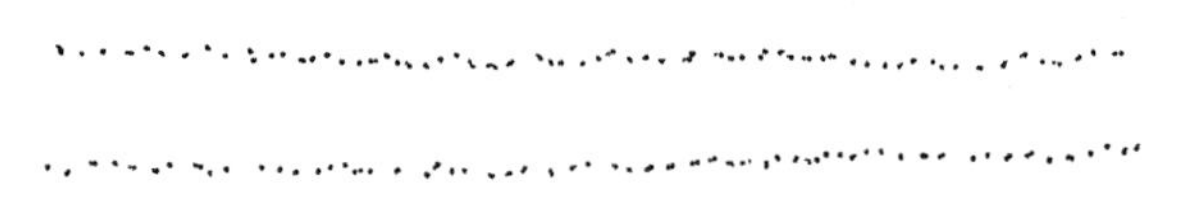

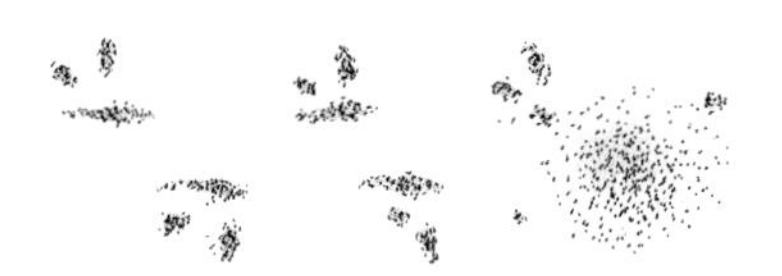

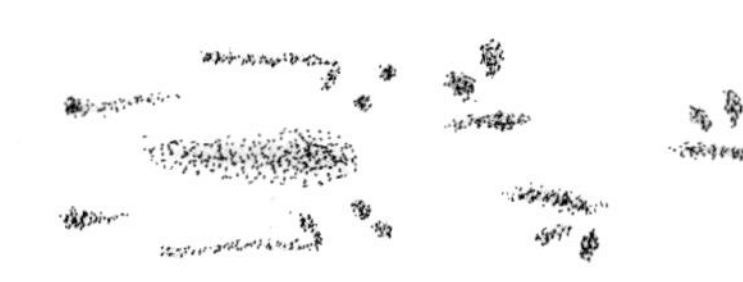

Grasshopper track pattern including a jump (two-thirds life size). Direction of travel is left to right.

can help you distinguish them from beetle trails. Their occasional long jumps are aided by their hind legs and unfurled wings. Hind tracks are innermost and oriented parallel to the direction of travel, often distinctly longer than the tracks of the other feet.

ANTLIONS

Order Neuroptera
Family Myrmeleontidae

Antlion larvae excavate a cone-shaped depression and bury themselves at its base. The cone traps ants and other small insects, making them easy prey. Traps vary in size, depending on the species, ¾–2 in. (1.9–5.1 cm) across, and are found in loose, sandy soils. In moist environments, look for them in drier microhabitats such as the overhanging base of a fallen log. Careful excavation may reveal the inhabitant.

MILLIPEDES

Class Diplopoda

A millipede's trail consists of two parallel lines comprising many small dots, each representing one of its many feet. Tracks are commonly found along the edges of mud puddles in forests west of the Cascade Crest.

Antlion trap

SIGNS ON TREES AND TREE BARK

BARK- AND WOOD-BORING BEETLES

Order Coleoptera

Bark- and wood-boring beetles play an important role in the structure of conifer forests east of the Cascades, a role which has been affected by humans through fire suppression and, according to a growing body of evidence, global climate change. These insects bore into the bark of dead or living trees and lay eggs in the cambium layer or sapwood. The eggs hatch and larvae tunnel into the inner bark or wood as they feed. These tunnels are called galleries.

Multitudes of beetles can tunnel into a tree, often killing it, especially if the tree is weakened by drought, crowding, or other issues. In areas where fire suppression has led to stands of evenly spaced ponderosa pine being replaced by dense forests, stressed trees have become susceptible to infestations. Warmer and drier summers have created favorable conditions for beetle infestations, and areas that had never before experienced outbreaks are now becoming susceptible.

A large beetle infestation can kill thousands of acres of trees—a natural occurrence for some species, such as in lodgepole pine, but historically rare for other species, such as whitebark pine. For whitebark pines, there is particular concern about how infestations will impact the high-elevation ecosystems in which they grow. This pine is a keystone in its habitat—it provides important food (large seeds) for many species of wildlife, and the trees serve as windbreaks that hold snow on ridgelines.

Bore holes and dust. Entrance holes created by adult beetles are apparent on the trunk of an infested tree. Pitch exuding from the holes is a defense mechanism of the tree. Fine boring dust can often be found during the warm months of the year around the base of an infected tree or sprinkled in the crevices of the tree's bark below bore holes.

Galleries. The passageways left behind by bark beetles are their most prolific and easily recognizable signs. The pattern of gallery construction and choice of host tree species is distinctive for each beetle species. You can find galleries by removing the outer bark of dead trees.

BARK BEETLES

Family Scolytidae

WESTERN PINE BEETLE

Dendroctonus brevicomis

PRIMARY HOST SPECIES Ponderosa pine. Serpentine tunnels move laterally and vertically, often crossing and recrossing themselves. Entrance holes on the outer bark are small, and red boring dust may appear at the tree's base.

MOUNTAIN PINE BEETLE

Dendroctonus ponderosae

HOST SPECIES All native pines (*Pinus* spp.).

A mountain pine beetle infestation killed this whitebark pine. Middle Rockies, Idaho.

Yellow spots of pitch reveal where mountain pine beetles have entered this living lodgepole pine. Similar signs can be found on numerous tree species. Middle Rockies, Idaho.

Entrance holes are often obvious and numerous on infected trees. A main tunnel is bored by the adult, parallel with the tree trunk, with smaller side tunnels roughly perpendicular to the trunk's axis.

DOUGLAS-FIR BEETLE

Dendroctonus pseudotsugae

HOST SPECIES Douglas-fir and occasionally western larch (*Larix occidentalis*). The main tunnel is parallel with the tree's trunk. Eggs are laid on alternate sides of the tunnel, and side tunnels are bored roughly perpendicular to the trunk.

SPRUCE BEETLE

Dendroctonus rufipennis

HOST SPECIES Engelmann spruce and lodgepole pine. Red boring dust may be found in bark crevices or at the base of an infested tree. A main tunnel often starts with a noticeable bend and then extends upward. Tunnels radiate from both sides and often form an interwoven fan shape. In winter, woodpeckers flake off the bark of infested spruce trees searching for larvae.

Spruce beetle (*Dendroctonus rufipennis*) galleries on a Engelmann spruce log. North Cascades, Washington.

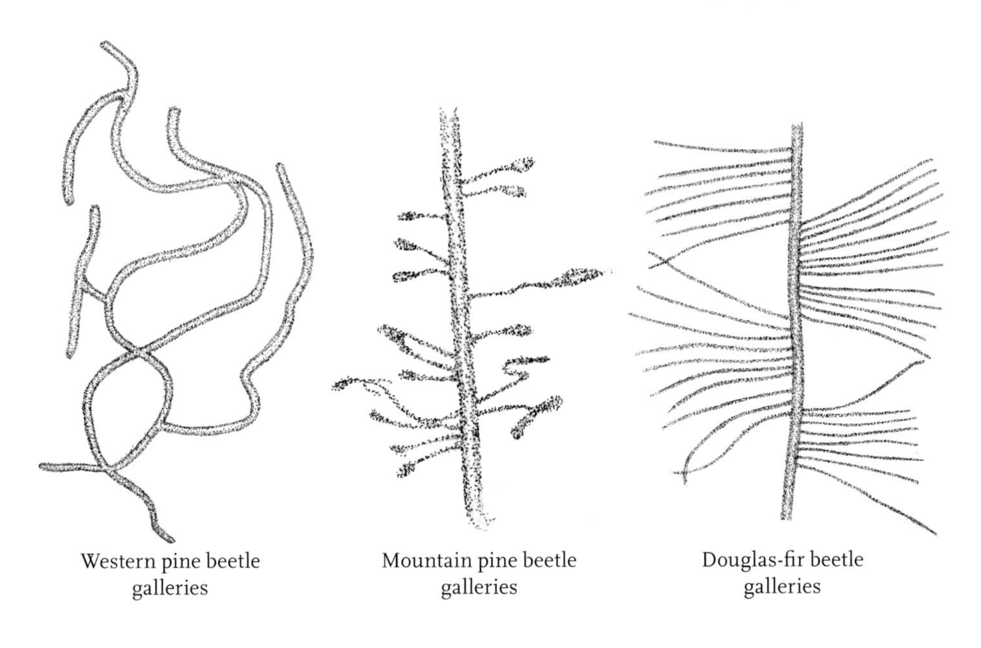

Western pine beetle galleries

Mountain pine beetle galleries

Douglas-fir beetle galleries

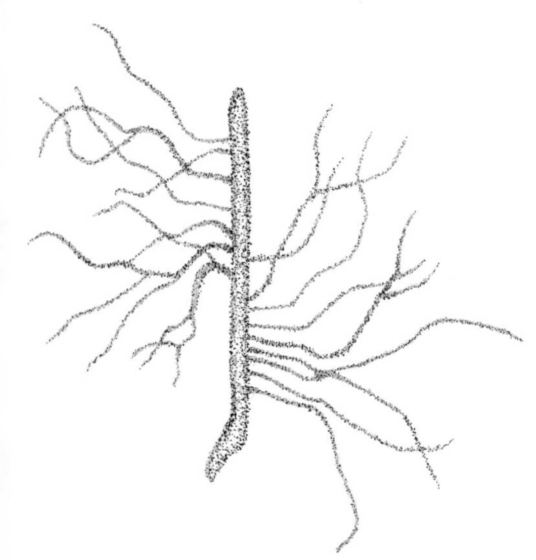

Spruce beetle galleries

FIR ENGRAVER BEETLE

Scolytus ventralis

HOST SPECIES True firs, Douglas-fir, and spruce (*Picea* spp.). Unlike tunnels of most bark beetles, the fir engraver beetle's main tunnel runs perpendicular to the trunk. Eggs are laid on both sides, and side tunnels run parallel to the trunk.

PINE ENGRAVER BEETLE

Ips spp.

HOST SPECIES Pines and occasionally Engelmann spruce. Galleries radiate from an enlarged central chamber created by a male beetle. The male mates with three or four females here, and each bores her own gallery up or down the trunk, leaving a Y- or H-shaped gallery.

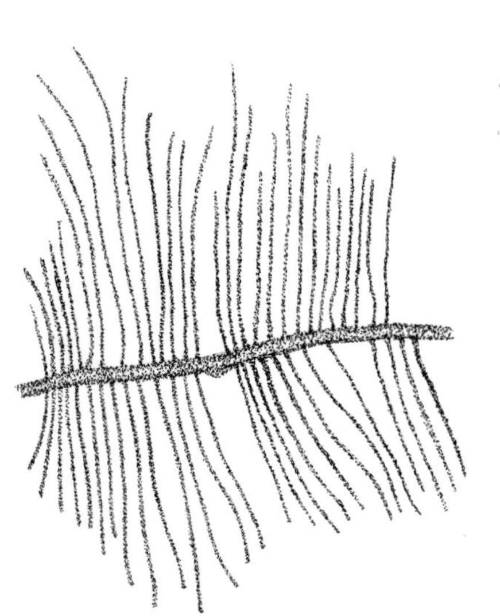

Fir engraver beetle galleries

Pine engraver beetle galleries

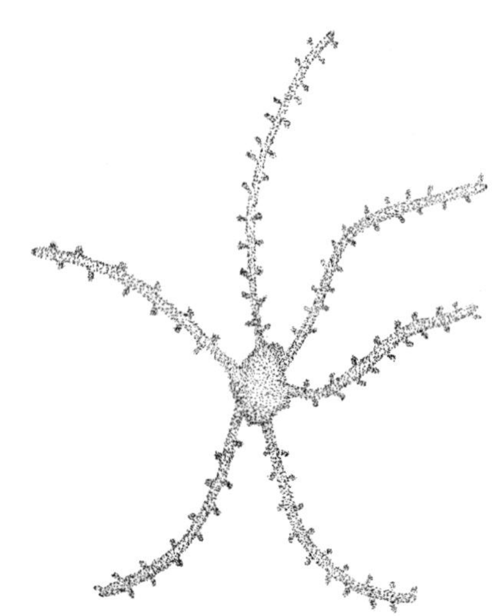

Western balsam bark beetle galleries

WESTERN BALSAM BARK BEETLE

Dryocoetes confuses

HOST SPECIES Subalpine firs (*Abies lasiocarpa*) and occasionally grand firs (*A. grandis*). This beetle's central chamber is similar to that of the pine engraver, but the secondary tunnels radiate in a spiral fashion from the central point.

Pine engraver beetle (*Ips emarginatus*) galleries mark a lodgepole pine. Woodpecker holes appear to the left of the galleries. Middle Rockies, Idaho.

WOOD-BORING BEETLES

Families Buprestidae and Cerambycidae

Coleoptera has two families of insects that feed on the cambium of western conifers and create galleries deep into a tree's heartwood. (Bark beetles feed only in the cambium layer.)

Flathead borers (family Buprestidae) produce flattened oval galleries that are often packed tightly with fine boring dust.

Roundhead borers (family Cerambycidae) produce round galleries that are more loosely packed with coarser material than those of flathead borers.

CATERPILLARS

Order Lepidoptera

WESTERN SPRUCE BUDWORM

Choristoneura occidentalis

HOST SPECIES Douglas-fir, true fir, Engelmann spruce, and western larch. Western spruce budworms lay their eggs on the tips of conifer branches. The hatched larvae feed on the terminal buds and young needles, and heavy infestations can kill the tree. Large infestations covering huge swaths of forest are not uncommon in our region. Infestations are easy to see from a distance

Spruce budworm larvae and feeding signs on the tip of a young Douglas-fir branch. East Cascades, Washington.

Snail and banana slug scat (shown here) can also be found on tree bark and is relatively common west of the Cascades. Northwest Coast, Oregon.

The pellets placed around this spider's hole are from a Nuttal's cottontail. Columbia Plateau, Oregon.

Engelmann spruce produce galls on branch tips in response to the presence and feeding activities of the Cooley spruce gall adelgid (*Adelges cooleyi*). North Cascades, Washington.

because branch tips of an infected tree are discolored.

OTHER SIGNS

Galls are produced by a plant in response to stimulation from an insect that lays its eggs under the bark or on the foliage. Galls come in all shapes, sizes, and colors and are often mistaken for fruit or cones. They rarely prove fatal to the host plant. Occasionally galls are opened by birds or mammals seeking the insect larvae inside. Dozens of different galls can appear on oak trees, and numerous galls appear on leaves and branches of willows. Insects that elicit galls include a variety of flies (order Diptera), wasps (order Hymenoptera), and leaf hoppers (order Homoptera).

TRACK MEASUREMENT GRAPHS

All measurements are from track sets found within our region.

For the sample sizes, a single record was defined by a discrete group of tracks that were clearly created by the same animal during the same travel event. Black bars represent the range.

For the hind track ranges, species for which the heel shows variably are represented by a gray bar that represents the track length with the heel showing (not included in the average). A two-step approach was used to calculate average track length values (represented by the central white bar for each species). Multiple measurements from the same track set were averaged to create one value per set, which were then averaged across sets to create a mean value for each species.

Small mammal hind track length range and average (inches)

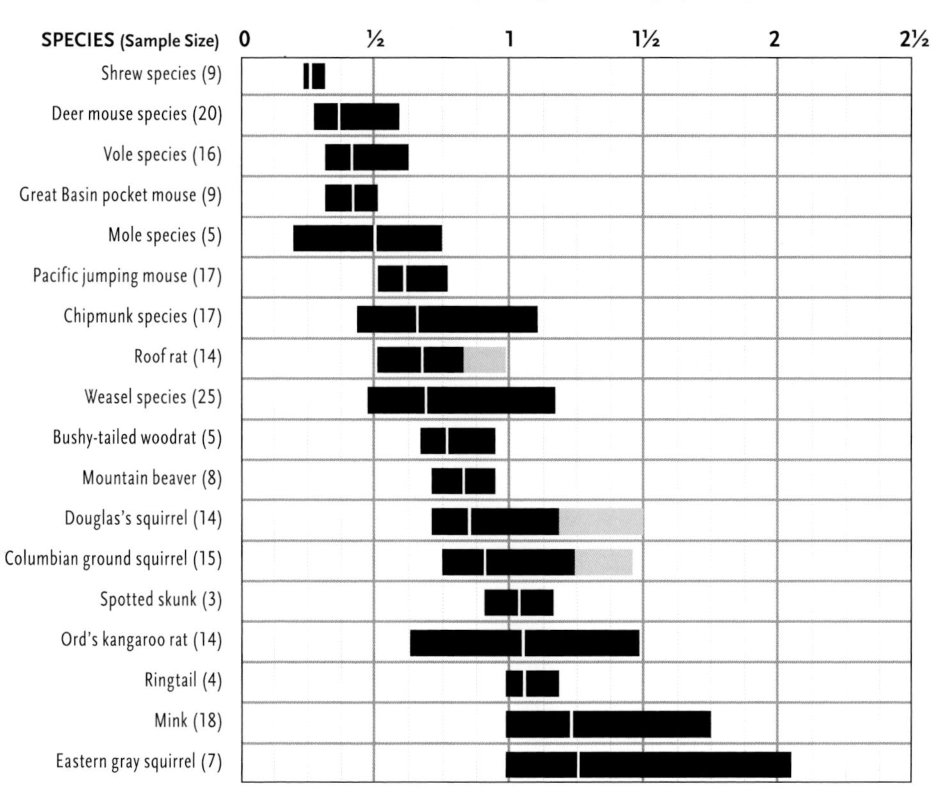

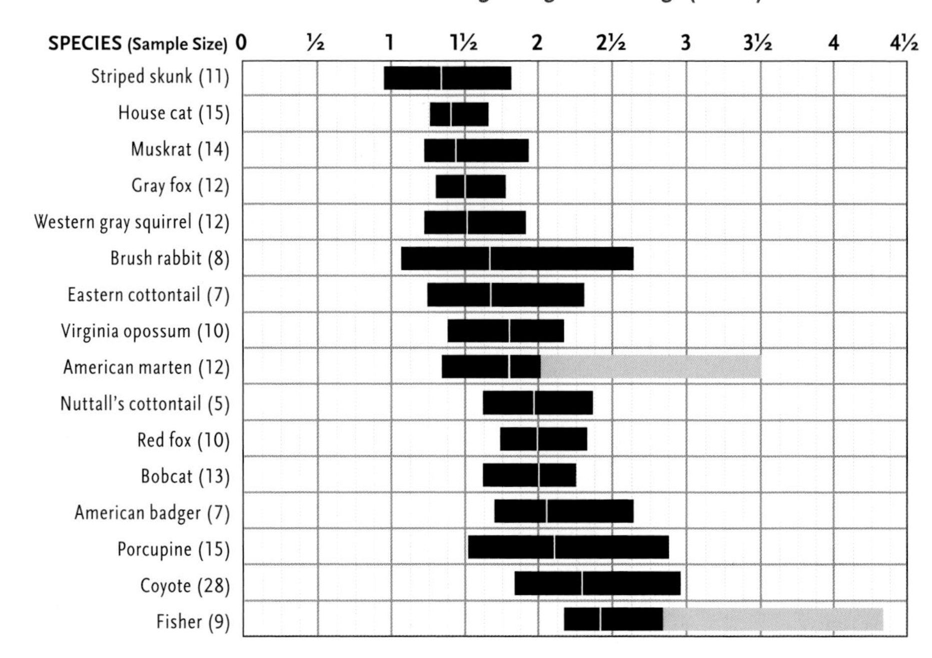
Medium mammal hind track length range and average (inches)
SPECIES (Sample Size)
0
½
1
1½
2
2½
3
3½
4
4½
Striped skunk (11)
House cat (15)
Muskrat (14)
Gray fox (12)
Western gray squirrel (12)
Brush rabbit (8)
Eastern cottontail (7)
Virginia opossum (10)
American marten (12)
Nuttall's cottontail (5)
Red fox (10)
Bobcat (13)
American badger (7)
Porcupine (15)
Coyote (28)
Fisher (9)

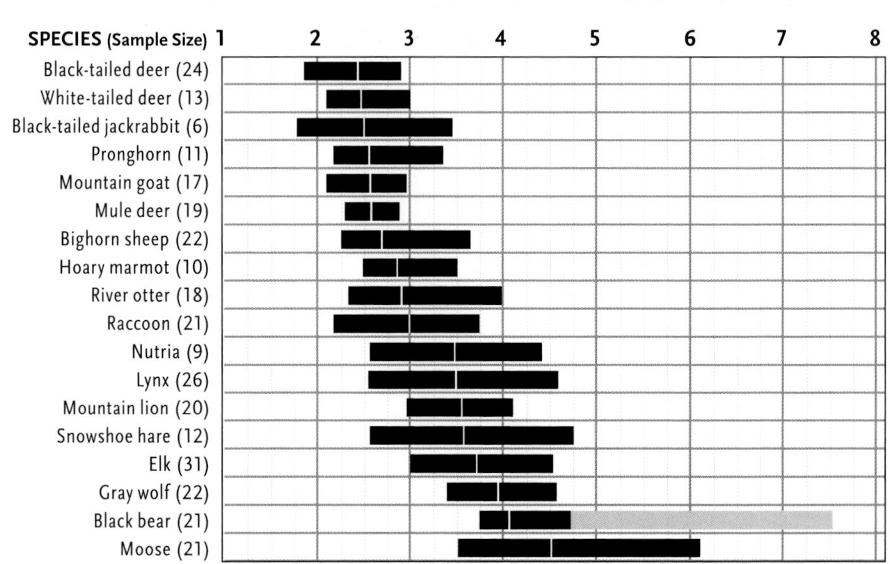
Large mammal hind track length range and average (inches)
SPECIES (Sample Size)
1
2
3
4
5
6
7
8
Black-tailed deer (24)
White-tailed deer (13)
Black-tailed jackrabbit (6)
Pronghorn (11)
Mountain goat (17)
Mule deer (19)
Bighorn sheep (22)
Hoary marmot (10)
River otter (18)
Raccoon (21)
Nutria (9)
Lynx (26)
Mountain lion (20)
Snowshoe hare (12)
Elk (31)
Gray wolf (22)
Black bear (21)
Moose (21)

Large mammal hind track length range and average (inches), continued

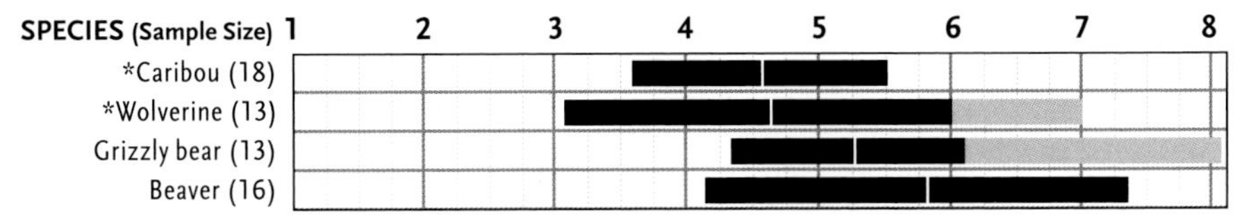

*For these species, the sample size is probably an overestimate of the number of individual animals included in the sample.

Bounding trail width range and average (inches)

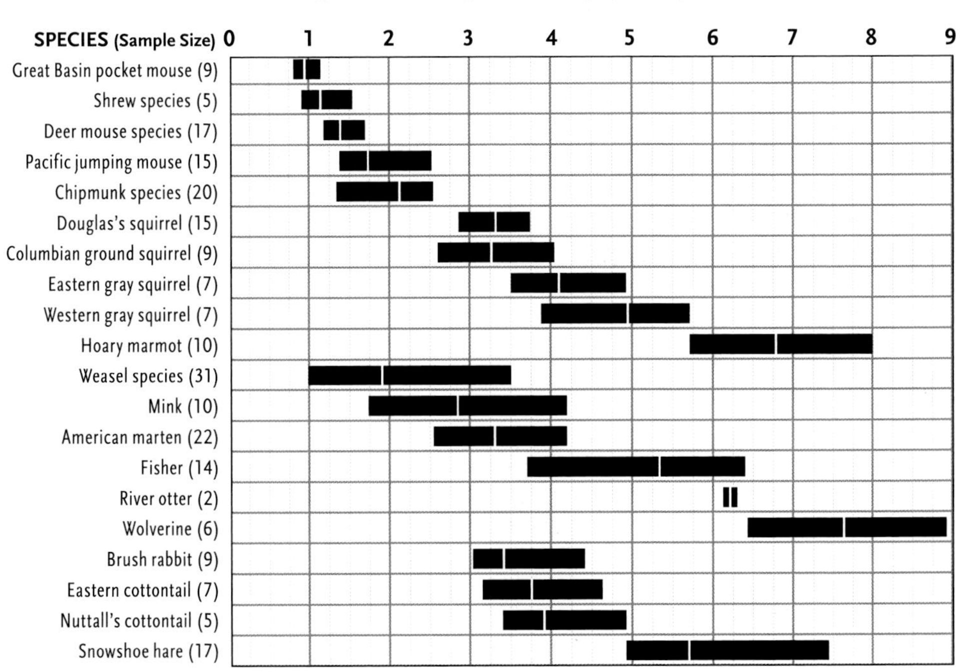

GLOSSARY

abiotic Nonliving parts of the natural world such as wind, rocks, and temperature.

altricial Young are born requiring a great deal of maternal care, often without hair, and with undeveloped thermoregulation, locomotion, vision, and hearing.

bound Form of quadruped locomotion that involves landing and pushing off with the hind feet in synchronicity.

cambium The layer of cells between the bark and wood of trees and shrubs that moves water and nutrients in the plant. Often referred to as "inner bark."

carnivore An animal that consumes animal tissue.

carrion The remains of dead animals, often in reference to a food source for other animals.

coprophagous An animal that reingests its own feces to extract more nutrients from the material, a normal behavior for several species of herbivorous mammals.

crepuscular Daily rhythm in which an animal is most active around dawn and dusk.

delayed implantation Reproductive cycle in which development of a fertilized egg is stopped within the female for a period of time, usually until a season most favorable for rearing young.

digitigrade Foot structure in which an animal bears all weight on its toes and palm pad. Heel rarely touches the ground.

direct register Track pattern in which hind foot lands directly on the track of the front foot, leaving what appears to be a single track (or nearly so).

diurnal Daily rhythm in which an animal is most active during daylight hours.

endemic A species whose distribution is limited to a defined geographic region.

estrus Physiological state in which a female is interested in breeding and is capable of conceiving.

extirpation Disappearance, or extinction, of a particular species from a defined and often isolated geographic region (the species still exists in other locations).

fossorial Adapted to digging and living beneath the earth's surface.

gait A defined sequence of foot movements by which an animal moves.

gallop Gait that involves the independent and asymmetric movement of each foot and one or two airborne phases.

generalist A species that is able to survive in many different ecological conditions.

heel Posterior portion of foot. May be hairless or entirely furred and may include one or more unfurred proximal pads.

herbivore An animal that consumes plant material.

home range The entire area of land across which an individual animal travels or uses for food collection, breeding, and rearing young.

keystone species A species that, because of its important role in an ecosystem, has a major impact on other species that is often disproportionately large in comparison to its abundance.

lope Gait in which all four feet move independently and asymmetrically; may have a single airborne phase.

metacarpal pad Fleshy, calloused pad(s) on the sole of an animal's foot, associated with the metacarpal (front) or metatarsal (hind) bones. Located between or immediately adjacent to toe pads, in the region referred to as the palm.

midden A collection of waste materials produced by feeding activities that accumulate at a habitually used feeding location.

natal Relating to birth.

niche Distinct ecological role that a particular animals fills within an ecosystem.

nocturnal Daily rhythm in which an animal is most active at night.

obligate Species that is restricted to a particular set of environmental conditions.

omnivore An animal whose diet includes both plant and animal tissue.

overstep Track pattern created when an animal's hind foot falls beyond where its front foot on the same side of the body landed.

pace Gait in which the front and hind feet on the same side of the body move in exact synchronicity. Rarely used by any of our region's native mammals.

palm The part of the foot proximal to the toes and distal to the heel. May include unfurred metacarpal pads or be entirely furred.

parturition The process of giving birth.

plantigrade Foot structure in which the animal bears its weight across the entire surface of the foot including its heel.

precocial Young that are born in a highly developed state, capable of a high degree of independence shortly after birth. They are furred, can walk shortly after birth, and thermoregulate well.

pronk Gait in which all four feet are employed in synchronicity to produce a jumping locomotion.

proximal pad Unfurred fleshy calloused structure on an animal's heel, often associated with carpal (front) or tarsal (hind) bones of foot.

riparian Ecological community defined by proximity to a body of water.

semiaquatic Animal that spends part but not all of its life in water.

sexual dimorphism Differences in the sizes and shapes of males and females within a single species.

sign Any piece of evidence of an animal's passage or activity observable in the field.

specialist Species with an ability to survive in a narrow set of ecological conditions, often with highly specialized adaptations to do so.

subnivean zone The space under snowpack.

territory The part of an animal's home range that is defended from other animals, often other members of its own species.

track An animal's footprint.

track pattern Repeating sequence of tracks on the ground produced by a distinct type of locomotion.

trot Gait in which the diagonally opposed front and hind feet move in exact synchronicity.

understep Track pattern created when animal's hind foot lands before its front foot track on the same side of the body.

unguligrade Foot structure in which the weight-bearing surface of an animal's feet is limited only to the tips of the toes, called a hoof.

walk Gait in which each foot is moved independently but symmetrically, with no airborne phase.

webbing Membrane of skin that attaches between the animal's toes. Webbing can be proximal (attaching close to where each toe connects to the rest of foot), mesial (attached about midway between the tips and terminus of the toes), or distal (attached at or nearly at the tip of the toes).

SELECTED BIBLIOGRAPHY

Aubry, K. B., K. S. McKelvey, and J. P. Copeland. 2007. Distribution and Broadscale Habitat Relations of the Wolverine in the Contiguous United States. *Journal of Wildlife Management* 71 (7): 2147–2158.

Bailey, Robert G. 1995. Description of the Ecoregions of the United States. Washington, D.C.: USDA Forest Service. Miscellaneous Publication 1391.

Bang, Preben, and Preben Dahlstrom. 2001. *Animal Tracks and Signs*. New York: Oxford University Press.

Bannick, Paul. 2008. *The Owl and the Woodpecker: Encounters with North America's Most Iconic Birds*. Seattle: The Mountaineers Books.

Brown, R. W., M. J. Lawrence, and J. Pope. 1984. *The Larousse Guide to Animal Tracks, Trails and Signs*. New York: Larousse & Co.

Brown, R. W., J. Ferguson, M. Lawrence, and D. Lees. 2003. *Tracks & Signs of the Birds of Britain & Europe*. 2d ed. London: Christopher Helm.

Brown, Tom Jr. 1983. *Tom Brown's Field Guide to Nature Observation and Tracking*. New York: Berkley Books.

Chadwick, Douglas H. 1983. *A Beast the Color of Winter: The Mountain Goat Observed*. Lincoln, Nebraska: University of Nebraska Press.

Corkan, C. C., and C. Thoms. 2006. *Amphibians of Oregon, Washington and British Columbia: A Field Identification Guide*. Vancouver, British Columbia, Canada: Lone Pine Publishing.

Cowan, I. M., and C. Guiguet. 1978. *The Mammals of British Columbia*. British Columbia Provincial Museum Department of Recreation and Conservation Handbook 11. Victoria, British Columbia, Canada: British Columbia Provincial Museum.

Csuti, B., T. O'Neil, M. Shaughnessy, E. Gaines, and J. Hak. 2001. *Atlas of Oregon Wildlife: Distribution, Habitat, and Natural History*. 2nd ed. Corvallis, Oregon: Oregon State University Press.

Darimont, C. T., P. C. Paquet, T. E. Reimchen, and V. Crichton. 2005. Range expansion by moose into coastal temperate rainforests of British Columbia, Canada. *Diversity and Distribution* 11: 235–239.

Elbroch, M., E. Marks, and C. D Boretos. 2001. *Bird Tracks & Sign: A Guide to North American Species*. Mechanicsburg, Pennsylvania: Stackpole Books.

Elbroch, Mark. 2003. *Mammal Tracks & Sign: A Guide to North American Species*. Mechanicsburg, Pennsylvania: Stackpole Books.

Haggard, P., and J. Haggard. 2006. *Insects of the Pacific Northwest*. Portland, Oregon: Timber Press.

Hagle, S. K., K. E. Gibson, and S. Tunnock. 2003. *A Field Guide to Diseases & Insect Pests of Northern & Central Rocky Mountain Conifers*. USDA Forest Service Report R1-03-08. http://www.fs.fed.us/r1-r4/spf/fhp/field_guide/index.htm. Accessed 31 October, 2008.

Halfpenny, James C. 1986. *A Field Guide to Mammal Tracking in North America*. Boulder, Colorado: Johnson Books.

———. 1999. *Tracking for Elk Hunters*. Video. Gardiner, Montana: A Naturalists World.

Harris, J. E., and C. V. Ogan, eds. 1997. *Mesocarnivores of Northern California: Biology, Management, and Survey Techniques, Workshop Manual*. Arcata, California: Humboldt State University, The Wildlife Society, California North Coast Chapter.

Hayssen, Virginia. 2008. Mammalian Species: Numbered Mammalian Species Accounts and Links to PDFs. http://www.science.smith.edu/departments/Biology/VHAYSSEN/msi/default.html.

Heffelfinger, Jim. 2006. *Deer of the Southwest: A Complete Guide to the Natural History, Biology, and Management of Southwestern Mule Deer and White-Tailed Deer.* College Station, Texas: Texas A & M University Press.

Hildebrand, Milton. 1989. The quadrupedal gaits of vertebrates. *Bioscience* 39: 766–775.

Howell, A. Brazier. 1944. *Speed in Animals.* Chicago: University of Chicago Press.

Idaho State University and Idaho Geological Survey. 2000. Digital Atlas of Idaho. http://imnh.isu.edu/digitalatlas/bio/mammal/mamfram.htm.

Johnson, David H., and T. A. O'Neil, eds. 2001. *Wildlife-Habitat Relationships in Oregon and Washington.* Corvallis, Oregon: Oregon State University Press.

Johnson, R. E., and K. M. Cassidy. 1997. Terrestrial Mammals of Washington State: Location Data and Predicted Distributions. Volume 3 in Washington State Gap Analysis Project Final Report (K. M. Cassidy, C. E. Grue, M. R. Smith, and K. M. Dvornich, eds.). Seattle: Washington Cooperative Fish and Wildlife Research Unit, University of Washington.

Kays, R. W., and D. E. Wilson. 2002. *Mammals of North America.* Princeton, New Jersey: Princeton University Press.

Lehmkuhl, J. F., K. D. Kistler, J. S. Begley, and J. Boulanger. 2006. Demography of northern flying squirrels informs ecosystem management of western interior forests. *Ecological Applications* 16 (2): 584–600.

Liebenberg, Louis. 1990a. *The Art of Tracking: The Origin of Science.* Claremount, South Africa: David Philip Publishers.

———. 1990b. *A Field Guide to Animal Tracks of Southern Africa.* Cape Town, South Africa: David Philip Publishers.

Magoun, A., and J. P. Copeland. 1998. Characteristics of wolverine reproductive den sites. *Journal of Wildlife Management* 62: 1313–1320.

Maser, Chris. 1998. *Mammals of the Pacific Northwest: From the Coast to the High Cascades.* Corvallis, Oregon: Oregon State University Press.

Merlin, Pinau. 1999. *A Field Guide to Desert Holes.* Tucson Arizona: Arizona Sonora Desert Museum Press.

Morse, Susan C. 2001. Is the lynx missing? *Northern Woodlands* Spring: 19.

———. 2004. When a wild cat kills a deer. *Northern Woodlands* Spring: 21.

———. 2007. Making sense of scent marking. *Northern Woodlands* Winter: 36–39.

Mote, P. W. 2003. Trends in temperature and precipitation in the Pacific Northwest during the twentieth century. *Northwest Science* 77 (4): 271–282.

Murie, O. J., and M. Elbroch. 2005. *The Peterson Field Guide to Animal Tracks.* 3d ed. Boston: Houghton Mifflin Harcourt.

Muybridge, Eadweard. 1957. *Animals in Motion.* New York: Dover Publications.

Nagorsen, David W. 1996. *Opossums, Shrews and Moles of British Columbia.* Royal British Columbia Museum Handbook. Vancouver British Columbia, Canada: University of British Columbia Press.

———. 2002. *An Identification Manual to the Small Mammals of British Columbia.* British Columbia, Canada: British Columbia Ministry of Sustainable Resource Management, Terrestrial Ecosystems Branch, British Columbia Ministry of Water, Land and Air Protection Ecosystems Branch, and Royal British Columbia Museum.

Reid, Fiona A. 2006. *A Field Guide to Mammals of North America.* 4th ed. New York: Houghton Mifflin.

Rezendes, Paul. 1999. *Tracking and the Art of Seeing: How to Read Animal Tracks and Sign.* 2d ed. New York: HarperCollins.

Ripple, W. J., and R. L. Beschta. 2004. Wolves and the ecology of fear: Can predation risk structure ecosystems? *BioScience* 54 (8): 755–766.

———. 2005. Linking wolves and plants: Aldo Leopold on trophic cascades. *Bioscience* 55 (7): 1–9.

Ruggiero, Leonard F., Keith B. Aubry, Steen W. Buskirik, L. Jack Lyon, and William J. Zielinski, eds. 1994. The Scientific Basis for Conserving Forest Carnivores: American Marten, Fisher, Lynx, and Wolverine in the Western United States. USDA Forest Service General Technical Report RM-GTR-254. Fort Collins, Colorado: USDA Forest Service, Rocky Mountain Forest and Range Experiment Station.

Sager, S. R., and L. A. K. Singh. 1991. Technique to distinguish sex of tiger (*Panthera tigris*) from pug-marks. *The Indian Forester* 117 (1): 24–28.

Schwartz, C. W., and E. R. Schwartz. 2001. *The Wild Mammals of Missouri*. 2d ed. Columbia, Missouri: University of Missouri Press.

Sheehan, S. T., and C. Galindo-Leal. 1997. Identifying coast moles, *Scapanus orarius*, and Townsend's moles, *Scapanus townsendii*, from tunnel and mound size. *Canadian Field Naturalist* 111 (3): 463–465.

Singleton, P. H., W. L. Gaines, and J. F. Lehmkuhl. 2002. Landscape permeability for large carnivores in Washington. USDA Forest Service Research Paper PNW-RP-549. Portland, Oregon: USDA Forest Service, Pacific Northwest Research Station.

St. John, Alan. 2002. *Reptiles of the Northwest*. Vancouver, British Columbia, Canada: Lone Pine Publishing.

St-Pierre, Caroline, Jean-Pierre Ouellet, France Dufresne, Audrey Chaput-Bardy, and François Hubert. 2006. Morphological and Molecular Discrimination of *Mustela erminea* (Ermines) and *M. frenata* (Long-tailed Weasels) in Eastern Canada. *Northeastern Naturalist* 13 (2): 143–152.

Stander, P. E., and D. T. Ghau. 1997. Tracking and the interpretation of spoor: A scientifically sound method in ecology. *Journal of Zoology* 242: 329–341.

Terres, John K. 1980. *The Audubon Society Encyclopedia of North American Birds*. New York: Wings Books.

Vaughan, T. A., J. M. Ryan, and N. J. Czaplewski. 2000. *Mammalogy*. 4th ed. Orlando, Florida: Harcourt College Publishers.

Verts, B. J., and L. N. Carraway. 1998. *Land Mammals of Oregon*. Berkeley, California: University of California Press.

Williams, D. W., and A. M. Liebhold. 2002. Climate change and the outbreak ranges of two North American bark beetles. *Agricultural and Forest Entomology* 4 (2): 87–99.

Wilson, D. E., and D. M. Reeder. 2005. *Mammal Species of the World*. 3rd ed. Baltimore, Maryland: Johns Hopkins Univrsity Press.

Worsham, Charles E. 1989. Revised 1992. Techniques of Tracking on Various Ground Covers. Unpublished manuscript.

Young, J., and T. Morgan. 2007. *Animal Tracking Basics*. Mechanicsburg, Pennsylvania: Stackpole Books.

Yuan, G., N. Hasler, R. Klette, and B. Rosenhahn. 2006. Understanding Tracks of Different Species of Rats. Communication and Information Technology Research Technical Report 187. Auckland, New Zealand: University of Auckland.

Zielinski, W. J., and R. L. Truex. 1995. Distinguishing tracks of marten and fisher at track-plate stations. *Journal of Wildlife Management* 59 (3): 571–579.

Zielinski, W. J., and T. E. Kucera. 1995. *American Marten, Fisher, Lynx and Wolverine: Survey Methods for their Detection*. (PSW-GTR-157) Berkeley, California: Pacific Southwest Research Station, U.S. Forest Service.

INDEX

STEVE SMITH

David Moskowitz, a professional wildlife tracker, photographer, and outdoor educator, has been studying wildlife and tracking in the Pacific Northwest since 1995. He has contributed his technical expertise to a variety of wildlife studies regionally and in the Canadian and U.S. Rocky Mountains, focusing on using tracking and other noninvasive methods to study wildlife ecology and promote conservation. David has worked on projects studying rare forest carnivores, wolves, elk, Caspian terns, desert plant ecology, and trophic cascades. He helped establish the Cascade Citizens Wildlife Monitoring Project, a citizen science effort to search for and monitor rare and sensitive wildlife in the Cascades and other Northwest wildlands.

David's extensive experience as an outdoor educator includes training mountaineering instructors for Outward Bound, leading wilderness expeditions throughout the western United States and in Alaska, teaching natural history seminars, and serving as the lead instructor for wildlife tracking programs at Wilderness Awareness School in Duvall, Washington. David holds a bachelor's degree in environmental studies and outdoor education from Prescott College. View his photography and find out about classes at www.davidmoskowitz.net.

HOW TO MEASURE TRACKS AND TRACK PATTERNS

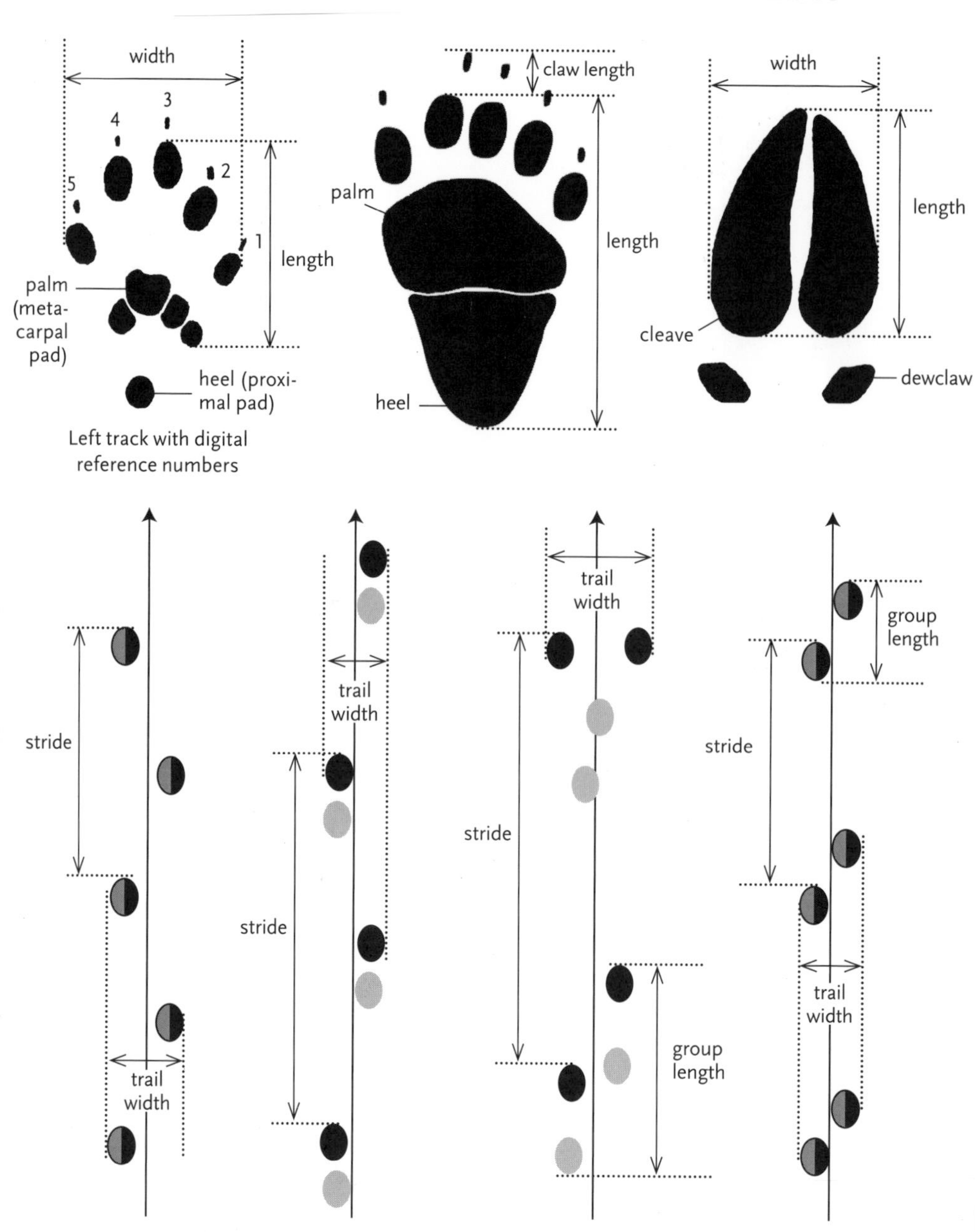

See also pages
49 and 349